NL ARMS

Netherlands Annual Review of Military Studies

Ministry of Defence

Series Information

The series aims to create an overview of scientific research in military sciences and it shows the extent to which researchers of the Faculty of Military Sciences (FMS) of the Netherlands Defence Academy (NLDA) contribute to this field of expertise.

Research at the Faculty is confined to military-relevant, multi-disciplinary areas and therefore deals with issues relating, among others, to:

- Command and control in military operations
- Leadership and ethics
- Military law and history
- Operational analysis
- Navigation
- Combat systems
- Military platforms
- Cyber warfare
- Exit Strategies
- Organizing for Safety and Security in Military Organizations

The series is peer-reviewed and reflects the diversity and quality of the Faculty of Military Sciences at its best. With NL ARMS the FMS hopes to serve as a trigger for more international comparative research on military sciences.

Editorial Office

F.E.J. de Pee-Vencken MA
Faculty of Military Sciences
Netherlands Defence Academy
P.O. Box 90 002
4800 PA Breda
The Netherlands

More information about this series at http://www.springer.com/series/13908

Paul A.L. Ducheine · Frans P.B. Osinga
Editors

NL ARMS
Netherlands Annual Review of Military Studies 2017

Winning Without Killing: The Strategic and Operational Utility of Non-Kinetic Capabilities in Crises

ASSER PRESS

Springer

Editors
Paul A.L. Ducheine
Faculty of Military Sciences
Netherlands Defence Academy
Breda
The Netherlands

and

Amsterdam Centre of International Law
University of Amsterdam
Amsterdam
The Netherlands

Frans P.B. Osinga
Faculty of Military Sciences
Netherlands Defence Academy
Breda
The Netherlands

ISSN 1387-8050 ISSN 2452-235X (electronic)
NL ARMS
ISBN 978-94-6265-188-3 ISBN 978-94-6265-189-0 (eBook)
DOI 10.1007/978-94-6265-189-0

Library of Congress Control Number: 2017940585

Published by T.M.C. ASSER PRESS, The Hague, The Netherlands www.asserpress.nl
Produced and distributed for T.M.C. ASSER PRESS by Springer-Verlag Berlin Heidelberg

Printed on acid-free paper

This T.M.C. ASSER PRESS imprint is published by Springer Nature
The registered company is Springer-Verlag GmbH Germany
The registered company address is: Heidelberger Platz 3, 14197 Berlin, Germany

Foreword

The prime task of the Netherlands Defence Academy's Faculty of Military Sciences is to educate officers of the four armed services—navy, army, air force and Royal Netherlands Marechaussee—thus contributing to the effectiveness of the Dutch armed forces.

The courses offered in the various bachelors, masters and other programs require up-to-date knowledge and understanding of contemporary crises, conflict and the (strategic) functions of armed forces. It is through the various research programs and Strategic Research Orientations that these knowledge and understanding are acquired by the Faculty's staff.

From my own experience as a Task Force Commander, modern conflict is volatile, complex, wicked and multi-layered, requiring more than just traditional military and kinetic capabilities to be effective. In general, the notion of a comprehensive approach and the concept of Diplomacy, Development and Defence (3D), proved beneficial to conduct contemporary operations where capabilities of a military and non-military nature are used in an orchestrated way, seeking synergy from the separate elements. In particular, looking back on operations in Uruzgan Province, the non-kinetic capabilities proved very useful in creating a safe and secure environment in a war-torn region.

This Netherlands Annual Review of Military Sciences or NL ARMS is dedicated to non-kinetic, non-military and—until now—sometimes unorthodox capabilities contributing to conflict resolution and crisis management in international relations. With this study, the researchers have sought to contribute to a more solid and grounded understanding of the potential benefits and limits of these capabilities, both at a strategic and at an operational level. Personally, I would have loved to have obtained this study prior to assuming command in Uruzgan. I am convinced that I could have used more than one insight provided, and I am also convinced that knowledge and understanding contribute to foresight and insight and thus to the effectiveness of armed forces in conflicts and crises.

Which is precisely the very purpose of the Faculty and the Defence Academy as a whole.

Major-General Nico Geerts
Commander
Netherlands Defence Academy
Breda, The Netherlands

Contents

About the Contributors

Editors

Brigadier-General Prof. Dr. Paul A.L. Ducheine is Professor for Cyber Operations and Cyber Security at the Netherlands Defence Academy and a Professor of the Law of Military Cyber Operations at the University of Amsterdam. Trained at the Netherlands Royal Military Academy, he joined the Engineer Corps in 1988 and switched to the Army Legal Service in 1998. He holds a Master of Science degree in Public Administration & Political Sciences as well as a Master degree and a Ph.D. in Law.

Air-Commodore Prof. Dr. Frans P.B. Osinga is Professor of Military Operational Art and Sciences, Chair of the War Studies Department at the Netherlands Defence Academy (Faculty of Military Sciences), and a member of the Faculty's Managing Board. Trained at the Royal Military Academy, he joined the Royal Netherlands Air Force in 1987 as an F-16 pilot. He defended his Ph.D. in 2005 on the relationshipbetween war, science and the strategic thinking of John Boyd.

Authors

Dr. F.H. Baudet is Associate Professor of Strategy and Military History at the Department of War Studies, Netherlands Defence Academy. He defended his Ph.D. in 2001 at the University of Utrecht on the Netherlands' (human rights) policy for Eastern Europe and Yugoslavia (1972–1989).

Prof. Dr. R.J.M. Beeres is Professor of Defence Economics, Department of Military Management Studies at the Faculty of Military Sciences, Netherlands Defence Academy and an Associate Professor at Nyenrode Business University. He obtained a Ph.D in administrative sciences at Radboud University Nijmegen.

Colonel Dr. R.M.M. Bertrand RA RC RO is Associate Professor of Accounting Control and Defence Economics, Department of Military Management Studies at the Faculty of Military Sciences, Netherlands Defence Academy. He defended his Ph.D. in 2016 at Maastricht University.

Prof. Dr. M.T.I.B. Bollen is Professor of Civil-Military Interaction and Chair of the Department of Military Management Studies and a member of the Managing Board of the Faculty of Military Sciences, Netherlands Defence Academy. She defended her Ph.D. in 2002 at Wageningen University. Professor Bollen is a Visiting Professor at the Baltic Defence College.

Lieutenant-Colonel G.L.C. van den Bosch, LL.M. MA is Assistant Professor of Military Law, Department of War Studies, Netherlands Defence Academy. He holds master degrees in Law and in Public Administration (Dutch Open University) and is preparing his Ph.D. thesis on the Law of Armed Conflict for Cyber Operations.

Colonel A.J.H. Bouwmeester, MSc MMAS is Associate Professor of Military Strategy and Land Warfare at the Department of War Studies, Netherlands Defence Academy. He holds a Master of Science degree in Public Administration & Political Sciences, is a graduate of the U.S. Army School of Advanced Military Studies and is preparing his Ph.D. thesis on Russia's information campaigns and deception.

Prof. Dr. Theo Brinkel is Assistant Professor of International Security Studies at the Department of War Studies, Netherlands Defence Academy and KVMO [Koninklijke Vereniging van Marineofficieren—Royal Association of Naval Officers] Professor of Civil Military Relations at the Faculty of Social Sciences of Leiden University. Theo Brinkel defended his doctoral thesis in 2006 on the subject of nation building and pluralism in South Africa. Previously he was a Member of Parliament for the Christian Democratic Party and a member of the Foreign Affairs and Defence Committees of the Dutch Parliament.

Dr. T.W. Brocades Zaalberg is Associate Professor of Intelligence and Security at the Department of War Studies, Netherlands Defence Academy and a University Lecturer in Contemporary Military History at Leiden University. He holds a Master of Arts degree from the University of Groningen. He is the author of Soldiers and Civil Power: Supporting or Substituting Civil Authorities in Modern Peace Operations, which earned him his Ph.D. in history from the University of Amsterdam in 2005.

Prof. Dr. P.C. van Fenema is Professor of Military Logistics, Department of Military Management Studies at the Faculty of Military Sciences, Netherlands Defence Academy. He obtained his Master of Science degree in Law and

Economics from Utrecht University, and his Ph.D. in Business Administration at the Erasmus University in 2002, both cum laude.

Commander (RNLN) Dr. M.D. Fink, LL.M. is Section Head of the Administrative Law Branch, Netherlands Ministry of Defence, and Senior Researcher Law of Armed Conflict and Peace Operations (ACIL/LACPO) at the Faculty of Law, University of Amsterdam. He holds a law degree from Leiden University and obtained his Ph.D. on the Legal Framework for Maritime Interception Operations at the University of Amsterdam in 2016.

Dr. F. Giumelli is Assistant Professor in International Relations, Department of International Relations and International Organization (IRIO), University of Groningen. He obtained a master degree in International Relations at Suffolk University, Boston (USA) and a Ph.D. in Political Science at the University of Florence in 2009.

Lieutenant J. van Haaster, LL.M. is a Ph.D. researcher for Cyber Warfare at the Department of War Studies, Netherlands Defence Academy and the Amsterdam Centre of International Law (ACIL).

Colonel R.H. van Harskamp is a Lecturer at the Department of War Studies, Netherlands Defence Academy. He commanded 42 Mechanized Infantry Battalion during deployment in Iraq (2004), and Task Force Uruzgan in Afghanistan (2008).

Dr. Berma Klein Goldewijk is Assistant Professor of International Security Studies at the Department of War Studies, Netherlands Defence Academy. She obtained a master degree (1984) and a Ph.D. degree (1991) on contextual methodology in international religious studies from Radboud University Nijmegen, both cum laude.

C.I.A. van Kuijck, LL.M. MA is a Researcher of non-kinetic influence at the Royal Netherlands Air Force's Innovation centre AIR, co-located at the Faculty of Military Sciences, Netherlands Defence Academy. She holds master degrees in International Law and Criminal Law from Tilburg University and a master degree in Military Strategic Studies from the Netherlands Defence Academy.

J.M.M. van Lieshout, MSc is Assistant Professor of Defence Economics, Department of Military Management Studies at the Faculty of Military Sciences, Netherlands Defence Academy. He obtained a Master degree in Social Sciences at Tilburg University.

B. Lubbers, MSc is Assistant Professor Operational Processes for Navigation Technology at the Department of Military Technical Studies, Netherlands Defence Academy. His research focusses on navigation integrity and navigation warfare.

Prof. Dr. P.J. Oonincx is Professor for Navigation Technology, Department of Military Technical Studies and Acting Dean of the Faculty Military Sciences, Netherlands Defence Academy. He obtained his Ph.D. degree from the University of Amsterdam in 2000 on mathematical signal processing for physics.

Dr. S.J.H. Rietjens, an engineer by training, is Associate Professor of Information Management, Department of Military Management Studies at the Faculty of Military Sciences, Netherlands Defence Academy. He obtained his master degree and Ph.D. on civil-military interaction during peace support operations from Twente University in 2006. His research focusses on effectiveness of (intelligence) operations.

Dr. M.G.D. Rothman is Associate Professor of International Security Studies at the Department of War Studies, Netherlands Defence Academy. He defended his Ph.D. in 2004 at Purdue University (IN, USA) on globalisation as a context for justice.

Prof. Dr. Eric Theunissen is Professor of Avionics at the Department of Military Technical Studies, Netherlands Defence Academy.

Colonel (RNLAF) Dr. J.E.D. Voetelink, LL.M. is Associate Professor of Military Law at the Department of War Studies, Netherlands Defence Academy and a senior researcher at the Amsterdam Centre of International Law (ACIL). He holds a degree in law from Tilburg University. He defended his Ph.D. thesis Status of Forces Agreements as an Impetus for Military Operational Law at the University of Amsterdam.

Dr. E.J. de Waard is Associate Professor of Strategic Organizing and Change, Department of Military Management Studies at the Faculty of Military Sciences, Netherlands Defence Academy. He holds a Master of Science degree from Radboud University Nijmegen and obtained his Ph.D. at the Erasmus University Rotterdam in 2010.

Captain (RNLN) N.A. Woudstra, LL.M. is Associate Professor of Maritime Strategy and Operations at the Department of War Studies, Netherlands Defence Academy. He holds a Master degree of Law and International Relations from the Dutch Open University.

Chapter 1
The Promise, Practice and Challenges of Non-Kinetic Instruments of Power

Frans P.B. Osinga

Contents

1.1 Rediscovering a Neglected Instrument

'The realm of strategy', to quote Lawrence Freedman, 'is one of bargaining and persuasion as well as threats and pressure, psychological as well as physical effects, and words as well as deeds. This is why strategy is the central political art. It is about getting more out of a situation than the starting balance of power would suggest. It is the art of creating power.'[1] Strategy, in other words, centres on achieving influence, the key theme of this book. Freedman's description matters because usually strategy is equated with the application of military force, battles, destruction and lives lost, in short, with warfare. This betrays a military bias and narrow perspective on strategy; strategy, as Clausewitz keeps reminding us of, is

[1] Freedman 2013, p. xii.

F.P.B. Osinga (✉)
Faculty of Military Sciences, Netherlands Defence Academy, Breda, The Netherlands
e-mail: FPB.Osinga@mindef.nl

P.A.L. Ducheine and F.P.B. Osinga (eds.), *Netherlands Annual Review of Military Studies 2017*, NL ARMS, DOI 10.1007/978-94-6265-189-0_1

governed by and set within *Politik*. Proper strategy includes non-military instruments of power. Indeed, diplomacy, and mediation and negotiation efforts are successful most of the time precisely because influence is created without using destructive violence. Over the past decade, and indeed the past few years, political events and the emergence of new military and political instruments combine to suggest that producing effects that achieve tactical, operational level, strategic or even political objectives *without* the need to resort for military force is now increasingly feasible.

In the diplomatic arena sanctions for instance, the subject of Giumelli's chapter, have become progressively more sophisticated. Following ineffective comprehensive sanctions implemented by the UN, targeted sanctions have become the defining feature since the end of the Cold War.[2] Financial warfare, waged by the US Treasury against organized criminal syndicates, networks of rogue regimes, and terrorist groups, has proven to be another effective new tool to constrict funding flows.[3] In the military arena new non-kinetic doctrines and operational concepts have seen the light of day which have been inspired and necessitated by Western experiences in Iraq, Afghanistan and, more recently, Syria and by Israeli experiences in its struggle against Hezbollah and Hamas. He US Counter-Insurgency manual of 2006 and the UK Stabilization Operations manual refer to non-kinetic approaches such as media operations, information operations, 'winning the hearts and minds', strategic communications, the use of strategic narratives, human terrain analysis, key-leader engagements and social network analyses.

Moreover, over the past decade Western militaries have embraced the idea that most security problems require a multi-pronged approach involving a concerted variety of military and non-military influence activities. Long lasting structural societal, economic and political change in fragile states can only be achieved when employing such a 'Comprehensive Approach', which integrates economic policies, financial aid, the construction of infrastructure, establishing a judicial system, and the creation of democratic institutions, as Brocades Zaalberg discusses in his chapter. NATO too has embraced these notions; the importance it now attaches to cyberwarfare, strategic communications and strategic narratives, for instance, is not only manifest in official summit communiques but also in the creation of two dedicated so called 'Centres of Excellence' that focus on those themes.

All attest to the importance of using non-destructive methods to influence target audiences and thereby engineer behavioural change among individuals and groups, and to guard that flank. Indeed, in US doctrine this awareness permeates the re-defined notion of combat power which is construed as the total means of destructive, constructive, and information capabilities that a military unit or formation can apply at a given time.[4] This reflects the fact that modern-day conflicts

---

[2] In addition to Chap. 6 in this volume by Francesco Giumelli, see for a recent excellent overview Biersteker et al. 2016.

[3] Zarate 2013.

[4] US Army 2013, p. 1-1.

can no longer be geographically defined nor can military success be measured merely in terms of terrain seized or the physical defeat of an opponent. Combat in turn involves more than physical engagement of an opponent; current and future conflicts require a broader take on the concept of combat in which both kinetic and non-kinetic approaches are integrally applied. The notion of 'influence' and information operations are now considered key combat capabilities; they integrate designated information-related capabilities in order to synchronize themes, messages, and actions with operations to inform United States and global audiences, influence foreign audiences, and affect adversary and enemy decision making, according to the US Army doctrine manual 3-13 titled 'Inform and Influence Operations'. In more abstract terms, what these influence activities seek to achieve is behavioural change.

While the doctrines are of recent date, this aspiration and the awareness of the importance of non-kinetic approaches in warfare are far from new of course. Winning without killing, the holy grail of strategists, was, to paraphrase Sun Tzu, always the epitome of strategic skills. Nor is the notion that information and decision processes are key variables in warfare novel. Clausewitz harped on the problem of friction for the commander. Taking his inspiration partly from Sun Tzu, Liddell Hart argued for the need to think in terms of paralyzing the opponent, creating mental and moral effects and not merely to aim for physical attrition in battle. Such insights also undergirded the so called Manoeuvrist Approach which has become the philosophical foundation of most Western doctrines since the 1980s. The Manoeuvrist Approach embraces the idea that commanders should aim to affect not just the physical domain and plan for physical destruction but also, and in particular, home in on the moral and mental domain of the enemy system, an insight entirely in line with the ancient teachings of Sun Tzu by the way. Manoeuvre and fires should be planned based on the cognitive effects one wanted to create. As one of the founding fathers of this doctrine and of the famous OODA loop—John Boyd—taught his audience, fear, distrust, uncertainty, information overload, ambiguity, rapid changes in the environment, informational isolation, all such effects were keys to paralyzing the enemy's decision making capabilities.[5] During the Cold War, with a glance to deterrence theory, one might argue that behavioural influence through mere threats of violence has long underpinned Western security policy. And, as Baudet argues in this volume, as an integral part of that policy, the West was quite actively engaging in propaganda activities.

After the Cold War, Desert Storm reminded analysts of the potency of considering information a key contested domain. Desert Storm, considered the first information war by some, coined the concept of Command and Control Warfare in recognition of the fact that the victory was also the result of a myriad of informational and kinetic actions that paralyzed Saddam Hussain's ability to command and control his armed forces.[6] And not long after that operation the seminal article

[5] Osinga 2006.

[6] Campen 1992.

'Cyberwar is Coming' was published, which reflected the idea that information and the cognitive domain were becoming perhaps the most decisive elements in war.[7] Knowledge had become a key weapon and target in war, both at the societal level and at the battlefield. As Gray noted, 'As a weapon, as a myth, as a metaphor, or as a force multiplier, as an edge, as a trope, as a factor, as an asset, information (and its handmaidens—computers to process it, multimedia to spread it, systems to represent it) has become the central sign of post-modernity. In war, information has always been important. Now it is the single most significant military factor',[8] not only for controlling the battlespace, controlling information at home is just as important.[9] The element of representation—war as spectacle—inspired Michael Ignatieff to comment on NATO's war in Kosovo—'a virtual war'—and the way such 'humanitarian' wars have been experienced by the Western publics—'spectator sport warfare'.[10] Joseph Nye subsequently concluded that the outcome of conflicts are shaped not merely by whose army wins but also by whose story wins.[11] As McMaster, then a colonel, currently the director of the US National Security Council, observed, 'Although combat in Afghanistan and Iraq continues to require the defeat of the enemy on physical battlegrounds, U.S. commanders have discovered that lasting success over terrorist and insurgent groups requires winning on the battleground of perception … Ideas are weapons in the information age.'[12] These insights have also gained currency in military thinking, for instance in the COIN-doctrine of NATO: 'Paradoxically, the reality is that what is believed is more important than what is true'.[13]

Indeed, power, according to several authors, not just military power, had to be redefined; it had now become the most important element of power because it is the most transferable. Groups, organizations, nation-states and even individuals can now influence policy at the systemic level by using information. Due to the explosion in technology, the power paradigm has vastly changed.[14] Real-time 24/7 news coverage of human suffering in Europe and Africa started to inspire local, regional and even global activists and NGOs and became a real factor behind launching humanitarian interventions. Part and parcel of this information revolution was the spread of Western ideas and values such as rule of law, individualism, democracy, freedom of speech, human rights, etc. into formerly communist countries. Accompanied by the introduction of capitalism and consumerism, as

---

[7] Arquilla and Ronfeldt 1993, 1997.

[8] Gray 1997.

[9] Gray 1997, pp. 38–40.

[10] Ignatieff 2000.

[11] Nye 2011, p. 19.

[12] McMaster 2009, pp. xi–xii.

[13] North Atlantic Treaty Organisation 2009.

[14] Armistead 2004, p. 13. Similarly see Arquilla and Ronfeldt 1997; Castells 2000; Toffler 1995; Mann 2012.

Rothman argues in his chapter, such soft power has proven to be quite a threatening force for dictatorial regimes.

Despite this renewed attention and novel tools, exerting influence effectively in practice using just non-kinetic instruments has proven very difficult for the West. This should not be a surprise, for even with the destructive application of military force, achieving desired political outcomes is certainly not guaranteed. Factors underlying that fact are similar for non-kinetic operations, but at least four problems seem to affect non-kinetic operations more than traditional kinetic ones: (1) asymmetry vis a vis savvy unrestrained opponents who beat the West in this game; (2) flaws in the strategic logic; (3) organizational, bureaucratic and political obstacles; (4) fundamental shortfalls in our understanding of affecting behavioural change.

1.2 When the Weak Become the Strong

The first problem is that, as many scholars have concluded, in particular authoritarian regimes such as Russia and violent non-state actors such as the Taliban, Hezbollah and ISIS have been successful in creating influence through non-kinetic means. One reason for this is the fact that they do not feel bound by international law. China for instance, as Woudstra explains in his chapter, employs a strategy of mere physical presence by creating a string of artificial islands in the South Chinese Sea and thus creating facts on the ground. Prohibited by international law this nevertheless serves to create a deterrence effect towards the US maritime power projection capabilities in the Pacific. Similarly, Brinkel and Bouwmeester in their respective chapter describe the Russian application of *maskirovka* in the Ukraine, the use of troll armies to create fake news and its use of twitter as a weapon. Allegations of Russian meddling the US presidential elections and ambitions to influence elections in Germany, France and the UK, reports of Russian funding of right-wing extremist political parties in Europe, cyberattacks on Eastern European infrastructure and German Bundestag as some other worrying recent examples of non-kinetic influence activities deep in Westerns societies that have come to be labelled as Hybrid Warfare.[15] Interestingly, Ducheine, van Haaster and van Harskamp illustrate the potential of cyber operations to create disruptive effects by describing not Western cases but two Russian successes.

Indeed, in particular the virtual arena has turned into a wonderful gift for authoritarian regimes and terrorist groups and into a vulnerable flank for Western militaries.[16] For Western militaries the reality since 1991, when the CNN effect

---

[15] Diamond 2016; Polyakova et al. 2016; Giles 2016; Munich Security Conference Report 2017; Meister and Puglierin 2015; Naja 2015, p. 2; Ambrosio 2016; Harding 2014; NATO Parliamentary Assembly 2015; Foster and Holehouse 2016.

[16] Betz 2008.

emerged, has been that the deep presence of media during military operations created an unprecedented level of transparency and hence accountability for the Western military. With each tactical incident immediately broadcast around the world often resulting in negative strategic spin-off, the perception within Western militaries of the CNN effect was that it resulted in more scrutiny and direct involvement from the political to the tactical level of operations, or as they called it, 'the 6000 mile screwdriver'. Both at the tactical and the strategic level, the presence of media resulted in an ever-energized legitimization process of military actions. Indeed, the concept 'lawfare', described by Voetelink, has been coined to signify the deliberate actions by opponents of the West to exploit and manipulate perceptions concerning legitimacy by spreading images—fake or real- that suggest collateral damage and civilian casualties due to Western military actions. One author introduced the term 'imagefare' to signify this trends where the virtual arena is more decisive than physical facts on the battlefield.[17] And in this arena, the weak become the strong.[18]

This trend, according to Lawrence Freedman in 2006, constituted nothing less than a transformation in strategic affairs.[19] Social media allows authoritarian regimes and violent non-state actors to create and disseminate a coherent strategic narrative for both internal recruitment purposes as well as for exploiting the Western CNN/YouTube effect. As Mike Vlahos notes 'War narrative does three essential things. First, it is the organizing framework for policy. Policy cannot exist without an interlocking foundation of 'truths' that people easily accept because they appear to be self-evident and undeniable. Second, this 'story' works as a framework precisely because it represents just such an existential vision. The 'truths' that it asserts are culturally impossible to disassemble or even criticize. Third, having presented a war logic that is beyond dispute, the narrative then serves practically as the anointed rhetorical handbook for how the war is to be argued and described'. Strategic narratives do not arise spontaneously but are deliberately constructed or reinforced out of the ideas and thoughts that are already current. They express a sense of identity and belonging and communicate a sense of cause, purpose and mission. The proliferation of social media allows terrorist groups, as Phares observed after 9/11, to portray their fight as a war of ideas, or even a 'cosmic war' and thus assign eschatological meaning to their historical cause.[20] Meanwhile, the innovative use of web-based media by groups such as Hezbollah and Hamas is well documented. Their sophisticated media campaign includes daily press releases, weekly and monthly magazines, video clips and full-length films. ISIS too displayed sophisticated skills in the virtual arena, which became evident for all to see in the horrific beheading video's that were distributed through social media.[21]

---

[17] Brooking and Singer 2016.

[18] McCauley 2015.

[19] Freedman 2006.

[20] Phares 2007; Juergensmeyer 2003, pp. 148–163.

[21] Among the abundance of studies on this, see for instance Farwell 2014; Ingram 2015.

The problem is that in the West narratives which are deliberately constructed by Government are almost immediately rejected for that reason whatever their inherent accuracy or falsity. The public is highly sensitized to 'spin', the media excels at revealing (and counter-spinning) it, and no narrative can long survive the perception that it is based, even in part, on a lie. Or, in the words of David Galula, 'The counterinsurgent is tied to his responsibilities and to his past, and for him, facts speak louder than words. He is judged on what he does, not on what he says. If he lies, cheats, exaggerates, and does not prove, he may achieve some temporary successes, but at the price of being discredited for good. And he cannot cheat much unless his political structures are monolithic, for the legitimate opposition in his own camp would soon disclose his every psychological manoeuver. For him, propaganda can be no more than a secondary weapon, valuable only if intended to inform and not to fool. A counterinsurgent can seldom cover bad or non-existent policy with propaganda'.[22] Insurgents on the other hand do not face such a dynamic: 'The insurgent, having no responsibility, is free to use every trick; if necessary, he can lie, cheat, exaggerate. He is not obliged to prove; he is judged by what he promises, not by what he does. Consequently, propaganda is a powerful weapon for him. With no positive policy but with good propaganda, the insurgent may still win.[23]

As such his propaganda differential is not new, the difference these days is that the arrival of social media and near real-time digital imagery means that the connection between the popular perception of the war and the physical battlefield is much more immediate and therefore volatile.[24] Indeed, social media makes for a huge advantage as research on civil wars, terrorism and authoritarian regimes suggests. Such actors enjoy the great advantage that through unrestrained media manipulation and inflammatory rhetoric they can construct political realities, create 'enemies of the state' and subsequently incite and legitimize violence.[25] Moreover, they can connect and reinforce these messages to harsh intimidation and outright violent repression. In their so-called night letter, the Taliban explained their cause—vanguards of the true Islamic faith, defenders against yet another Christian invasion of a Muslim country—reminding everyone of their duty to support them, and threatened those who would not and those suspected of supporting ISAF with severe punishment. ISIS published a manual—The Management of Savagery—and glossy magazines glorifying their jihad and the orgy of violence they unleashed upon the unlucky people living in the regions they conquered.[26] But above all, social media have played an essential role in the jihadists' operational strategy in Syria and Iraq, and beyond. Like Al Qaeda, ISIS use the Internet—YouTube and especially Twitter—for organization and planning, proselytizing and entertainment,

[22] Galula 2005, pp. 14–15.

[23] Galula 2005, pp. 14–15.

[24] Kurth Cronin 2006.

[25] Kaufman 2006; Jackson and Dexter 2014.

[26] Friis 2015.

to educate the believers and to spread the images of terror towards apostates and infidels in the Middle-East, Europe and the US.[27] As Osama bin Laden wrote in 2002: 'It is obvious that the media war in this century is one of the strongest methods; in fact, its ratio may reach 90 percent of the total preparation for the battles'.

1.3 Flaws in the Strategic Logic

Another problem in using non-kinetic tools successfully is a failing to understand the strategic dynamic of a crisis and the limits of non-kinetic tools. Comprehensive sanctions, while successful as a signalling instrument, often failed to produce tangible policy changes in the target regime simply because those sanctions failed to produce costs that hurt the regime. Instead the population was bearing the brunt of the effects of sanctions. Moreover, authoritarian regimes could exploit the sanctions to tighten control over the population by centralizing distribution of commodities that had become scarce due to the sanctions. Behind this failure was a failure to disaggregate the opponent's political and military system.[28] Only through a painful learning process and much intelligence efforts combined with vigorous scholarly analysis has the UN been able to develop specific sanctions that targeted specific individual and groups within the political, industrial and military elites.

A similar failure to understand the stakes involved, the objectives and value patterns of the leaders of the warring factions, the vicious dynamics of civil wars, and overestimation of the perception of one's own credibility and influence have all contributed to faltering efforts to stop the violence in civil wars through negotiations, deterrence and coercive diplomacy. During, for instance, the Balkan crises, the stakes involved and the costs were simply too high for parties to give into NATO demands, while the threats communicated by NATO were considered bearable or lacking credibility in light of the fragility of the Western coalition. The threat of force and the costs that may accrue from attacks may also be in doubt and defensive measures and counter-measures such as threatening with counter-strikes, hostage taking of peacekeepers may offer a straw to grasp at for the target regime.[29] As a result, ensuring the credibility of threats often requires an actual demonstration of the will to apply military force.

And sometimes there is simply not a strategic link between the political objective, the threat that is communicated and the military capabilities that are

27 Klausen 2015.

28 Kirshner 1997; Brooks 2002.

29 Byman and Waxman 1999, 2000; Posen 2000.

available to effectuate the threat. This applies to regimes that will await not only their political but also their physical end if they concede to demands of the deterrer. Threats of nuclear attack are not credible when the crisis does not concern vital interests or concern humanitarian values. The religious motivation of some types of terrorists makes deterrence an irrelevant or very limited tool. Countering radicalization, for instance through delegitimizing the terrorist group using its own narrative foundation (for instance its adherence to the Islamic faith) has proven to be very complex and largely frustrating.[30] This has been the experience of Israel vis-à-vis Hamas and Hezbollah. Deterrence can still be applied and non-kinetic instruments (such as delegitimization), publishing of lists of known leaders of such groups, raising physical barriers to mitigate the chance of success of terrorist attacks all have role to play in mitigating the risk of terrorism. Yet deterrence in such cases needed to be re-conceptualized as Wilner and Rid have demonstrated: deterrence-the threat of using force to deter certain types of behaviour—is aimed to restrict certain types of violence. Deterrence does not fail when one terrorist attacks occurs. Like fighting crime, deterrence succeeds if it limits the risks of terrorism to an extent society can bear and function normally. It involves a communication process in which norms of unacceptable behaviour are established, which must occasionally be enforced through arrests or kinetic operations such as targeted killings.[31]

Such a strategic disconnect—or at least a lack of knowledge concerning the strategic logic, is also part of the current debate on cyber weapons. In their chapter, Ducheine, van Haaster and van Harskamp offer timely suggestions on how to employ cyber weapons. Those are valuable for cyber warfare, thus far, while a major concern for Western societies is still surrounded by questions to what extent cyber weapons can be harnessed by the West in an instrumental fashion. There is no shortage of incidents of cyber vandalism, hacking, violations of privacy, data-theft, disruption of governmental and industrial processes and espionage, and those certainly need attention and defensive countermeasures. The allegations of Russian cyber activities to influence the US elections in 2016 are a case in point.[32] Such disruption of the integrity and normal functioning of Western societal institutions is a prominent risk identified by various Western intelligence and state security agencies.[33] Nevertheless, analysts struggle to decide whether cyber weapons can produce predictable and controllable results that can be translated into tangible military and political influence which in turn would suggest these weapons may serve as deterrent and military offensive instrument.[34]

[30] For an argument to at least make the effort, see Long and Wilner 2014.

[31] Wilner 2011, 2013; Adamsky 2017; Rid 2012.

[32] For an extensive reconstruction of these activities, see Osnos et al. 2017.

[33] See for instance National Cyber Security Centre 2016.

[34] Rid 2011; Libicki 2009.

1.4 Organizational, Bureaucratic and Political Obstacles

A third problem is of a practical nature. In the fight against terrorist groups and their leaders in Iraq and Afghanistan US special forces, the CIA and various other governmental intelligence agencies learned the benefits of seamless cooperation and sharing intelligence. This facilitated rapid fusion and analysis of various intelligence sources, which enabled quick exploitation of that intelligence in the form of raids and shadow operations, which in turn fuelled the process for subsequent counter-terrorist operations. This proved pivotal for success against fleeting targets such as individual leaders, ideologues, bomb-making experts and financiers. Setting up this process and structure, which has obtained the acronym F3EAD (for Find, Fix, Finish, Exploit, Analyse and Disseminate) required dedicated and high-energy efforts of SOF commanders, who had to convince the other intelligence organizations, military and non-military, of the merits of this model.[35] Interorganizational cooperation does not come naturally.

Indeed, organizational issues, apart from the enormity of the challenges of state building,[36] has often affected the implementation of the Comprehensive Approach, as Brocades Zaalberg observes. While in principle most experts lauded the military efforts to engage NGOs and governmental departments within their country to help the state building effort in Afghanistan, such organizations often did not consider it a high priority, were reluctant to cooperate and coordinate closely with the military, or were simply not organized and resourced for such lengthy expeditionary operations. In addition, various organizations had different perspectives concerning priorities specific problems should be addressed. Different planning and operating procedures too affected cooperation.[37] Similarly, strategic communications and other information operations, despite concerted efforts by several NATO countries during the ISAF mission in Afghanistan, have produced little tangible effects. Strong counter-messaging by the Taliban was one factor but another cause was the fact that each participating country defined the motives and objectives in its communications differently. There was no coherent message emanating from Brussels or Kabul.[38] These problems were often compounded by the fact that in most Western militaries media operations and strategic communications sections without the headquarters were under-resourced, and not receiving quite the high level attention these functions required. Similar organizational issues plagued the UN mission in Mali. Even with a proper intelligence operations, as Rietjens, de Waard and van Fenema in their analysis on the UN mission in Mali demonstrate, bureaucratic and political obstacles within the local UN organization—a reluctance or downright refusal to act upon the intelligence—may degrade the impact of intelligence such operations potentially might have.

---

[35] Lamb and Munsing 2011; Gomez 2011.

[36] See Paris and Sisk 2009.

[37] See de Coning and Friis 2011.

[38] Ringsmose and Børgesen 2011; Dimitriu 2011.

1.5 Limited Understanding of Engineering Behavioural Change

A final reason that is related to the previous one is precisely the topic of much study underlying recent doctrine manuals: understanding what drives human behaviour, or rather the lack thereof. All relevant recently published studies on military influence activities acknowledge that any success derives from a deep understanding of the drivers of behaviour of individuals, groups, and leaders in a particular operational environment. However, recognizing the importance of influence and the cognitive domain does not automatically answer the question how to influence behaviour. Despite the flurry of publications on the importance of influence-associated concepts, engineering behavioural change—those micro-foundations of strategy—has remained relatively underexplored. Most strategic theories dwell on it only implicitly or assign economic rationality to strategic actors. And that is problematic as the following brief and incomplete discussion suggests.

1.5.1 The Limits of Economic Rationality

Research has clearly demonstrated that the economic rationality assumption underlying most strategic theories and theories of behaviour is deeply flawed.[39] The famous but simple rational actor model explains behaviour by suggesting that such a decision revolves around maximizing value in choosing among options, each with clearly perceived costs and benefits for the decision maker. Rational choice theory traditionally assumes that individuals make behavioural decisions based on a calculation of the expected costs and benefits of behaviour. Underlying this model are several assumptions, three of which are of particular interest in military settings as they directly suggest that limited utility of this model in operational environments in which militaries engage with different cultures: the target audience has a stable set of preferences (which can be known to the military planner); the target audience have access to all the relevant information bearing on the decision (they know all the available options, know the benefits and costs); and they are fully able to process this information in order to reach the optimal (utility maximising) decision. In reality this is never the case and the concept of 'bounded rationality' explains how, even when individuals are pursuing utility, their decision making processes are 'bounded' by psychological and environmental constraints. Thus personal abilities and situational factors (including how choices are presented, and also the context in which the decision is made, for instance under time pressure) limit people's capacity for deliberation.

[39] This draws from an excellent wide-ranging overview of models and theories by Darnton 2008, pp. 5–9; as well as Larson et al. 2009, pp. 11–17.

1.5.2 Emotion: Fear, Uncertainty, Honour, Habits, Values

Work in the fields of evolutionary psychology, cognition, and group dynamics adds that significant departures from the rational actor model can be caused by the fundamental structure of the brain, brain chemistry (the influence of a high base testosterone level), impairment of the brain (due to damage or drugs), misperception, real-world treatment of risk, and dysfunctions in group decision making (notably group-think). As Payne recently commented in the context of deterrence theory, high testosterone levels in men may induce risk taking behaviour in a competitive high stakes environment.[40]

Culture and the social environment too can have a direct impact on perception and thus behaviour. The social environment includes the norms, values, codes of conduct, habits, routines, identity, one's role in a social environment, etc. Whether an individual is willing to display a specific behaviour depends in part on the dominant attitudes and norms, the relative importance of attitudes and norms, beliefs about what others may think, and beliefs about the risk it poses to one's position vis a vis others.[41] Religion or ideology for instance may result in mistaken threat perceptions, leading states to perceive dangers where none exist or ignore threats with an objective reality. Significant ideological differences between two states can result in miscommunication, both in words and in actions (force deployments and exercises, for example) intended to convey intent. Ideology may also cause a state to fall victim to 'mirror-imaging' and assume that an adversary's behaviour is as predictable as its own, because the adversary is motivated by the same or similar logic, values, and objectives.[42] Religious beliefs, peer pressures, peer respect and lack of prospects are also factors involved in radicalizing young men and preparing them for suicide attacks. In such an emotionally charged context, non-kinetic deradicalization efforts are bound to be ineffective.

1.5.3 Risk Propensity

Risk framing, and risk acceptance/avoidance in particular has gained much interest. Prospect theory (the study of how decisions under uncertainty actually are made, rather than, as with expected utility theory, how they should be made) indicates that leaders may deal with risk in ways at odds with the rational actor model. Leaders may take more risks to maintain their international positions, reputations, and domestic political support than they do to enhance those positions. Leaders also tend not to accommodate to losses (in territory, reputation, and domestic political

[40] Payne 2011, p. 402.

[41] This is a brief summary of a series of Expectancy Value related models that add the impact of the social environment to economic models of behaviour. See Darnton 2008, pp. 11–26.

[42] Payne 2011, pp. 404–405.

support), but to take excessive risks to recover them. In short, risks, honour, history, framing of situations in terms of losses or gains, all affect the weighing of options.[43]

1.5.4 Short Cuts, Biases and Heuristics

Indeed, although we think we do, we hardly behave in accordance with rational thought and conscious decision processes. In practice, human behaviour is much more based on automatic and rule based responses, and influenced by emotions more than we realise. Fear often invites automatics responses. Stress due to threat and time pressure affects the cognitive functions dramatically by limiting the number of options under consideration, by reasoning by stereotyping and analogy, and by inducing a reduction in cue awareness for incoming information combined with an aversion to ambiguity. Moreover, in order to efficiently make decisions, people rely on past experiences for filtering and interpreting incoming information and coming to a decision. Incoming information is often also filtered based on whether is confirms existing perceptions, stereotypes and beliefs. This is what these is called the 'echo-chamber of social media'; people monitor websites and social media that reflect their own preconceptions and worldviews, preventing nuanced, alternative or contradicting messages and evidence to ever reach and potentially influence them. Such cognitive dynamics result in a fascinating set of biases and heuristics.[44] Combined these too explain decisions that seem counter to the expectations of the rational actor model and makes any prediction of future behaviour based on influence activities suspect.

1.6 Progress, Caution, and Necessity

These four factors in no way diminish the value of exploring and exploiting non-kinetic instruments. A variety of surveys have benefited doctrine development in this regard. A recent extensive overview of theories and models concerning individual and group behaviour states that effective influence operations include the following elements: (1) they are aimed at achieving specific desired objectives and effects, typically a change in a key attitude, belief, preferred policy, or behaviour; (2) They are directed toward key target audiences, whether an individual, a decision making group, a military unit, a population subgroup, or the mass public of a nation; (3) They are mindful of audience characteristics, including pre-existing attitudes and beliefs that may condition an audience's willingness to be influenced;

[43] Levy 1997. See also his treatment of decision making theories in Levy and Thompson 2010, Chapters 3–7.

[44] See Levy and Thompson 2010, pp. 133–150. See also Jervis et al. 1985.

(4) They rely upon messages with compelling message characteristics, i.e., those whose content, format, cognitive and emotional appeal, and other characteristics will most resonate with the audience.[45]

Based on such research, the process of Target Audience Analysis, for instance, has been developed and refined, as described by van Kuijt. Such a micro-marketing analysis can highlight stories, narratives, themes, and messages that strike a responsive chord with the intended audience. Each aspect of the analysis is articulated in language that resonates with the audience and respects religious, cultural, psychological, linguistic, and psychological factors that affect the audience and shape not only their behaviour but also their attitudes. Messages designed to shape behaviour should appeal to emotions and be rooted in values, which are critical to human decision-making and behaviour.[46] Such insights, that inform US Army Manual FM 3-13[47] and borrow from mass marketing and political campaigning, are certainly useful, as Beerens et al. demonstrate in their analysis of influence activities by groups like ISIS.[48]

Still, despite this progress, some caution is warranted. Countering the narratives of religious terrorist groups for instance—counter-marketing—has become a lively subject of study. Some see potential, but also practical and methodological obstacles. For a start, systematic, theory-based guidelines for the execution of such strategies are largely absent from the literature.[49] As Jackson has argued, while marketing is useful as an analytical tool, it does not easily translate into an influence instrument in conflict situations. Not only is the product 'we' try to sell not a commercial commodity, Western narratives are simply not in demand. Moreover, the marketing model suggests that the efficacy of information operations will depend not simply on the message and its delivery (promotion) but on the behaviour the sender seeks to induce (the product), the costs of that behaviour (the price), and the opportunities available for such behaviour (the placement). When the cultural differences between the sender and the receiver are vast, and the desired behaviour potentially costly for the receiver and alien to his social environment, the well-intended marketing effort is likely to fail.[50]

Moreover, while there is an abundance of theories and models that seek to explain individual-level attitudes, persuasive communications, and behavioural change, and although several appear to be relatively generalized and empirically robust, even the best of these have relatively modest explanatory power. As a RAND study concluded cautiously, planners should be mindful of the practical implications that result from the heterogeneity of the scholarly literature, the absence of a larger meta-theory of influence, and the situational or context

---

[45] Larson et al. 2009, p. xix. See also Tatham and Giles 2015.

[46] Farwell and Arakelian 2016, p. 74.

[47] US Army 2013.

[48] See also Helmus et al. 2007.

[49] Braddock and Horgan 2016, p. 400.

[50] Jackson 2016.

dependence of persuasion efforts: The various theories provide only starting points for planners and operators; and even after detailed analysis, vast uncertainties are likely to remain regarding the efficacy of various alternative approaches to communication of messages to target audiences. And, as it states: 'the required level of intellectual and analytic effort for such endeavours can be substantial. The effort may in many cases not only outstrip the capabilities of planning staffs but also may require more intensive inputs and effort than their results perhaps merit'.[51]

Yet precisely because of such methodological obstacles, theoretical voids, organizational issues and real world strategic problems vis-à-vis terrorist groups and authoritarian regimes, there is a requirement to continue exploring the challenges of effectively employing non-kinetic instruments and trying to win without killing. More empirical studies and theory development and testing are required, at least from a military perspective, if for no other reason than to understand the dynamics of the virtual contest that is currently ongoing within and against Western liberal societies.

References

Adamsky D (2017) From Israel with Deterrence: Strategic Culture, Intra-war Coercion and Brute Force. Security Studies 26(1):157–184

Ambrosio T (2016) The rhetoric of irredentism: The Russian Federation's perception management campaign and the annexation of Crimea. Small Wars & Insurgencies 27(3):467–490

Armistead L (2004) Information Operations, Warfare and the Hard Reality of Soft Power. Brassey's, Washington DC

Arquilla J, Ronfeldt D (1993) Cyberwar is coming! Comparative Strategy 12(2):141–165

Arquilla J, Ronfeldt D (eds) (1997) In Athena's Camp - Preparing for Conflict in the Information Age. RAND, Santa Monica

Betz D (2008) The virtual dimension of contemporary insurgency and counterinsurgency. Small Wars and Insurgencies 19(4):513–543

Biersteker T, Eckert S, Tourinho M (2016) Targeted Sanctions - The Impact and Effectiveness of United Nations Action. Cambridge University Press, Cambridge

Braddock K, Horgan J (2016) Towards a Guide for Constructing and Disseminating Counternarratives to Reduce Support for Terrorism. Studies in Conflict & Terrorism 39 (5):381–404

Brooking ET, Singer PW (2016) War Goes Viral, How social media is being weaponized across the world. The Atlantic (Nov 2016) http://www.theatlantic.com/magazine/archive/2016/11/war-goes-viral/501125/ Accessed 1 March 2017

Brooks R (2002) Sanctions and Regime Type: What Works, and When? Security Studies 11(4):1–50

Byman D, Waxman M (1999) Defeating US Coercion. Survival 41(2):107–120

Byman D, Waxman M (2000) Kosovo and the Great Air Power Debate. International Security 24 (4):5–38

Campen AD (ed) (1992) The First Information War: The Story of Communications, Computers and Intelligence Systems. AFCEA International Press, Fairfax, VA

Castells M (2000) The Rise of the Network Society - The Information Age: Economy, Society and Culture. Volume 1, 2nd edn. Blackwell, Malden

[51] Larson et al. 2009, p. xx.

Cronin AK (2006) Cyber-mobilization: The New Levee en Masse. Parameters 36(2):77–87
Darnton A (2008) GSR Behaviour Change Knowledge Review. University of Westminster, http://www.gsr.gov.uk. Accessed 23 July 2008
de Coning C, Friis K (2011) Coherence and Coordination - The Limits of the Comprehensive Approach. Journal of International Peacekeeping 15: 243–272
Diamond L (2016) Russia and the Threat to Liberal Democracy. The Atlantic (9 Dec 2016) http://www.theatlantic.com/international/archive/2016/12/russia-liberal-democracy/510011/ Accessed 1 March 2017
Dimitriu GR (2011) Winning the story war: Strategic communication and the conflict in Afghanistan. Public Relations Review 38:195–207
Farwell JP (2014) The Media Strategy of ISIS. Survival 56(6):49–55
Farwell JP, Arakelian DJ (2016) Using Information in Contemporary War. Parameters 46(3):71–86
Foster P, Holehouse M (2016) Russia accused of clandestine funding of European parties as US conducts major review of Vladimir Putin's strategy. The Telegraph (16 January 2016)
Freedman L (2006) The Transformation of Strategic Affairs. Adelphi Paper no. 379 (March). IISS, Oxford University Press, Oxford
Freedman L (2013) Strategy - A History. Oxford University Press, Oxford
Friis SM (2015) Beyond anything we have ever seen: beheading videos and the visibility of violence in the war against ISIS. International Affairs 91(4):725–746
Galula D (2005) Counterinsurgency Warfare: Theory and Practice. Praeger, New York
Giles K (2016) Russia's 'New' Tools for Confronting the West; Continuity and Innovation in Moscow's Exercise of Power. Chatham House, RIIA, London
Gomez J (2011) The Targeting Process: D3A and F3EAD. Small Wars Journal, 16 July 2011
Gray CH (1997) Postmodern War, the New Politics of Conflict. Routledge, London
Harding L (2014) We Should Beware Russia's Links with Europe's Right. The Guardian (8 December 2014)
Helmus T, Paul C, Glenn R (2007) Enlisting Madison Avenue, the marketing approach to earning popular support in theaters of operation. RAND, Santa Monica
Ignatieff M (2000) Virtual War: Kosovo and Beyond. Penguin, London
Ingram HJ (2015) The strategic logic of Islamic State information operations. Australian Journal of International Affairs. 69(6):729–752
Jackson CF (2016) Information Is Not a Weapons System. Journal of Strategic Studies 39(5–6):820–846
Jackson R, Dexter H (2014) The Social Construction of Organized Political Violence: An Analytical Framework. Civil Wars 16(1):1–23
Jervis R, Lebow RN, Stein JG (eds) (1985) Psychology of Deterrence. Johns Hopkins University Press, Baltimore
Juergensmeyer M (2003) Terror in the Name of God. 3rd edn. University of California Press, Berkeley
Kaufman S (2006) Symbolic Politics or Rational Choice? Testing Theories of Extreme Ethnic Violence. International Security 30(4):45–86
Kirshner J (1997) The Microfoundations of Economic Sanctions. Security Studies 6(3):32–64
Klausen J (2015) Tweeting the Jihad: Social Media Networks of Western Foreign Fighters in Syria and Iraq. Studies in Conflict & Terrorism, 38(1):1–22
Lamb C, Munsing E (2011) Secret Weapon: High Value Target Teams as an Organizational Innovation. NDU Press, Washington DC
Larson EV, Darilek RE et al (2009) Foundations of Effective Influence Operations: A Framework for Enhancing Army Capabilities. RAND, Santa Monica
Levy J (1997) Prospect theory, rational choice, and international relations. International Studies Quarterly 41(March):87–112
Levy J, Thompson W (2010) Causes of War. Wiley-Blackwell, Oxford
Libicki M (2009) Cyberdeterrence and cyberwar. RAND, Santa Monica
Long JM, Wilner AS (2014) Delegitimizing al-Qaida: Defeating an 'Army Whose Men Love Death'. International Security 39(1):126–164

Mann M (2012) The Sources of Social Power: Volume 4, Globalizations, 1945-2011. Cambridge University Press, Cambridge

McCauley T (2015) The war of ideas on the Internet: An asymmetric conflict in which the strong become weak. Dynamics of Asymmetric Conflict 8(1):79–90

McMaster HR (2009) Foreword. In: David GJ, McKeldin TR (eds) Ideas as Weapons: Influence and Perception in Modern Warfare. Potomac Books, Washington DC

Meister S, Puglierin J (2015) Perception and Exploitation: Russia's Non-Military Influence in Europe. DGAP kompakt Nr.10 (October 2015) http://www.dgap.org/en/article/getFullPDF/27185 Accessed 1 March 2017

Munich Security Conference Report 2017, Post-Truth, Post-West, Post-Order? (February 2017)

Naja B (2015) Understanding propaganda and disinformation. At a glance, European Parliamentary Research Service (13 November 2015) http://www.epthinktank.eu/2015/12/01/russia-manipulation-of-information-and-the-eus-response/ Accessed 1 March 2017

NATO (2009) Allied joint doctrine for counterinsurgency (COIN) AJP 3.4.4 (Ratification Draft 1) Brussels

NATO Parliamentary Assembly (2015) Annual report. The battle for the hearts and minds: countering propaganda attacks against the Euro-Atlantic Community. http://www.nato-pa.int/default.asp?SHORTCUT=4012 Accessed 7 March 2017

National Cyber Security Centre (2016) Cyber Security Assessment Netherlands 2016. Ministry of Security and Justice, The Hague (The Netherlands) http://www.ncsc.nl/english/current-topics/Cyber+Security+Assessment+Netherlands/cyber-security-assessment-netherlands-2016.html Accessed 10 March 2017

Nye J (2011) The Future of Power. Public Affairs Books, New York

Osinga F (2006) Science, strategy and war, the strategic theory of John Boyd. Routledge, Abingdon

Osnos E, Remnick D, Yaffa J (2017) Trump, Putin, and the New Cold War. The New Yorker (7 March 2017)

Paris R, Sisk T (2009) The Dilemmas of Statebuilding. Routledge, Abingdon

Payne KB (2011) Understanding Deterrence. Comparative Strategy 30(5):393–427

Phares W (2007) The War of Ideas, Jihadism against Democracy. Palgrave McMillan, New York

Polyakova A et al (2016) The Kremlin's Trojan Horses. Atlantic Council, Washington DC (November) http://www.atlanticcouncil.org/publications/reports/kremlin-trojan-horses Accessed 1 March 2017

Posen B (2000) Serbia's Political Military-Strategy. International Security 24(4):39–84

Rid T (2011) Cyber War Will Not Take Place. Journal of Strategic Studies 35(1):5–32

Rid T (2012) Deterrence beyond the State: The Israeli Experience. Contemporary Security Policy 33(1):124–147

Ringsmose J, Børgesen BK (2011) Shaping public attitudes towards the deployment of military power: NATO, Afghanistan and the use of strategic narratives. European Security 20(4):505–528

Tatham S, Giles K (2015) Training Humans for the Human Domain. US Army War College Press, Carlisle, PA

Toffler A (1995) War and Anti-War. Warner Books, New York

US Army (2013) Field Manual FM 3-13 Inform and Influence Activities. Headquarters Department of the Army, Washington DC (January)

Wilner AS (2011) Deterring the Undeterrable: Coercion, Denial, and Delegitimization in Counterterrorism. Journal of Strategic Studies 34(1):3–37

Wilner A (2013) Fencing in Warfare: Threats, Punishment, and Intra-war Deterrence in Counterterrorism. Security Studies 22(4):740–772

Zarate J (2013) Treasury's War - The Unleashing of a New Era of Financial Warfare. Public Affairs, New York

Author Biography

Air-Commodore Professor Dr. F.P.B. Osinga is Professor of Military Operational Art and Sciences, Chair of the War Studies Department at the Netherlands Defence Academy (Faculty of Military Sciences), and a member of the Faculty's Managing Board. Trained at the Royal Military Academy, he joined the Royal Netherlands Air Force in 1987 as an F-16 pilot. He defended his Ph.D. in 2005 on the relationship between war, science and the strategic thinking of John Boyd.

Chapter 2
The Resilient Mind-Set and Deterrence

Theo Brinkel

> Je suis rentré chez moi à pied, traversant la moitié de Bruxelles, longeant le quartier des institutions européennes – cette forteresse lugubre, entourée de taudis. Le lendemain, je suis allé voir un imam à Zaventem. Et le surlendemain – le lundi de Pâques – en présence d'une dizaine de témoins, j'ai prononcé la formule rituelle de conversion à l'islam.
>
> (Michel Houellebecq, Soumission. Flammarion 2015, 257).

Abstract Finding themselves in an ideological competition with Russian nationalism and extreme Islamic jihadism, European societies need to mobilise their moral resources to provide themselves with the sense of purpose necessary to face down and neutralise the influence of current threats. The ambition of this chapter is to explore if and how Western societies can mentally arm themselves against such threats and whether the concept of resilience can play a role in this. Two options present themselves. The isolationist option aims at closing off society from outward threats and maintaining the status quo. The other option sees resilience as renewal. This one lends itself as a useful interpretative framework to further study of the resilience of the home front as a contribution to deterrence. It seeks to influence the international context in such a way that the core values of society are thought anew, modernised or made to work in a new context.

Keywords Deterrence · EU · Hybrid warfare · ISIS · Jihadism · NATO · Populism · Putin · Resilience · Trump · Values

Contents

T. Brinkel (✉)
Faculty of Military Sciences, Netherlands Defence Academy, Breda, The Netherlands
e-mail: TBFM.Brinkel@mindef.nl

P.A.L. Ducheine and F.P.B. Osinga (eds.), *Netherlands Annual Review of Military Studies 2017*, NL ARMS, DOI 10.1007/978-94-6265-189-0_2

2.1 Introduction

"We, the Heads of State and Government of the North Atlantic Alliance, stand united in our resolve to maintain and further develop our individual and collective capacity to resist any form of armed attack. In this context, we are today making a commitment to continue to enhance our resilience against the full spectrum of threats, including hybrid threats, from any direction. Resilience is an essential basis for credible deterrence and defence and effective fulfilment of the Alliance's core tasks."[1] Thus declared the heads of state and government of NATO at the Summit in Warsaw in 2016. They used a fairly new word in securityspeak: resilience.

Over the last few years, the word resilience has emerged in debates on security in the West in the context of ISIS attacks on European soil and of threats originating in Russia. The downing of civil airliner MH 17 in the summer of 2014 over the territory of Ukraine, with 298 people on board, among whom 196 Dutch nationals, shows how a regional conflict can have direct effects in countries that seemingly had nothing to do with it. Similarly, ISIS, operating in Iraq and Syria, is hurting Western societies, not only because of the terrorist attacks in European cities, but also because individuals in Western societies are attracted to the battle ISIS claims to be fighting. Furthermore, war in the Middle East and North Africa causes thousands of people to flee and apply for asylum in Europe, putting another kind of pressure on societies there.

In the meantime, public opinion in the West sailed its own course. In 2016 Dutch voters said no in a referendum on the question whether the Netherlands should ratify the Trade Agreement between the EU and Ukraine. A little later the United Kingdom voted to leave the European Union. And in the United States of America Donald Trump was elected as the new president on a platform of putting "America first". In these cases, the outcome seems to have been influenced by the sentiment among broad groups in society to seek security in turning away from international developments. It has been branded as a wave of populist politics.[2]

On a deeper level, mainstream Western leaders find themselves in a new ideological competition. On one side, there is populism, with its main example in Vladimir Putin, the president of the Russian Federation. On the other side, there is the amalgamation of jihadist groups and individuals, of which currently ISIS is the

[1] NATO 2016.

[2] There are numerous sources underlining this. One example may be *Foreign Affairs* November/December 2016.

main proponent.[3] And thirdly, there is the traditional Western world that claims to stand for values of freedom, democracy and the rule of law. Western governments find themselves steering between the Scylla of increased pressure from Russia and jihadism on the one hand and the Charybdis of the desire to keep increasingly isolationist voters satisfied. They find it hard to answer these ideological challenges with the need for unity and solidarity that the Atlantic Alliance and the European Union are supposed to uphold.

Many authors have tried to understand the currents at work in these developments. In his reaction to the terrorist attacks on 13 November 2015 in Paris, in the Dutch journal *Trouw* Philippe Bilger, former advocate general in France, said that we can't win the battle (against terrorism, *ThB*) with the technical solutions that are now constantly debated. I know for sure, he said, that we need to offer an alternative to a belief that commends death. We need something more solid, something that is useful in peacetime as well. We need to look for the conviction that our idea of living together is as strong as their sick ideas.[4] This is the very problem that this chapter wishes to explore.

The ambition is to find if and how Western societies can mentally arm themselves against such kinds of ideological threats and whether the concept of resilience can play a role in this. The provisional assumption is that a resilient society enhances overall mental deterrence against hybrid or terrorist attacks. In NATO parlance, deterrence is seen as the ability to discourage an opponent from taking an unwelcome action. This can be done through deterrence by punishment (i.e. the threat of retaliation) or through deterrence by denial (blocking the opponents ambitions). Especially in view of hybrid warfare, deterrence cannot be limited to military force only. As will be discussed below, hybrid warfare as employed by Russia involves a set of instruments, most of which in fact are non-military in nature.

Unquestionably, the presence of NATO's Very High Readiness Joint Task Force in countries under threat, such as the Baltic Republics, is essential to deterrence. But deterring hybrid war will subsequently also require non-kinetic means, such as increased resilience of cyber networks, diversification of energy supplies, and strategic communications. Such measures make up "deterrence-by-resilience"; together, they are meant to dissuade the opponent by demonstrating that his threat is futile.[5]

This idea is as follows: Finding themselves in an ideological competition, European societies need to mobilise their moral resources to provide the people with the sense of purpose necessary to face down and neutralise the effect and influence of current threats. This raises questions such as: How are these threats

---

[3] Several names are used to designate this terrorist group: Islamic State, Islamic State in Iraq and Syria, Islamic State in Iraq and the Levant. Daesh is an Arabic acronym that is used in France as well. For the sake of simplicity, this chapter uses the acronym ISIS. See amongst others: ICG 2014.

[4] Bilger 2015.

[5] NATO 2015.

relevant to our societies? And how can we deter them? Can the answer be found in the notion of resilience? How can resilience as a concept be used to enhance deterrence?

This chapter will first try to map the problem at hand and will attempt to indicate in what way Westernn societies have become vulnerable to ideological challenges resulting from hybrid warfare and terrorism that originate in other parts of the world. Next, this chapter will investigate if and how the values on which Western societies are based are losing attractiveness. Then, the emergence of the resilience debate will be described, as well as the way theoretical elements of this debate can be applied to current security issues. It should be noted here that this chapter does not deal with engineering aspects of resilience, e.g. the protection of critical infrastructure, cyber capabilities and other physical aspects of defence, but primarily with non-kinetic notions of morale, the mind-set of the society. The chapter will mainly have an agenda-setting character, signalling the importance of the subject and setting out markers for policy development. These will be presented in the concluding paragraphs of this contribution.

2.2 Home Front Under Threat

Resilience can be seen as a strategy for dealing with threats and shocks. This paragraph will look at threats that sometimes express themselves as shocks or disasters and that have been captured with the term hybrid warfare. Today, they mostly originate from along the Eastern and Southern borders of Europe. The term hybrid warfare—which will be dealt with extensively elsewhere in this volume—has been applied to the Russian aggression towards Ukraine in 2014 and the attacks of ISIS over the last three years in Paris, Brussels, Nice, Berlin or Istanbul. In these cases, the resilience of Western societies came under threat—indirectly as well as directly. In both cases the "lethality of state conflict" was mixed with "the fanatical and protracted fervour of irregular warfare".[6]

The most cited Russian source on hybrid warfare is an article written by general Valeri Gerasimov, Chief of Staff of the Russian armed forces. As is described in the chapter by Han Bouwmeester elsewhere in this volume, Gerasimov's article appeared in February 2013 under the title "The Value of Science in Prediction". His article contains an analysis of how the West operated in the last 25 years, but it also reveals the current thinking in the Russian armed forces. Gerasimov noted how non-military action was far more effective than the use of force, especially in the field of information, psychological warfare and communication.[7] Gerasimov himself uses the term non-linear warfare and stresses the importance of information warfare, which is aimed at influencing the morale of the society of the opponent.

[6] Lasonjarias and Larsen 2015, p. 3.
[7] Perry 2015, p. 6.

Hybrid warfare can be defined as "… the true combination and blending of various means of conflict, both regular and unconventional, dominating the physical and psychological battlefield with information and media control, using every possible means to reduce one's exposure".[8] Activities are aimed to mislead, cause confusion through psychological warfare, propaganda and media manipulation and cyber attacks. The purpose is to test the political will power and credibility of Western deterrence.[9] In hybrid warfare there is not always a clearly demarcated war theatre. Frequently irregular troops are involved and the battle is often waged among the population. These conflicts are also played out in the virtual domain and through information and disinformation in so-called social media.[10]

In the case of Russia, the opponent is a state. However, the Russian state uses non-state proxies, such as ethnic Russian nationalist insurgents.[11] In the case of Syria, ISIS has been defined as a terrorist organisation, although it currently has features of a state and defines itself as an Islamic State. One may also take the example of Hezbollah, which is a political party that takes part in Lebanese politics, a religious social movement, as well as an armed group that in the view of Western governments is considered a terrorist organisation. The distinction between state and non-state actors has thus become blurred.

2.3 Vulnerability of the West

The terrorist attacks in Europe or the downing of the MH 17 show that our own societies are directly affected. However, insecurity is not the only consequence of such acts. With each shock, the vulnerability of Western society is highlighted. Underneath the incidents, more structural developments can be discerned, such as a decline of social cohesiveness in Western societies. As a result of these phenomena, the morale of Western societies has become involved.

Many analysts have raised this issue. In his seminal work on long-term social changes, "Bowling Alone", Robert Putnam sees a connection between declining social capital and civic disengagement. When, he argues, social capital loses in quality we will be more and more inclined to free-rider behaviour and neglect our civic duties that make possible the functioning of democracy.[12] In the Netherlands Kim Putters argues that as a society, we are losing a shared commitment that is based upon shared prosperity and freedom, together with trust in political leaders who, and the institutions which, are supposed to make that happen.

---

[8] Lasonjarias and Larsen 2015, p. 3.

[9] Osinga 2016b, p. 18.

[10] Fjäder 2014, p. 117.

[11] Ruiz Palmer 2015, pp. 62–63.

[12] Putnam 2000, p. 347.

Putters' analysis is based on trends in Dutch society, but these developments can be seen in other Western countries as well. There are growing divisions between rich and poor, the highly educated and the less educated, young and old, different ethnic backgrounds and the stronger and the more vulnerable social groups. Especially among the poorer, less educated and older parts of the population, there is an increased feeling of insecurity, because of the economic and financial crisis, the influx of refugees and other immigrants, terrorist attacks, and insecurity in the labour market.[13] To this, Fareed Zakaria added the factors of the openness and therefore vulnerability of Western economies and societies to the global market and the resulting loss of traditional jobs as well as high public debts and as a result a limited room for manoeuvre for government spending.[14]

According to Cas Mudde, mainstream political parties appeared to be powerless in the face of two developments: the transfer of power to supranational entities such as the EU and the IMF as well as the transfer of power from elected politicians to non-elected functionaries such as bankers and lawyers.[15] Concepts of belief, truth, the preparedness to bring sacrifices for the common good have weakened to such an extent that we feel threatened by people who are prepared to fight and die for an ideal.[16] Such trends have made many in the public in the West more impressionable to alternative influences of populists and radicals.

One of the alternatives is presented by the rise of populist parties and politicians in many Western countries, which at the time of the writing of this chapter culminated in the election of Donald Trump as president of the United States. The character of populist politics may differ per country, but all its politicians perceive themselves as the mouthpiece of the common man. They express distrust of mainstream politics and share a nationalistic, isolationist or even xenophobic political view and an incredible bad taste in general. According to Michael Kazin, Donald Trump, during his election campaign saw himself as the spokesman for "working families", "our middle class", or "the American people".[17]

Populist leaders appear to feel ideologically close to the Russian president Putin or with his party, United Russia. Several publications have stressed the affinity of populist politics in the West with Putin's political preferences. According to the German weekly *Der Spiegel*, for example, Russia has spun a tight network of right-wing leaders and movements in Europe: Marine Le Pen of the Front National in France, Mateusz Pikorski in Poland, Viktor Orbán the prime minister of Hungary, the British UK Independence Party and the extreme right-wing *Alternative für Deutschland* (AfD) in Germany. According to Olga Oliker,

[13] Putters 2016.

[14] Zakaria 2016, pp. 10–12.

[15] Mudde 2016, pp. 25–30.

[16] Durodié 2005a, pp. 42–43.

[17] Kazin 2016, pp. 17–24.

populism differs from Putinism, but … "it is notable that so many of those who seek to take Western democracies in a less pluralistic direction see Russia as a model".[18]

By supporting these politicians, parties, and movements, Russia hopes to create a wedge between Europe and the United States and to weaken cooperation in the West in NATO and the EU. The least that can be said about these populist movements is that they fit into Putin's international ambitions.[19] "It may well end in tears," the Huffington Post wrote, "but the union between Putin and the populists is no marriage of convenience. It is true love."[20] To give but one example: Right after his election as prime-minister of Greece, Alexis Tsipras called for lifting the EU economic sanctions against Russia that had been introduced after the Russian annexation of the Crimea in 2014.[21]

This populist movement has been identified above as one of the ideological competitors. The other one is the apparently much smaller group of sympathisers with jihadist extremism. In radical Islamist circles, it is argued that Western societies and secular and humanistic values have a corrupting influence on Muslims. Such views are regularly expressed in radical jihadist periodicals such as ISIS' *Dabicq* or Al-Qaida's *Inspire*. Furedi found that for many young people in Western Europe it has become easy to embrace Islamic counter-narratives that criticise the Enlightenment and applaud "jihadist identity politics".[22]

With their attacks, terrorists hope that feelings of insecurity and powerlessness among the general public are increased. The more this happens, Furedi argues, the greater the prestige of the attackers. Before ISIS became an attractive option, there has been a long-term development among young people in the West searching for a meaningful alternative.[23] Western societies have not been able to sufficiently integrate these groups on the basis of its fundamental values.

Finding a remedy is complicated because democracy presupposes openness towards alternative ideologies and political convictions. In this line of thought, Henri Beunders finds that there is no dominant morality in Western societies anymore. Postmodernist thinking made everything relative and truth was seen as a linguistic construction of the powerful.[24] The only moral guidelines a liberal democracy is able to produce, Francis Fukuyama argues, are the universal obligations of tolerance and mutual respect. When democracies were relatively homogeneous, this was not so much of a problem. However, societies today have become much more diverse than they had been fifty years ago. Relativism undermines democracy when the very political values upon which the political system is founded are questioned as well.[25]

---

[18] Oliker 2017, p. 17.

[19] Bidder 2016. See also Bidder and Witrock 2014.

[20] Baggini 2016.

[21] Jones et al. 2015.

[22] Furedi 2015d.

[23] Furedi 2015a, d.

[24] Beunders 2016.

[25] Fukuyama 1999, pp. 281–282.

To sum up, Western societies have become vulnerable to shocks and underlying trends. Both the populist and the jihadist tendencies present radical alternatives to mainstream Western politics. Populist parties dominate the political debate. Mainstream parties find difficulty in voicing a convincing alternative. Western vulnerability can on the other hand partly be situated in the existence of marginalised groups in the societies that are impressionable for radicalisation or ideological mobilisation. In reaction, many governments say they feel the need to strengthen their resilience. Morale, in the sense of a resilient mind-set, plays an essential role in this. This is where the discussion of the concept of resilience is situated in this chapter.

2.4 The Emergence of Resilience

In an attempt to deal with these developments, debates have started in various relevant publications to include resilience within security strategies. With a bias towards the Netherlands, the following section refers to some recent papers that deal with resilience. The 2015 Dutch Clingendael Monitor, for instance, deals with the vulnerability of open, democratic and plural Western societies. They are said to be more vulnerable to threats of returning jihadists, undermining propaganda, economic pressure and cyber attacks. The Monitor notes that it is difficult to find the right answers to these developments.[26] However, the Monitor leaves the question how societies can be made more resilient unanswered.

A report by the Hague Centre for Strategic Studies on resilience building makes some sound remarks on the subject. It says that resilience essentially makes up the immunity system of a society against conflicts. With that, the report points at people and social structures that help prevent conflicts from breaking out and that spring back once they have started.[27] The report, however, concentrates on societies of countries where stabilisation operations are taking place. The focus in this chapter is on domestic resilience in the sense of the mind-set of Western societies in response to external threats. This specific field of research is relatively underexplored.

Domestic resilience is only implicitly referred to in the Netherlands' government policy white paper on International Security of 2014 (*Beleidsbrief Internationale Veiligheid*), where the remark is made that nowadays our convictions regarding freedom of religion and speech, democracy and rule of law are ever more openly questioned.[28] The concept of resilience is discussed more extensively in the Dutch national Counter-terrorism Strategy. There, resilience is predominantly seen as an instrument in preventing fear of and support for terrorism. Resilience is discussed in the field of communication with the general public as well as in strengthening the

---

[26] Rood et al. 2015.

[27] HCSS 2016, p. 7.

[28] NL Ministry of Foreign Affairs 2014b, p. 5.

mind-set among groups where potential supporters for radicalisation and extremist or jihadist ideologies can be found. According to the Counter-terrorism Strategy, suspect behaviour will be monitored and counteracted, and cooperation will be sought with educational institutions, associations, groups in society, and professionals. Terrorist websites will be removed and violent extremist narratives will be counteracted.[29]

The Global Strategy document of the European Union is one of the few policy papers that has tackled the moral dimension of resilience. It builds upon the premise that we "… live in times of existential crisis, within and beyond the European Union". It reasons that resilience makes up an important element of the EU strategy and it sees values as central to that. The document says: "To engage responsibly with the world, credibility is vital. The EU's credibility hinges on our unity, on our many achievements, our enduring power of attraction, the effectiveness and consistency of our policies, and adherence to our values." These are said to be the values that were at the basis of the EU's foundation, i.e. respect for and defence of human rights, fundamental freedoms and the rule of law. Living up to these values internally and externally is seen as the most crucial of all measures, and as "the strongest antidote we have against violent extremism".[30]

As was seen at the beginning of this chapter, the 2016 NATO Summit in Warsaw recognised the importance of resilience, mainly in the sense of civil preparedness, the sense of strengthening continuity of government, continuity of essential services and security of critical civilian infrastructure.[31] In a short piece on the same subject, NATO's Jamie Shea argues that as the world has become more globalised, complex and confrontational, resilience has become a main topic among NATO's leaders. Shea sees two challenges: one is to make sure that troops and materiel can be deployed rapidly to every relevant hot-spot within alliance territory, the other is the ability to anticipate, identify, mitigate and recover from hybrid attacks in such a way that the social, political and military cohesion is least affected.[32]

What the above-mentioned documents have in common is that the importance of moral resilience, the mind-set of the home front, is mentioned, but generally as a sideline that does not (yet) merit research on its own term. The EU Global Strategy paper is an important exception. Before we turn our attention to the mind-set of the home front, in the following sections we will try to learn more about the term resilience.

[29] NL Ministry of Security and Justice 2016, pp. 13–14.

[30] EEAS 2016.

[31] NATO 2016.

[32] Shea 2016.

2.5 Resilience Theory and Security

The emergence of the concept of resilience in security policies runs parallel with the rise of the academic debate on the concept. In the field of security studies, this discussion has come up in relation to the hybrid character of warfare. Hybrid threats and attacks do not occur along the border between two opponents only, but can strike at the heart of the societies involved as well. Hybrid warfare is a challenge to traditional distinctions between war and peace, military and non-military instruments, between conventional and irregular warfare, and between perceptions of security at home and security abroad.

Resilience research started in ecology studies and psychology. The latter was mainly oriented at the capacity of individuals to recover from traumas and distress. The focus in the former was on concern for the decay of ecosystems and the capacity of ecosystems to recover from disturbances and shocks. From there, research broadened to the capacity of economic systems to recover from crises, to disaster management of societies to withstand and recover from disasters such as earthquakes. Recently, resilience as a concept has been extended to the way societies deal with (the threat of) terrorist attacks.[33]

Rodin has made a useful definition of resilience in a general sense. According to her resilience is "…the capacity of any entity—an individual, a community, an organization, or a natural system—to prepare for disruptions, to recover from shocks and stresses, and to adapt and grow from a disruptive experience".[34] Martin-Breen and Anderies define a system as a "collection of agents and relationships among those agents". A system may be a village, a region, a nation or the whole world.[35] In Rodin's definition three types of resilience can be discerned: resilience as (1) the capacity to recover, as (2) the capacity to adapt and as (3) the capacity to grow. We will have a brief look at these three types of resilience.

Resilience as the capacity to recover, also called coping or engineering resilience, is aimed at the ability of a social system to survive after a shock and return to its pre-existing state. According to this type of resilience a social system is capable of returning to a situation of stability after a temporary disturbance, or to continue functioning in a situation of instability. Another term used in this category is robustness.[36] Much of the appreciation of resilience in government white papers or in NATO-related publications belongs to this category. Recommendations in this sphere are the ability of basic infrastructure to bounce back after a shock, the level of preparedness of society against terrorist attacks, counter-terrorist measures or emergency plans. An example of engineering resilience is the way cyber security is discussed in the Dutch policy paper on Defence Cyber Strategy. There, digital resilience is discussed in various paragraphs, but within the context of making the

[33] Frerks 2015, p. 490.

[34] Rodin 2015, p. 3.

[35] Martin-Breen and Anderies 2011.

[36] Keck and Sakdapolrak 2013, pp. 6–7; Rodin 2015, p. 50; also Fjäder 2014, p. 12.

digital Defence network safe against attacks and capable of functioning as intended after a crisis has taken place.[37]

This concept of resilience runs parallel to Bourbeau's categorisation of resilience as maintenance: a way of constructing a shock with the intention to use resources and energies to maintain the status quo. Measures that serve the continuation of the status quo are formulated in such a way that the status quo is legitimized. Shocks are constructed as dangerous. According to Bourbeau an example of this can be found in the way the public debate on migration is framed in many Western countries.[38]

The second type of resilience has been labelled adaptive resilience. It describes the ability of a system to adapt to changing future threats, to absorb shocks and reorganise in order to preserve the same essential structure, function and identity. This category takes change into account whilst keeping the functionality of a social system, such as food, water, housing, health-care, intact. In times of stability, changes may be gradual; in times of crisis they may be rapid.[39] Giegerich also takes this approach when he defines resilience as the ability of societies to manage threats and risks, to adapt and to recover without losing the ability to provide basic functions and services.[40]

The third type of resilience is related to the capacity to grow, often described as the transformative capacities of complex adaptive systems. According to Martin-Breen and Anderies this extends beyond the system itself and envisages the ability of a system to change the landscape of stability and create a fundamentally new one when ecologic, economic or social structures have become untenable.[41] Therefore, according to Osinga, in complex adaptive systems, resilience is best defined as the ability to withstand, recover from, and reorganize in response to crises.[42] What changes under crisis is the understanding of normality. In the words

---

[37] NL Ministry of Defence 2015.

[38] Bourbeau 2015b, pp. 177–178. What Bourbeau is looking for is to theorize resilience in order to get a better understanding of some aspects of securitization theories. Resilience can help to develop hypotheses that purport to get a better understanding of how societies adapt to shocks, and to compare how societies adapted in time or how different societies adapt differently in different places. This can help with the analysis of how security performances are received, evaluated and approved by audiences. From a constructivist perspective, resilience can help provide the literature with the socio-political use of memory and myth. Constructivism sees a shock or a trauma as an interpretative moment, where governments can use the opportunity to dictate a certain narrative about the origin, the meaning or the implications. Resilience as maintenance means that reactions are informed by the intent to maintain the status quo and it means that reactions are dominated by rigidity and anxiety. When resilience is presented in this manner, it may lead to securitization. For instance "(t)he social construction of refugee movements and mass migration as a significant disturbance necessitating a resilient strategy has led, ultimately, to the securitization of migration." See: Bourbeau and Vuori 2015, p. 389.

[39] Keck and Sakdapolrak 2013, p. 7; Martin-Breen and Anderies 2011, pp. 6–7.

[40] Giegerich 2016, p. 69.

[41] Martin-Breen and Anderies 2011, p. 38.

[42] Osinga 2016a, p. 46.

of Bourbeau, this opens up the focus on the possibility of renewal of self-organisation, rearrangement and reorientation.[43]

This third approach can therefore also be seen as what Bourbeau in another context calls resilience as renewal.[44] A crisis can, thus interpreted, be used as an opportunity to transform basic policy assumptions and reshape social structures. Bourbeau took the migration crisis in Europe as an example. A society that chooses resilience as renewal, he reasons, can use the crisis as a trigger to reform its understanding of migration, diversity in society and integration and then develop a new policy in the field of citizenship and refugees.[45] Such new understandings do not emerge in a vacuum, but build on experiences from the past, on collective memories and social history. In this way, in the words of Bourbeau, a crisis can be seen as an "interpretative moment".[46]

As the resilience debate has come up in the context of hybrid warfare, where communication, narratives and images play a crucial role, these points deserve some further exploration. Bourbeau's findings can be situated in securitization theories, which deal with perceptions, justifications and the framing of security. Current government policies dealing with resilience are informed by the perception that we are dealing with threats against which a total defence is impossible. Nevertheless, Coaffee and Wood argue, governments can influence the way risks are dealt with.[47] According to these authors, after the al-Qaida attacks on New York and London, a new protective and regulative state has emerged, which is expressed in terms of resilience rhetoric and which involves the spread of responsibilities for security over all layers of government, private organisations and citizens themselves. This rhetoric has replaced conceptions of emergency planning and civil protection from the Cold War era.[48]

In the words of Bourbeau, the "cultural regime" informs the way a society adapts to shocks.[49] The way people handle an emergency or a disaster is shaped by the cultural narrative that forms expectations and that sensitizes people more for one problem than for another. Furedi points at reactions to the notorious London smog of December 1952. Although 4000 people were killed in it, nobody at that time interpreted this as a disaster. He notes that "…perceptions regarding causation are shaped by a cultural script that seeks to endow events, especially extreme ones, with meaning."[50]

This is where the mind-set, the morale, of a society, comes in: it affects resilience to security threats and openness to the future. Bourbeau's interpretation of

---

43 Bourbeau 2015b, p. 175.

44 Bourbeau 2015a, pp. 384–385.

45 Bourbeau 2015b, pp. 181–182.

46 Bourbeau 2015a, p. 375.

47 Coaffee and Wood 2006, p. 505.

48 Coaffee and Wood 2006, p. 514.

49 Bourbeau 2015a, p. 382.

50 Furedi 2007, p. 484.

resilience as renewal lends itself as a useful interpretative framework to further study of the resilience of the home front as a contribution to deterrence. It incorporates elements such as the ambition to actively influence the context in which the social system is supposed to operate. It also allows for taking aspects such as cultural narrative, interpretation, or memory into account. That would make it relevant to understanding Western government policies that use the concept of resilience to deal with hybrid threats.

2.6 Resilience Theory and Deterrence

Now that we discussed the concept of resilience, we turn to the relevance of resilience in relation to the deterrence of hybrid threats. We have seen that deterrence is the ability to discourage an opponent from taking an unwelcome action. And we will specifically look into the question how that can be enhanced through resilience of the mind-set. Two options present themselves. One aims at closing off society from outward threats and maintaining the status quo. This can be called the isolationist option and more or less follows the logic of Bourbeau's coping resilience. The other option is in line with the concept of resilience as renewal. The choice between the two options is influenced heavily by the interpretation of the vulnerability of society in the face of risks.

The isolationist option is an interpretation of threats and risks that fits Bourbeau's concept of resilience as coping, which would mean that resilience is primarily seen as a defensive instrument against risks that come from outside and threaten us. This means that the state today has to cope with insecurity in the sense of lack of knowledge about the future, insecurity in the sense of socio-economic uncertainty and insecurity in the sense of terrorist threats. According to Ulrich Beck, on the one hand these undermine the power of the state, on the other hand these confirm the power of the state beyond democratic legitimacy.[51] Governments are increasingly inclined to limit civil rights and freedoms in order to protect their societies against risks, which, in practice, means more control at airports, more cameras, more interception of Internet data. This only results in giving up the free and open society governments are supposed to uphold, whereas the terrorist threat is in no way abolished.[52]

Faced with the choice between freedom and security, Beck argues, "… the vast majority of human beings seem to prioritize security, even if that means civil liberties are cut back or even suffocated."[53] Coaffee and Wood are even more critical in that they add that governments deliberately appropriate the terrorist threat and create a climate of fear in order to justify the building of technologically driven

[51] Beck 2006, pp. 343–345.

[52] Beck 2006, p. 330.

[53] Beck 2006, p. 431.

control strategies. Thus the state of civil liberties in the West comes under pressure.[54] And the need for security is in principle insatiable. With ever-tighter measures the feeling of insecurity only increases. The idea settles that the authorities have lost control over security and are on the defensive in the fight against terrorists. To the sympathisers of jihadism this only shows how powerful they are, which further encourages other “hoaxers, loners and cranks”.[55]

The alternative may be found in the other option, which sees resilience as renewal, as future oriented and proactive. This option would be in line with Beck’s plea to tap new sources of legitimacy such as global civil society and to strive for a cosmopolitan form of government that is a counterweight to international capital and that can create a post-national order on the basis of interdependence.[56] This option seeks to influence the international context in such a way that the core values and interests are thought anew, modernised or made to work in a new context. Such a reinvigorated orientation in the world around us starts with charting who we are, what as a society we stand for and how this can be brought forward effectively in the world that surrounds us.

This is a widely shared analysis. Resilience is only possible if, as a society, we know what we find worthwhile to defend and where we want to be heading. Durodié is convinced that currently society does not have clarity about where it stands. This position undermines all technical solutions to make society more resilient. He feels that resilience basically is not a new technology but a cultural attitude that stems from a self-confident commitment to the society and clarity about its meaning and purpose.[57] On this basis, deterrence can be enhanced in a context of hybrid warfare.

Often the resilience debate is about technical solutions and infrastructure. But public debate about common values and objectives is seen by many authors as a contribution to social capital and therefore, trust. Rodin points to the so-called softer characteristics of resilience, such as policy, norms and behaviour, and identity.[58] Real resilience depends on who we want to be with each other. In the words of Durodié, that is what orients society to a sense of purpose and what is needed to win the hearts and minds of our own population. People are prepared to bring sacrifices if the purpose is important enough to live for.[59]

Identity comes to life in communities. Keck and Sakdapolrak see social capital as a central factor in building and maintaining social resilience.[60] According to

[54] Coaffee and Wood 2006, p. 515.

[55] Durodié 2005a, p. 45.

[56] Beck 2006, p. 343.

[57] Durodié 2004. See also: Durodié 2005a, p. 52.

[58] Rodin 2015, p. 63.

[59] Durodié 2005b, pp. 17–19.

[60] Keck and Sakdapolrak 2013, p. 11.

Rodin, communities react in ways that provide more than just the sum of individual connections. Members of communities feel committed to the groups they are part of, such as neighbourhoods or voluntary organisations. What results, is social capital, "the glue that holds people together" and that leads to commitment, common values and opinions, a common purpose and a feeling of identity.[61] Or, to quote Putnam: "Communities with strong social networks and grassroots associations are better at confronting unexpected crises than communities that lack such civic resources. […] Social capital, the evidence increasingly suggests, strengthens our better, more expansive selves. The performance of our democratic institutions depends in measurable ways upon social capital."[62]

The existence of such communities makes a society more resilient. What you get are relatively autonomous enterprises in which participants feel committed to a common purpose. The state needs to recognise the rights and freedoms of the communities and organisations in society and not impose efficiency from above. That would mean that a resilient society is not one that is centrally coordinated from above. The state should work with social capital, foster it, and not lead it. If one of the organisations—or even the state itself—breaks down, others can take over some of its functions.[63]

Thus, Martin-Breen and Anderies argue, a society comes about with multiple strategies and multiple mechanisms that can perform similar functions. Efficiency is not the most important criterion.[64] From a military operational perspective, Osinga argues, the more variables are available, the more dynamic and flexible a system can be. The key is self-organization. What makes an organization resilient are patterns of varying forms of interdependence. If one pattern of interdependency is eliminated, other areas can take over their functions.[65] This is recognized by the EU Global Strategy document, where it says that the resilience of society can be enhanced by deepening relations with the civil society. The EU proposes to seek better contact with cultural organizations, religious communities, social partners and human rights defenders and will defend the role of civil society abroad.[66]

The paragraphs above show how the mind-set can be resilient and enhance deterrence: common values and a sense of purpose, a feeling of community and commitment of its participants. A society that beliefs in what it stands for, is a resilient society.

---

[61] Rodin 2015, pp. 63–64, 193–194.

[62] Putnam 2000, p. 349.

[63] Martin-Breen and Anderies 2011, pp. 18, 25, 46.

[64] Martin-Breen and Anderies 2011, p. 10.

[65] Osinga 2016a, p. 49.

[66] EEAS 2016.

2.7 Conclusions

What we tried to do in this chapter is to explore if and how Western societies can mentally arm themselves against the mental aspects of hybrid threats and whether the concept of resilience can play a role in this. We have seen that Western societies experience threats from two sources: pressure stemming from Russia on its Eastern frontier and terrorist attacks from the Middle Eastern region. These threats are of a kinetic and a non-kinetic character. Given the focus of this volume, this chapter has dealt with the latter type of threat. In the face of this, governments feel the need to strengthen resilience. The importance of moral resilience, the mind-set of the home front, is mentioned, but generally as a sideline that did not (yet) merit much research on its own term. The EU Global Strategy paper is an important exception.

In the discussion of resilience theories we have seen that, under certain conditions, a resilient society enhances overall mental deterrence against hybrid or terrorist attacks. Two alternatives were presented. The first was resilience as maintenance, a way of constructing a shock with the intention to use resources and energies to maintain the status quo. Its basic attitude was to close the society off from the world outside. The second alternative was resilience as renewal. It incorporates elements such as the ambition to actively influence the context in which the social system is supposed to operate. It also allows for taking aspects such as cultural narrative, interpretation, or memory into account. It was suggested that by stressing common values, a sense of purpose, a feeling of community and commitment, the mind-set of Western societies could be more resilient.

Pleas for a government narrative and propaganda campaigns always run the risk of getting a manipulative character if stressing values is just an instrument and not felt as an end in itself. One might suggest that shocks in the form of propaganda campaigns can be a starting point for discussions in society about values in order to build trust in the nation and in the European Union. One might argue that information, moral, psychological and ideological measures aimed at the society as a whole, and well-chosen political and military leaders are the first phase in the defence against hybrid attacks resulting in the emergence of a resilient collective mind-set.[67] Such arguments don't work however, if the public doesn't perceive them as authentic.

Sometimes, the message is confusing. On the one hand, Western governments can be seen as supportive of processes of democracy, respect for human rights and the fostering of the rule of law, such as during the transformation period of the countries in Central and Eastern Europe that asked for and were granted membership of the EU and NATO. On the other hand, governments of the United Kingdom or the Netherlands were hardly seen fighting for a yes-vote in the referendum campaigns on the membership of the European Union or the EU trade agreement with Ukraine. Western governments stood by, doing little to help the population when over a period of five years when Aleppo was utterly destroyed.

---

[67] Bilban 2015, p. 11.

The EU can stress its values in its policy papers, but should also practise them when faced with emission fraud by the Volkswagen car manufacturers.

Resilience requires focused activity and consistency. In this era of alternative facts, post-truth, deliberate manipulation of the media by populist politicians and Russia, a passive attitude will cede the battleground to strong consistent counter-narratives that resonate easily in the echo chambers of social media. And, no matter how well constructed the narrative may be, if you don't mean it and if you don't act accordingly it's all to no avail.

References

Adger WN (2000) Social and ecological resilience: are they related? Progress in Human Geography 24(3):347–364

Baggini J (2016) Why Putin and the Populists Love Each Other Deeply. Huffington Post (14 December 2016) http://www.huffingtonpost.com/entry/putin-trump-populists_us_58506581e4b0ee009eb41e14 Accessed 4 January 2017

Barber BB (1996) Jihad versus McWorld; How Globalism and Tribalism are Reshaping the World. Ballantine Books, New York

Beck U (2006) Living in a World Risk Society; A Hobhouse Memorial Public Lecture given on Wednesday 15 February 2006 at the London School of Economics. Economy and Society 35 (3): 329–345

Beunders H (2016) Het beste wapen? Schrijf het betere verhaal. Nationaal Coördinator Terrorismebestrijding en Veiligheid (ed) Magazine Nationale Veiligheid en Crisisbeheersing 14(5–6):11–13

Berzins J (2014) Russia's New Generation Warfare in Ukraine: Implications for Latvian Defense Policy, Policy Paper No 2. Center for Security and Strategic Research, National Defence Academy of Latvia

Bidder B, Witrock P (2014) Putin und die Populisten; Das rechte Netz des Kreml. Spiegel Online (24 November 2014) http://www.spiegel.de/politik/ausland/afd-und-front-national-putin-umwirbt-europas-rechtspopulisten-a-1004746-druck.html Accessed 19 December 2016

Bidder B (2016) Populisten Netzwerk in Europa; Russlands Rechte Freunde. Spiegel Online (4 February 2016) http://www.spiegel.de/politik/ausland/russland-wladimir-putins-rechtsextreme-freunde-in-europa-a-1075461-druck.html Accessed 19 December 2016

Bilban C (2015) Resilience: Silver Bullet in Challenging Hybrid Warfare? 210098 SE M4 a: Resilienz in den Internationalen Beziehungen (WS) L V-Leitung: Mag. Dr. Jan Pospisil

Bilger P (2015) Trouw (17 November 2015)

Bourbeau P (2015a) Resilience and International Politics: Premises, Debates, Agenda. International Studies Review 17(3):374–395

Bourbeau P (2015b) Resiliencism and Security Studies; Initiating a dialogue. In: Balzaq T (ed) Contesting Security: Strategies and Logics. Routledge, London/New York pp 173–188

Bourbeau P, Vuori JA (2015) Security, resilience and desecuritization: multidirectional moves and dynamics. Critical Studies on Security 3(3):253–268

Chandler D (2012) Resilience and human security: The post-interventionist paradigm. Security Dialogue 43(3):213

Coaffee J, Wood DM (2006) Security is Coming Home: Rethinking Scale and Constructing Resilience in the Global Urban Response to Terrorist Risk. International Relations 20(4):503–17

Cohen N (2016) Leavers are angry, for their lies will return to haunt them. The Guardian (18 December 2016). http://www.theguardian.com/commentisfree/2016/dec/18/brexit-leavers-fear-their-lies-will-haunt-them Accessed 19 December 2016

Davis J (2015) Continued Evolution of Hybrid Threats: The Russian Hybrid Threat Construct and the Need for Innovation. The Three Swords Magazine (28):19–25
Durodié B (2004) Cultural influences on resilience and security. Monitor (September) http://www.durodie.net/pdf/CulturalInfluences.pdf Accessed 7 December 2015
Durodié B (2005a) Cultural Precursors and Psychological Consequences of Contemporary Western Responses to Acts of Terror. In: Wessely S, Krasnov VN (eds) Psychological Responses to the New Terrorism: A NATO-Russia Dialogue. IOS Press, Amsterdam, pp 37–53
Durodié B (2005b) The Limitations of Risk Management: Dealing with Disasters and Building Social Resilience. Politik 8(1):14–21
EEAS (2015) Food-for-thought paper "Countering Hybrid Threats". Working Document of the European External Action Service. Crisis Management and Planning Directorate (EEAS(2015) 731, 13 May) http://www.statewatch.org/news/2015/may/eeas-csdp-hybrid-threats-8887-15.pdf Accessed 19 November 2015
EEAS (2016) Shared Vision, Common Action: A Stronger Europe. A Global Strategy for the European Union's Foreign and Security Policy. EEAS, Brussels (June 2016) http://eeas.europa.eu/top_stories/pdf/eugs_review_web.pdf Accessed 7 July 2016
Fjäder C (2014) The Nation-State, National Security and Resilience in the Age of Globalisation. Resilience 2(2):114–129
Frerks G (2015) Help or Hindrance? The Contribution of the Resilience Approach to Risk Governance. In: Urbano FP (ed) Risk Governance: The Articulation of Hazard, Politics and Ecology. Springer, Dordrecht, pp 489–494
Fukuyama F (1995) Trust; The Social Virtues and the Creation of Prosperity. Hamilton, London
Fukuyama F (1999) The Great Disruption; Human Nature and the Reconstitution of Social Order. The Free Press, New York
Furedi F (2007) The changing meaning of disaster. Area 39(4):482–489
Furedi F (2015a) Exploding the Myth of Radicalisation. Spiked (16 June 2015) http://www.spiked-online.com/newsite/article/exploding-the-myth-of-radicalisation/17075#.WL5nI3_dV8E. Accessed 1 March 2017
Furedi (2015b) Ten Years After 7/7 They Still Don't Get It. Spiked (6 July 2015). http://www.spiked-online.com/newsite/article/ten-years-after-7-7-they-still-dont-get-it/17146#.WL5mG3_dV8E. Accessed 1 March 2017
Furedi F (2015c) Islamic State Propaganda Buys into Western Culture of Victimhood. The Australian (25 July 2015) bit.ly/2mQR1VO. Accessed 1 March 2017
Furedi F (2015d) After Paris, We Must Refuse to be Terrorised. Spiked (16 November 2015). http://www.spiked-online.com/newsite/article/after-paris-we-must-refuse-to-be-terrorised/17634#.WL5lp3_dV8E. Accessed 1 March 2017
Giegerich B (2016) Hybrid Warfare and the Changing Character of Conflict. Connections QJ 15 (2):65–72
HCSS (2016) Si vis pacem para utique pacem; Individual Empowerment, Societal Resilience and the Armed Forces. HCSS StratMon 2016. The Hague Centre for Strategic Studies, The Hague
ICG (2014) Rigged Cars and Barrel Bombs: Aleppo and the State of the Syrian War. Report 155 (9 Sep) http://www.crisisgroup.org/en/regions/middle-east-north-africa/syria-lebanon/syria/155-rigged-cars-and-barrel-bombs-aleppo-and-the-state-of-the-syrian-war.aspx. Accessed 1 March 2017
Jacobs A, Lasconjarias G (2015) NATO's Hybrid Flanks: Handling Unconventional Warfare in the South and the East. In: Lasconjarias G, Larsen JA (eds) Response to Hybrid Threats. NATO Defence College Forum Papers Series, Rome, pp 257–276
Johnson D (2015) Russia's Approach to Conflict: Implications for NATO's Deterrence and Defence. In: Lasconjarias G, Larsen JA (eds) Response to Hybrid Threats. NATO Defence College Forum Papers Series, Rome, pp 137–160
Jones S, Hope K, Weaver C (2015) Alarm bells ring over Syriza's Russian links. Financial Times (28 January 2015)
Joseph J (2013) Resilience as embedded neoliberalism; a governmentality approach. Resilience 1 (1):38–52

Kazin M (2016) Trump and American Populism: Old Wine, New Bottles. Foreign Affairs (Nov/Dec):17–24

Keck M, Sakdapolrak P (2013) What is Social Resilience? Lessons Learned and Ways Forward. Erdkunde 67(1):5–19

Kennan GF (1948) On Organizing Political Warfare (April 30, 1948) Wilson Center Digital Archive, http://digitalarchive.wilsoncenter.org/document/114320.pdf?v=941dc9ee5c6e51333ea9ebbbc9104e8c Accessed 7 November 2016

Lasonjarias G, Larsen AJ (2015) Introduction: A New Way of Warfare. In: Lasconjarias G, Larsen JA (eds) Response to Hybrid Threats. NATO Defence College Forum Papers Series, Rome, pp 1–13

Martin-Breen P, Anderies JM (2011) Background Paper - Resilience: A Literature Review. The Bellagio Initiative Ed.; The Future of Philanthropy and Development in the Pursuit of Human Wellbeing (Draft November 2011) http://opendocs.ids.ac.uk/opendocs/bitstream/handle/123456789/3692/Bellagio-Rockefeller%20bp.pdf?sequence=1 Accessed 1 March 2017

McDermott R, Reisinger H, Smith-Windsor B (2015) Cold War Déjà Vu? NATO, Russia and the Crisis in Ukraine. Lasconjarias G, Larsen JA (eds) Response to Hybrid Threats. NATO Defence College Forum Papers Series, Rome, pp 41–50

Mudde C (2016) Europe's Populist Surge: A Long Time in the Making. Foreign Affairs (Nov/Dec):25–30

NATO (2015) Deterrence: what it can (and cannot) do. NATO Review. http://www.nato.int/docu/review/2015/also-in-2015/deterrence-russia-military/EN/index.htm Accessed 1 March 2017

NATO (2016) Commitment to Enhance Resilience, Issued by the Heads of State and Government participating in the meeting of the North Atlantic Council in Warsaw, 8–9 July 2016 (8 July 2016) http://www.nato.int/cps/en/natohq/official_texts_133180.htm Accessed 7 November 2016

Nicolini M, Janda J (2016) In the Area or Out of Business: Building Resilience to Hybrid Attacks. The Polish Quarterly of International Affairs (1):77–87

NL Ministry of Defence (2015) Updated Defence Cyber Strategy (Dutch: Actualisering Defensie Cyber Strategie, 23 February 2015) Parliamentary Papers (Kamerstukken) II 2014-15, 33 321, no. 5, The Hague

NL Ministry of Foreign Affairs (2014a) (Dutch) Leidraad geïntegreerde benadering. De Nederlandse visie op een samenhangende inzet op veiligheid en stabiliteit in fragiele staten en conflictgebieden (Den Haag, 11 juli 2014) http://www.rijksoverheid.nl/documenten/kamerstukken/2014/07/11/kamerbrief-inzake-leidraad-geintegreerde-benadering Accessed 9 November 2015

NL Ministry of Foreign Affairs (2014b) Turbulent Times in Unstable Surroundings. Policy letter on international security (14 November 2014) http://www.government.nl/binaries/government/documents/parliamentary-documents/2015/01/16/turbulent-times-in-unstable-surroundings/turbulent-times-in-unstable-surroundings.pdf Accessed 1 March 2017

NL Ministry of Home Affairs (2007) (Dutch) Strategie Nationale Veiligheid (Den Haag, 14 mei 2007) http://www.rijksoverheid.nl/documenten/kamerstukken/2007/05/14/strategie-nationale-veiligheid. Accessed 2 November 2015

NL Ministry of Security and Justice (2016) (Dutch) Nationale Contraterrorismestrategie 2016 – 2020 (July) Nationaal Coördinator Terrorismebestrijding en Veiligheid, Den Haag

Oliker O (2017) Putinism, Populism and the Defence of Liberal Democracy. Survival 59(1):7–24

Osinga F (2016a) Organizing for Insecurity and Chaos: Resilience and Modern Military Theory. In: Beeres R, Bakx G et al (eds) Organizing for Safety and Security in Military Organizations-Netherlands Annual Review of Military Studies. TMC Asser Press, The Hague, pp 43–73

Osinga F (2016b) Hybrid Warfare en de uitdaging van de nieuwe geopolitieke realiteit. Nationaal Coördinator Terrorismebestrijding en Veiligheid (ed) Magazine Nationale Veiligheid en Crisisbeheersing 14(5–6):16–21

Perry B (2015) Non-Linear Warfare in Ukraine; The Critical Role of Information Operations and Special Operations. Small Wars Journal (14 August 14) http://smallwarsjournal.com/jrnl/art/non-linear-warfare-in-ukraine-the-critical-role-of-information-operations-and-special-opera Accessed 1 March 2017

Praks H (2015) Hybrid or Not: Deterring and Defeating Russia's Ways of Warfare in the Baltics – The Case of Estonia. In: Lasconjarias G, Larsen JA (eds) Response to Hybrid Threats. NATO Defence College Forum Papers Series, Rome, pp 219–241

Putnam P (2000) Bowling Alone; The Collapse and Revival of American Community. Simon and Schuster Paperbacks, New York

Putters K (2016) Land tussen hoop en vrees: over het politiek en maatschappelijk vertrouwen van Nederlanders; Drees-lezing (17 November 2016) http://www.willemdrees.nl/wp-content/uploads/2016/11/Willem-Dreeslezing-2016-door-Kim-Putters.pdf Accessed 1 March 2017

Reisinger H, Golts A (2015). Russia's Hybrid Warfare: Waging War below the Radar of Traditional Collective Defence. In: Lasconjarias G, Larsen JA (eds) Response to Hybrid Threats. NATO Defence College Forum Papers Series, Rome, pp 113–136

Rodin J (2015) The Resilience Dividend; Managing Disruption, Avoiding Disaster, and Growing Stronger in an Unpredictable World. Profile Books, London

Rood J, van der Putten FP, Meijnders M (2015) Een wereld zonder orde? Clingendael Monitor 2015. Netherlands Institute of International Relations Clingendael, The Hague

Ruiz Palmer DA (2015) Back to the Future? Russia's Hybrid Warfare, Revolutions in Military Affairs, and Cold War Comparisons. In: Lasconjarias G, Larsen JA (eds) Response to Hybrid Threats. NATO Defence College Forum Papers Series, Rome, pp 49–71

Samaan JL (2015) An End-of-Time Utopia: Understanding the Narrative of the Islamic State. In: Lasconjarias G, Larsen JA (eds) Response to Hybrid Threats. NATO Defence College Forum Papers Series, Rome, pp 243–253

Shea J (2016) Resilience: a core element of collective defence. NATO Review (no date) http://www.nato.int/docu/Review/2016/Also-in-2016/nato-defence-cyber-resilience/EN/index.htm Accessed 12 July 2016

Smith R (2005) The Utility of Force: The Art of War in the Modern World. Random House, New York

Vest R (2015) Baltic COIN: Using a counterinsurgency model to counter Russian hybrid warfare in the Baltics. Joint Military Operations Department (18 May 2015), US Naval War College, Newport RI

Werther GFA (2014) Resilience: Its Conceptual Links to Creating Society-Specific Forecasts about Emerging Change. Armed Forces and Society 40(3):428–451

Wijk R (2016) Hybride dreigingen. Nationaal Coördinator Terrorismebestrijding en Veiligheid (ed) Magazine Nationale Veiligheid en Crisisbeheersing 14(5–6):14–15

Zakaria F (2016) Populism on the March: Why the West is in Trouble. Foreign Affairs (Nov/Dec):9–15

Author Biography

Professor Dr. Theo Brinkel is Assistant Professor of International Security Studies at the Department of War Studies, Netherlands Defence Academy and KVMO [Koninklijke Vereniging van Marineofficieren – Royal Association of Naval Officers] Professor of Civil Military Relations at the Faculty of Social Sciences of Leiden University. Theo Brinkel defended his doctoral thesis in 2006 on the subject of nation building and pluralism in South Africa. Previously he was a Member of Parliament for the Christian Democratic Party and a member of the Foreign Affairs and Defence Committees of the Dutch Parliament.

Chapter 3
On the Instrumentality of Soft Power; or Putin Against Democracy Promotion

Maarten Rothman

Abstract This chapter links democracy promotion to soft power in order to highlight how the spread of democracy has changed the balance of power between the West and Russia. It considers Putin's speech to the 2015 United Nations General Assembly as his intellectual response to democracy promotion and democratization. In the last four decades, democratization added significantly to the number of states on friendly terms with the West, in particular extending its sphere of influence in Eastern Europe. In response Putin invokes realism as a denial of the agency of pro-democracy protesters. Contrary to both Putin and soft power theory, this chapter argues that the agents of change are individuals and non-governmental organizations who operate largely independent of state power. While they produce outcomes conducive to Western interests, Western states cannot control their activities without irreparable damage to their own societies' core values.

Keywords Putin · Cold War · Democracy promotion · Democratization · Third Wave · Revolution · Soft power · Realism · Liberalism · Agency · Activism · Civil society

Contents

M. Rothman (✉)
Faculty of Military Sciences, Netherlands Defence Academy, Breda, The Netherlands
e-mail: mgdrothman@hotmail.com

P.A.L. Ducheine and F.P.B. Osinga (eds.), *Netherlands Annual Review of Military Studies 2017*, NL ARMS, DOI 10.1007/978-94-6265-189-0_3

3.1 Introduction

On September 28, 2015, at the UN General Assembly in New York, the president of the Russian Federation, Vladimir Putin, asked the impossible: for the West to stop exporting revolutions.[1] His examples range from Iraq in 2003 to Libya and Syria in 2011 to Ukraine in 2014. It is a strange list but Putin's argument does contain some interesting points, which help shed light on the always tricky subjects of state power and agency.

At issue is an alleged policy of democracy promotion, which results in an extension of Western influence. From a Western perspective, democracy promotion can be considered as a form of soft power. This term was coined by Joseph Nye to describe "when one country gets other countries to want what it wants".[2] It is clear that such an outcome is not to Putin's liking. Over the past few years much has been written about Putin's aggressive actions in Ukraine and Syria, his meddling in the 2016 US election and his support for right-wing populists in Europe, and his repression of dissent at home. The impression given by such stories is that Putin is an active threat to Western interests and to global peace and stability. This chapter reverses the perspective and instead considers how certain Western activities are constructed as threats by Putin. My purpose is not to excuse Putin's actions but rather to analyse the balance of power between the sides by correcting what appears to me as an oversight, namely Western pundits' neglect of potentially the strongest "weapon" in our arsenal. This is the attraction our model of government, democracy, exerts on people everywhere.

Linking democracy promotion to soft power highlights how the spread of democracy has already changed the balance of power between the West and Russia in recent decades and how it may continue to do so in the future. This approach has an important drawback, however. Both Nye and Putin focus on outcomes, conceived in terms of state power. Neither pays much attention to the agents of democratization, which are overwhelmingly non-state actors. In Putin's case this is deliberate; I will show that he has carefully constructed his argument so that it denies agency to non-state actors, appealing in particular to realist power politics. It is not easy, however, to gainsay his argument when the Western conception of soft power also reduces autonomous agents to mere channels of government influence. I will also show, therefore, that private individuals and non-governmental organizations play a crucial role in spreading democracy. Unlike most weapons, Western governments do not control who wields it and against whom.

The first part of this chapter lays out the stakes: over the last forty years democratization has affected a large number of states in all parts of the world, adding significantly to the number of democratic states and to the number of states on friendly terms with the West. The second part discusses Putin's speech to the General Assembly as his intellectual response to these facts on the ground. The

[1] Putin 2015. All quotes are from the version published by the 'Office of the President of Russia'.
[2] Nye 1990, p. 166.

third part then looks at Putin's invocation of realism as a denial of the agency of pro-democracy protesters. By contrast, soft power gets the process of change partly right, though, as the fourth part explains, only partly. In the fifth part we will finally be looking at the process of democratic revolution, concluding that it unpredictable to such an extent that it makes its instrumentalization exceedingly difficult. The West can still win adherents by inspiring democratization elsewhere but it can't very well direct that inspiration and by its very nature it can't put the instrument aside.

3.2 Democratization and the Balance of Power

Putin's problem is that the Maidan protests and the Arab Spring are part of a worldwide wave of democratization. Academics debate whether this is the third or fourth of such waves, or perhaps the fifth. The focus of the debate is on the mechanism of and limits to contagion between countries. All parties agree that the first wave started with the French Revolution in 1789 and ended shortly after the First World War, and that the second wave consisted of the democratization of formerly fascist countries after the Second World War. The third wave started with Portugal's Carnation Revolution in 1974 and encompassed the overthrow of military dictatorships in the Mediterranean and Latin America. The question is whether the fall of the Berlin Wall in 1989 started a new fourth wave or should still be reckoned part of the third wave.[3] It is clear that Latin American democrats looked to Iberian examples, so they can be grouped together. The Philippines in 1986 are also often classified with them, and from there the inclusion of South Korea and Taiwan makes sense. But did the Eastern Europeans take inspiration from the Latins or was their democratization an autonomous process? How about the various African transitions in the late 1980s and early 1990s? The Eastern Europeans, in turn, probably inspired the overthrow of Milosevic in 2000 and the Colour Revolutions of 2003–2005, at least those in former Soviet states. Lebanon's Cedar Revolution of 2005 is also sometimes counted as a colour revolution but may also be seen as the first in a separate, fifth wave of Middle Eastern democratization. The number of waves probably does not matter to Putin; what does matter is that pro-democracy movements have now occurred in every cultural-religious region on earth.

Democratization has been extraordinarily successful in terms of the number of countries affected. There have been similar waves inspired by other ideologies in the past, most prominently fascism in the Interbellum and communism between 1917 and 1989, but neither has achieved the same number of regime changes.[4] Not all are dictators overthrown, some are reforms initiated from inside the regime (e.g. Spain after the death of Franco) and many attempts remain incomplete, resulting in

[3] Doorenspleet 2005. See also Huntington 1991.

[4] Hale 2013.

hybrid regimes, or fail altogether. The numbers are nevertheless impressive. Freedom House counts 67 cases of successful democratization between 1972 and 2005, 50 of them (over 70%) driven by popular movements; that is one and a half regime change per year on average.[5] Though some were later reversed, the number of free states rose from 43 to 88 over the same period, roughly one per year. Kinsman 2011 counts 60 revolutions between 1974 and 2011, not counting regime-initiated transitions.[6] Neither study counts failed protests. Polity IV puts the number of democracies at 33 in 1974 and 89 in 2009, with the percentage of the world population living in a democracy reaching 50% in the mid-1990s.[7]

Democratization added significantly to the number of states on friendly terms with the West. While many of the regimes overthrown in the 1970s and 1980s were Western allies or stooges, the political calculus gradually but fundamentally changed during the same period. After Nixon's fall from grace and the accession of Carter, the United States pushed for the inclusion of human rights in accords with the Soviet bloc. Regime change in Portugal, Spain, Greece and (intermittently) Turkey allowed NATO to rebrand itself as a democratic alliance, while the European Community (later EU) had been exclusively democratic from the beginning. Reagan's idealism was more muscular but he nevertheless kept to the same policy and acquiesced in the reform or overthrow of various allied military dictatorships in Latin America and East Asia. By the end of the Cold War the West had made big strides in matching its foreign policy to its democratic ideals.[8] By consequence it managed to retain friendly ties with formerly dictatorial allies and extend them to formerly autocratic enemies.

Democratization thus translates into Western influence—in the case of Eastern Europe, indeed into an extension of the Western sphere of influence. Some nuances are in order, such as the preference of pro-democracy protesters in many countries for specifically European social and political arrangements sharply distinguished from the US model and their reluctance to tie domestic reform to foreign policy realignment. Western powers, including the US, engaged in democracy promotion not only out of direct self-interest but also for the sake of peace, believing (justifiably) that democracies are more pacific, at least in their relations with each other.[9] However, accession to EU and NATO did follow within a few years after 1989 throughout Eastern Europe. Nor is it only Putin who made the connection, Western leaders too have often made the link, even discussing possible Ukrainian and Georgian membership after those countries' Colour Revolutions in 2004 and 2005.

---

[5] Karatnycky 2005.

[6] Kinsman 2011.

[7] Cited in Roser 2016.

[8] Baudet 2008. As Traub 2017 points out, US democracy promotion dates back to Woodrow Wilson but the US nevertheless supported dictators throughout the Cold War; the present author follows Baudet's argument that the change in policy ties into the Helsinki process. See Chap. 4 by Baudet in this Volume.

[9] For a recent summary of the debates surrounding democratic peace theory, see Hayes 2011. For an overview of the empirical record, see Dafoe 2011.

With all the concerns about democracy at home, captured in Fareed Zakaria's concept of "illiberal democracy"[10] (and aided by Putin's meddling in the 2016 US election),[11] it doesn't come natural to many Westerners to think of democracy as a rising tide—but it is. Zakaria's worry is that democracy is increasingly decoupled from constitutional liberties, creating systems in which rulers are selected or confirmed by elections but ruling without the safeguards of civil rights. Recent examples are Hungary under Orbán and Turkey under Erdogan. But Zakaria also argues that "illiberal democracies gain legitimacy, and thus strength, from the fact that they are reasonably democratic".[12] In other words, illiberal democracy (as the name implies) is democratic. Democracy continues to win converts, as Zakaria acknowledges right at the start of both of his exposés. Indeed, that is precisely how the majority of illiberal democracies arise. A small number of established democracies have elected leaders with an authoritarian style but they have not, so far at least, changed their countries' constitutions. Very few strongmen have tried to abolish elections, preferring instead to hollow them out by such means as dominating the media and harassing opponents. Even slides such as these are, however, low in number compared to the rising tide of democracy. While datasets that measure democracy on a binary scale don't always capture this, index scales do; and those (for example, Freedom House and Polity IV) also show democracy rising. In fact, all of them have the same trend line.[13]

It is true that democracy's largest gains happened in the 1980s and 1990s, and that democratic revolutions after 2005 have been more often reversed. However, the decline in the rate of democratization is relative, not absolute, and it does not negate its historical significance. There is every reason to inquire into the causes of democratization's success in Europe, the Americas and along the Pacific Rim and its relative failure elsewhere,[14] but that should not blind us to the fact that democracy has proved attractive everywhere. Even the widely maligned Arab Spring yielded one, fledgling but still functioning, democracy in Tunisia, as well as liberalizing constitutional reforms in Jordan, Morocco and Oman.[15]

Moreover, the number of uprisings has not declined but remains stable at around one and a half per year. From Putin's perspective, a failed revolution is still a threat. Even if a thermidorian reaction reverses the revolution, it is likely to install a different strongman in place of the old regime. Sisi may not be any better than Mubarak but that is of little comfort to the deposed dictator. Even in Russia, there have been pro-democracy protests; though they were quickly suppressed, they indicate that Putin's position is weaker than that of Western countries who have only occasionally seen a protest in a third country fail but were never directly

[10] Zakaria 1997, 2003.

[11] E.g. Diamond 2016; Polyakova et al. 2016; Munich Security Report 2017.

[12] Zakaria 1997, p. 42.

[13] Wilhelmsen 2006.

[14] E.g. Huntington 1991; Way 2011.

[15] Rothman 2011a.

challenged on their own turf. The civil wars in Syria and Ukraine are humanitarian disasters about which (and about their spillover) Western publics are understandably worried but they are also geopolitical setbacks for Russia, which has lost influence over four-fifths of Ukraine and seen his Syrian allies totter; the military and financial cost of propping up Assad and seizing the Crimea are considerable.[16] No wonder then, that Putin asked the West to stop propagating democracy.

3.3 A New Cold War?

Putin's speech to the General Assembly may be considered his intellectual response to democratization. There are a number of interesting points in this speech: his condemnation of Soviet foreign policy, for example, though Putin appears to quickly balance it by the subtly nationalist claim that the international order was inaugurated at Yalta, in Russia, by the great powers, including Russia. But I will focus on the overarching theme, laid out at the beginning of the speech in a passage that is worth quoting in full:

> We should all remember the lessons of the past. For example, we remember examples from our Soviet past, when the Soviet Union exported social experiments, pushing for changes in other countries for ideological reasons, and this often led to tragic consequences and caused degradation instead of progress.
>
> It seems, however, that instead of learning from other people's mistakes, some prefer to repeat them and continue to export revolutions, only now these are "democratic" revolutions. Just look at the situation in the Middle East and Northern Africa already mentioned by the previous speaker. Of course, political and social problems have been piling up for a long time in this region, and people there wanted change. But what was the actual outcome? Instead of bringing about reforms, aggressive intervention rashly destroyed government institutions and the local way of life. Instead of democracy and progress, there is now violence, poverty, social disasters and total disregard for human rights, including even the right to life.
>
> I am urged to ask those who created this situation: do you at least realize now what you have done? But I am afraid that this question will remain unanswered, because they have never abandoned their policy, which is based on arrogance, exceptionalism and impunity.[17]

A fine piece of the rhetorical art, the speech first positions the speaker as an elder statesman, sadder and wiser, who tried to advise his impetuous Western colleagues about the dangers of unintended consequences. A council of prudence does not seem out of place after the adventures of the Bush administration; Obama's own maxim, "Don't Do Stupid Shit",[18] contains the same implicit criticism, and Putin's ascription of the rise of ISIS to the US invasion and occupation of Iraq between

---

[16] Tsygankov 2015. Though Tsygankov only discusses the conflict in Ukraine, the point can easily be extended to Syria.

[17] Putin 2015.

[18] Allen 2014.

2003 and 2011 is commonplace by now. The ethos of responsibility Putin promotes also seems uncontroversial. The devil is in the details.

Putin insists on acting in accordance with the tried and tested principles of international order, which means deferring to the UN Security Council. That this august body sometimes fails to come to agreement is as intended, in Putin's words resolutions "either pass or they don't." Indeed, the Council's refusal to endorse the removal of Saddam Hussein from Iraq can be viewed as the wiser course. On the other hand, its inaction with regard to the Rwandan genocide in 1994 and ethnic cleansing in Kosovo in 1999 was widely regarded as a failure and led directly to attempts to reform both its composition and its procedures; the former has stalled but both the Security Council and the General Assembly passed the latter in the form of the Responsibility to Protect program. Putin does not mention any of this. With respect to the intervention in Libya in 2011, he remarks only that it violated its mandate but does not address the mandate itself.

Putin also insists on deference to sitting governments, such as "Syrian authorities and government forces who valiantly fight terrorists on the ground." Military support to Assad as Syria's "legitimate government" is allowed; support for opposition groups and especially military intervention is not. So far, so hypocritical; this is after all the same man who annexed a province of neighbouring country, orchestrated a civil war in another province, and keeps troops in several other neighbouring states against their wishes. The inconsistency harms his credibility, of course—it must be assumed that Putin knows this, considering the fondness of the Russian media for pointing out real and alleged inconsistencies between Western words and actions. He is not aiming to convince us anyway. The strong air of told-you-so and the vicious description of the character of Western leaders make that clear.[19]

There are plenty of governments around the world which dislike Western interventionism, which might look to Russia as their champion. Certain Western critics of their governments' foreign policies share the contempt for Western politicians' sanctimonious chatter; after all, this line of criticism has roots in the Interbellum West, in the likes of E.H. Carr and Carl Schmitt.[20] But Putin couches his argument in a direct appeal to Western leaders, whom I assume therefore also to be part of the intended audience.

In respect of insulting speeches, Putin has form: the ostensible justification for Russian intervention in South Ossetia in 2008 was a direct copy of NATO's statement on Kosovo in 1999, very little changed except the names.[21] The purpose of that speech (delivered by placeholder president Medvedev) could not have been to convince anyone of the justice of Russia's cause, but instead aimed to

[19] An alternative explanation (Larson and Shevchenko 2014) is that Russian policy is motivated by resentment, which also subconsciously affects the language used by its leaders. Considering the scrutiny routinely given to big speeches by leaders of major states, I assume that Putin's tone is intentional.

[20] Carr 2001; Schmitt 2007.

[21] Schouten 2010.

demonstrate that NATO's claims were of the same order of transparent lies. This speech fits that pattern, its intent is not to set out a convincing moral argument but to deny the conviction of Putin's enemies.

The antagonism has a second function: to paint a picture of a world divided into two camps opposing each other at every turn. Our anger at Putin only makes that picture more convincing. The bipolar world order made prudence a necessity, an essential survival skill. It is the kind of world in which there is no room for idealist adventures. The speech is thus also a performative enactment of a kind of new cold war.[22]

3.4 Realist Framing

Why the Cold War nostalgia? Russia's status as one of the two powers that really mattered in the world is one part of it. The other is realism. This theory or approach was dominant during the Cold War among both scholars and politicians. It drew on historical example and diplomatic practice (and later on rational choice methodology) to derive lessons for the present: that states are the most important actors, that their power and interest are paramount in international politics, that peace is the result of a precarious military balance between opposing alliances, and that the best counsel for statesmen is prudence. Putin draws a straight line between the US invasion of Iraq, the intervention in Libya, the Syrian civil war and the situation in Ukraine: all of them are the result of so-called Western idealism. At first sight, his argument that an excess of utopianism and a lack of clear-eyed analysis are to blame for the failures of Western foreign policy, echoes decades of realist criticism beginning with E.H. Carr.

Of course, these alleged Western adventures are not all of the same kind. The first is an armed invasion aimed at regime change unsanctioned by the United Nations. The second was a UN-sanctioned intervention to protect the lives of civilians that adopted regime change, according to its backers, as a means to prevent a recurrence. The third was not an armed operation at all, nor the fourth. The reasons to abstain were political considerations about the prospects of a clash with Russia and the complexity of the task of rebuilding Syria—even after Assad's use of chemical weapons against population centres, the US preferred not to enforce its "redline" and instead allowed Putin to mediate (which did not prevent further chemical attacks). The fourth example, too, does not show military intervention on the part of the West but rather its reluctance to involve itself too directly in a struggle against Russian soldiers, thinly disguised as local militias and volunteers. This is precisely the kind of prudence Putin seems to be advocating.

For Putin the Iraq War is a convenient example of the West attempting to spread democracy by military force but the other examples show that this is not at all

---

[22] Sakwa 2013.

necessary; in fact, it was counterproductive. Putin's comment about Libya is also misleading as it draws attention to what he considers only the second Western intervention, the more obvious one but not the most important one. The core of Putin's complaint is not military intervention, it is that democracy keep spreading.

As we have already seen, Putin considers Western idealism a sanctimonious lie. Therefore, the imprudence cannot be the result of an excess of optimism. Instead, the cause is another, equally human failing, greed:

> Nevertheless, NATO has kept on expanding, together with its military infrastructure. Next, the post-Soviet states were forced to face a false choice between joining the West and carrying on with the East. Sooner or later, this logic of confrontation was bound to spark off a major geopolitical crisis. And that is exactly what happened in Ukraine, where the people's widespread frustration with the government was used for instigating a coup d'état from abroad. This has triggered a civil war.[23]

The passage features some interestingly contorted language. Note the shift from the active to the passive form: the prudential statesman is far away now; instead Putin appears as the unintended consequence of Western action. Not wanting to admit his culpability, he hides his own agency. It is not the only disappearance act: the protesters in Maidan square appear only as circumstance, enabling the only party with an active role, NATO, to make a power grab. In this way Putin turns revolution into coup d'état. Agency and motive go hand in hand: Putin sees great powers fighting over spheres of influence, not a people, or part of one, struggling for a chance of economic improvement against a kleptocratic regime.

The contradictions between the charges of reckless utopianism and imperialist expansion do not seem to bother Putin. But they should bother the serious analyst. Unfortunately this particular story appeals to a segment of the Western foreign policy elites. Articles by Henry Kissinger and John Mearsheimer give support to Putin's view that great power rivalry is the root cause of the conflict.[24] Realists are divided on policy, some arguing that the best response is confrontation, others appeasement. Either way, they have already acquiesced in the denial of agency to the people of Ukraine. That is both ethically dubious and descriptively wrong.

3.5 Democratization as Soft Power

Soft power at least represents the process of change better than realism. Joseph Nye's concept appreciates that changes in the balance of power need not be produced by the traditional tools of state power. Its definition—"the ability to get preferred outcomes through the co-optive means of agenda setting, persuasion, and

---

[23] Putin 2015.

[24] Kissinger 2014; Mearsheimer 2014. Tsygankov 2015 pays more attention to domestic factors, specifically to historical and cultural ties between Russia and Ukraine; his critique of the Russian imperialism thesis also negates one half of the realist explanation.

attraction"[25]—has two parts: means and outcomes. While outcomes are conceived in statist terms, as in realism, the means by which they are achieved are more varied and allow some space for non-statist actors.

Nye's central argument is that the US has policy instruments at its disposal that are often more effective than military power or economic sanctions. In *The Paradox of American Power*, Nye warned against the Bush Administration's tendency to unilateralism, correctly predicting that the 2003 US invasion of Iraq and attendant abuses would undermine the credibility of the US and thereby its capacity to persuade.[26] In his later writings, however, Nye moved away from the opposition between soft and hard power towards a prescriptive preference for a combination of the two, unimaginatively named "smart power".[27] The evolution of the concept within Nye's own works is one reason for confusion but not the only one.

The confusion stems in large part from choices Nye has made from the beginning and has stuck to throughout. Wanting his concept to be policy relevant,[28] he made soft power agent-centred, state-centred, and even US-centred. Agent-centred because it is focused on the capabilities of the actor exercising it; state-centred because that actor is a state; and US-centred because that state is specifically the US. Critics argue that by mixing theory with prescription, Nye ends up with a flawed concept.[29] Several critics have pointed out that historical, cultural or religious ties or shared civilizational values may be as effective as, or more so than, the supposedly universal values promoted by the US.[30] Economic prosperity is another basis for attractiveness on which some non-Western powers may be competitive, though the case is much easier to make for China than for Russia.[31]

More fundamentally, how soft power works is arguably better understood from the perspective of the receiver.[32] How are countries or their inhabitants persuaded or attracted? What exactly is the difference between the two? As Steven Lukes points out, Nye makes no distinction between "different ways in which soft power can co-opt, attract and entice".[33] Janice Bially Mattern adds that Nye treats attraction sometimes as a condition for soft power and sometimes as its outcome.[34] She proposes instead a theory of attraction as constructed through communicative exchange involving verbal fighting, hence her claim that soft power isn't so soft.[35]

---

[25] Nye 2011, p. 16. This is the most commonly reproduced definition; variants, both by Nye and by others, usually specify means and outcomes.

[26] Nye 2002.

[27] Nye 2011, p. 20.

[28] Nye 2011, p. 12.

[29] Wilson 2015, p. 290.

[30] Tsygankov 2013. See also Wilson 2015.

[31] Bogomolov and Lytvynenko 2012.

[32] Lukes 2005; Mattern 2005.

[33] Lukes 2005, p. 487; see also Lebow 2005.

[34] Mattern 2005, p. 591.

[35] Mattern 2005, p. 596.

The underlying problem, according to Lukes and Mattern, is that Nye confuses attributional, relational and structural dimensions of power.[36]

The issue is further complicated by the relative neglect of non-state actors. Nye treats NGOs, companies, social media networks mostly as transmission belts[37] between states and only rarely as actors in their own right. To be sure, Nye's neglect does not compare to Putin's denouncements of NGOs as agents of US intelligence but his focus on the benefits accruing to states, and to the US in particular, does play into Putin's hand. There is little doubt that soft power generally, and democratization in particular, has led to outcomes preferred by Western states. But that was not the primary intent of most of the actors involved in the process, whose priorities clearly lay with the benefits for their own people. As autonomous actors, they frequently initiated protests against regimes with whom the US had come to a pragmatic understanding and at inconvenient times.

3.6 How Democracy Spreads

Who could have foreseen that the decision not to sign a treaty of economic association between Ukraine and the European Union would motivate tens of thousands Ukrainians to take to the street and to stick to their protests for four cold winter months? For that matter, who could have foreseen that the police harassment and subsequent self-immolation of Mohamed Bouazizi, a small-time market salesman in a provincial town in Tunisia, would lead to widespread protests in 18 countries, the overthrow of three governments, constitutional reforms in three more countries, and civil war in two more? The question is important because the ability to influence the course of events depends at least in part on the ability of the actor to predict it. Revolution itself poses a collective action problem, where the efficacy and even the safety of the participants depends on their number.[38] Some triggers are fairly predictable, such as eruptions of anger after blatant election fraud.[39] In such cases, participants can be relatively confident that others will also come out to protest. Contagion effects, the spreading of revolution from one country to another, are also expected by now.[40] But that does not tell us when the contagion starts; the trigger for mass protests in the majority of cases could not be confidently foreseen.

The structural conditions for the Arab Spring are much easier to analyse than the trigger. Though experts admitted overestimating the stability of autocratic regimes in the region, they knew that leaders in Tunisia, Libya and Egypt were preparing for

[36] Lukes 2005; Mattern 2005.
[37] The metaphor is from Keohane and Nye 1977.
[38] Tilly 1978.
[39] Tucker 2007.
[40] Hale 2013. See also Snow and Benford 1992; Saideman 2012.

succession and that splits within the ranks of the regime were possible.[41] They also knew that many people were fed up with repression and daily corruption and that they could see democracies just across the Mediterranean which provided mechanisms to address these issues, not to mention delivered better economic results. They knew that democratic social movements had sprung up in similar conditions in other countries[42] and that on at least one occasion revolution had spread to the Arab Middle East. And yet, they could not predict the timing or the extent of the Arab Spring, just as they had not predicted the fall of the Berlin Wall. The problem is not that structural causes are unknown but that it is very hard to know when they will reach a tipping point. Existing regimes are resilient.[43] They have endured for decades under more or less the same conditions, they outlasted several earlier revolutionary waves (not all of them democratic),[44] so why should this one be different?

One possible answer lies in coordination. The role of social media in recent democracy protests has perhaps been overemphasized but it remains true that direct communication between potential protesters diminishes the collective action problem.[45] Add to this pre-existing (usually clandestine) pro-democracy groups and organizations who better prepared for the event than spontaneous protesters. In both the Colour Revolutions and the Arab Spring, activist groups coordinated signs, banners and press contacts, set up time schedules so that the protest site would not be left unattended during the night, trained protesters in nonviolent techniques to resist police pressure, and so on. They also formed a cadre of committed protesters prepared to see it through (sometimes for several months) when initial protests did not yield immediate results. Though activists could not know in advance when their skills would called upon, they influenced the course of events by promoting press coverage abroad and by resisting counter-protest activities by the authorities.

Outside actors could and did in turn exert influence by training the activists. While the overwhelming majority of activists is domestic, some of them did attend courses in other countries and foreign activists in some cases came over to advise them. Putin is right to see the Western influence in the Ukraine revolutions: an institute at the University of Belgrade, sometimes in cooperation with other Western NGOs, teaches nonviolent protest techniques, lessons learned from the Serbian revolution to the present. Alumni of the Center for Applied Non-Violent Action and Strategy were present in Georgia in 2004, Ukraine in 2004 and 2014 and Egypt in 2011, carrying placards with its distinctive fist logo embedded in their own signs.[46] Western states could curtail activities of this sort but, as teaching

[41] Goldstone 2011.

[42] Hale 2013.

[43] Goldstone 2011; Hess 2016.

[44] Rothman 2011b.

[45] Hussain and Howard 2013; Little 2016.

[46] Center for Applied Non-Violent Action and Strategy (no date).

nonviolent protest techniques in the West and on the internet is protected free speech, that would be unlikely to stop their spread entirely.

Putin's claim that the CIA was behind the Ukrainian revolution is controversial, to say the least. It also misses two crucial points. First, activists are self-selected. They are members of the public who take the trouble, and run the associated risks, of organizing in preparation for a popular revolt. Activist groups play a key role sustaining social movements but they do not get far without popular support. They do not determine the timing of revolutions nor, again, their extent. Activists have limited influence over the course of events after the revolution as well, as the takeover of the Egyptian revolution by the Muslim Brotherhood and the political squabbling by the heirs of the Orange Revolution both show. The presence of a (small) number of activists who received training by Western NGOs does not, therefore, prove that the revolution is not genuine; to the contrary, it shows that discontent runs deep enough that members of the public have organized to remedy their situation.

Second, trigger events are mostly homegrown. The only exceptions are revolutionary contagion and, in 1989, the withdrawal of external support by the Soviet Union for Eastern Europe's oppressive regimes which cleared the way for domestic discontent to out.[47] One may argue that Western pressure on the Soviet Union was behind this withdrawal but the West did not instigate revolutions directly. In Ukraine democracy promotion by Western states and by the EU between 1991 and 2005 failed to deliver political change.[48] Though Western fact-checking strengthened the hand of the protestors in the 2004 Orange Revolution and may have deterred violent repression[49] (unlike in Syria), it was constrained to act on a domestic trigger: election fraud. The Arab Spring provides another indication, in the form of widespread scepticism about the consequences of the democratic revolution in Egypt. The Egyptian regime had been viewed as a stable factor in a volatile region and its overthrow was seen as a security risk, particularly to Israel. While Western leaders did express support for democracy, they were quickly resigned to the reestablishment of the military regime a year later. Clearly, most would have preferred gradual reform over revolution, if only to avoid the tough choice between democratic ideals and established security policy. The conclusion thus must be that democracy promotion is a blunt instrument, which its beneficiaries are incapable of timing precisely or directing to specific countries and which, more importantly, they cannot decide not to use.

[47] Hale 2013, p. 342.

[48] Solonenko 2009.

[49] McFaul 2007.

3.7 Conclusion

As Susan Strange pointed out three decades ago, stories of the decline of the West are common in international politics but usually grossly exaggerated.[50] Not much has changed. Western pundits are pessimistic about the West's ability to resist what they view as a resurgent Russia. The reality today is otherwise: Putin is on the defensive, fighting rearguard actions against actual and potential regime changes that favour the West. Democratization has already doubled the number of democratic states over the past four decades and, though the rate of successful democratization has slowed, there is no indication that it will stop altogether. The West's strategic position has improved enormously since the end of the Cold War, while Russia is struggling to hold on in Syria and parts of Ukraine.

On the other hand, research on democratic revolutions shows that their timing or the extent depends mostly on domestic trigger events. It is beyond the capacity of Western states, even the US, to control events on the local scale, such as the self-immolation of a provincial Tunisian market vendor. The West also cannot control whom it inspires. Among the underlying causes is a role for the West as an example of a better practice, something it does simply by existing. Following Mattern, I assume that the attractiveness of democracy results in part from the West's ability to shape the terms of the discourse. Practically, it is clear that pro-democracy protesters cite economic performance, including a relative lack of corruption, as often as political freedoms. The key attribute of democracy in this respect is the ability to replace a dysfunctional or ineffective government. The West teaches this principle not only through public diplomacy but also through its own practice, which is imperfect but still clearly outmatches the performance of non-democratic regimes.[51] Since the West cannot simply put a stop to its democratic practice, and certainly would not do so only to satisfy foreign critics, and since it is impossible to cut all channels of communications from our side without ending our democracy,[52] there is no way to fulfil Putin's demand.

References

Allen M (2014) Playbook. Politico (1 June 2014) http://www.politico.com/tipsheets/playbook/2014/06/brunch-edition-forget-leading-from-behind-and-singles-doubles-west-wing-has-its-own-obama-doctrine-shorthand-hagel-in-afghanistan-lehane-fabiani-update-212543 Accessed 6 Sep 2016

Baudet F (2008) It was cold war and we wanted to win: Détente, human rights and the CSCE. In: Wenger A, Mastny V, Nuenlist C (eds) Origins of the European Security System: The Helsinki Process Revisited, 1965-75. Abingdon, London, pp 183–198

[50] Strange 1987.

[51] Roser 2016.

[52] It is possible to do so from the other side, even without foregoing the benefits of the internet altogether. See Rod and Weidmann 2015.

Bohomolov O, Lytvynenko OV (2012) A ghost in the mirror: Russian soft power in Ukraine. Chatham House, London

Carr EH (2001) The twenty years' crisis, 1919–1939: An introduction to the study of international relations. Perennial, New York

Center for Applied Non-Violent Action and Strategy (no date) http://www.canvasopedia.org/about-us Accessed 8 Jan 2017

Dafoe A (2011) Statistical critiques of the democratic peace: Caveat emptor. American Journal of Political Science 55(2):247–262

Diamond L (2016) Russia and the Threat to Liberal Democracy. The Atlantic (9 December 2016) http://www.theatlantic.com/international/archive/2016/12/russia-liberal-democracy/510011/ Accessed 4 March 2017

Doorenspleet R (2005) Democratic transitions: Exploring the structural sources of the fourth wave. Lynne Rienner Publishers, Boulder CO

Goldstone J (2011) Understanding the Revolutions of 2011 - Resilience and Weakness in Middle Eastern Autocracies. Foreign Affairs 90(3):8–16

Hale HE (2013) Regime Change Cascades: What We Have Learned from the 1848 Revolutions to the 2011 Arab Uprisings. Annual Review of Political Science 16(1):331–353

Hayes J (2011) Review Article: the Democratic Peace and the new Evolution of an old Idea. European Journal of International Relations 18(4):767–791

Hess S (2016) Sources of Authoritarian Resilience in Regional Protest Waves: The Post-Communist Colour Revolutions and 2011 Arab Uprisings. Government and Opposition 51(1):1–29. DOI:https://doi.org/10.1017/gov.2014.39

Huntington SP (1991) The third wave: Democratization in the late twentieth century. University of Oklahoma Press, Norman OK

Hussain MM, Howard PN (2013) What Best Explains Successful Protest Cascades? ICTs and the Fuzzy Causes of the Arab Spring. International Studies Review 15(1):48–66

Karatnycky A (2005) How freedom is won: from civic resistance to durable democracy. Freedom House, New York. http://freedomhouse.org/sites/default/files/How%20Freedom%20is%20Won.pdf Accessed 4 March 2017

Keohane RO, Nye JS (1977) Power and interdependence: World politics in transition. Little, Brown, Boston

Kinsman J (2011) Democracy rising: Tunisia and Egypt, when idealists got it right. Policy Options Montreal 32(4):37–43

Kissinger H (2014) To settle the Ukraine crisis, start at the end. Washington Post (5 March 2014)

Larson DW, Shevchenko A (2014) Russia says no: Power, status, and emotions in foreign policy. Communist and Post-Communist Studies, 47(3–4):269–279

Lebow RN (2005) Power, persuasion and justice. Millennium - Journal of International Studies 333(3):551–581

Little AT (2016) Communication Technology and Protest. The Journal of Politics 78(1):152–166

Lukes S (2005) Power and the Battle for Hearts and Minds. Millennium - Journal of International Studies 33(3):477–493

Mattern J (2005) Why 'Soft Power' Isn't So Soft: Representational Force and the Sociolinguistic Construction of Attraction in World Politics. Millennium - Journal of International Studies 33 (3):583–612

McFaul M (2007) Ukraine Imports Democracy: External Influences on the Orange Revolution. International Security 32(2):45–83

Mearsheimer JJ (2014) Why the Ukraine crisis is the West's fault: The liberal delusions that provoked Putin. Foreign Affairs, 93(5):77–89

Munich Security Report (2017) Post-Truth, Post-West, Post-Order? Munich Security Conference Report 2017 http://www.securityconference.de/en/discussion/munich-security-report/munich-security-report-2017/ Accessed 4 March 2017

Nye JS (1990) Bound to lead: the changing nature of American power. Basic Books, New York

Nye JS (2002) The paradox of American power: Why the world's only superpower can't go it alone. Oxford University Press, Oxford

Nye JS (2011) Power and foreign policy. Journal of Political Power 4(1):9–24
Polyakova, A et al (2016) The Kremlin's Trojan Horses. Atlantic Council, Washington DC
Putin V (2015) 70th session of the UN General Assembly. http://en.kremlin.ru/events/president/news/50385 Accessed 6 Sep 2016
Rod EG, Weidmann NB (2015) Empowering activists or autocrats? The Internet in authoritarian regimes. Journal of Peace Research 52(3):338–351
Roser M (2016) Democracy. http://OurWorldInData.org.ourworldindata.org/democracy/OurWorldInData.rg.ourworldindata.org/democracy/Accessed 20 Dec 2016
Rothman M (2011a) De Arabische Lente als historische gebeurtenis. Vrede en Veiligheid 40(2–3):209–227
Rothman M (2011b) Al-Tahrir vs. Bin Laden: De botsing van ideologieën in het Midden-Oosten van 9/11 tot de Arabische Lente. In: Osinga F, Soeters J, van Rossum W (eds) Nine Eleven - Tien Jaar Later. Boom, Amsterdam, pp 235–248
Saideman SM (2012) When Conflict Spreads: Arab Spring and the Limits of Diffusion. International Interactions 38(5):713–722
Sakwa R (2013) The cold peace: Russo-Western relations as a mimetic cold war. Cambridge Review of International Affairs 26(1):203–224
Schmitt C (2007) The concept of the political. University of Chicago Press, Chicago
Schouten P (2010) Zuid-Ossetië gespiegeld aan Kosovo: Een onderzoek naar de Russische vergelijking tussen de militaire interventies in Kosovo en Zuid-Ossetië. Bachelorthesis NLDA, Breda
Snow DA, Benford RD (1992) Master frames and cycles of protest. In: Morris AD, McClurg Mueller C (eds) Frontiers in social movement theory. Yale University Press, pp 133–155
Solonenko I (2009) External democracy promotion in Ukraine: the role of the European Union. Democratization 16(4)709–731
Strange S (1987) The persistent myth of lost hegemony. International Organization 41(4):551–574
Tilly C (1978) From mobilization to revolution. Addison-Wesley Pub Co, Reading MA
Traub J (2017) Donald Trump: Making the World Safe for Dictators. Foreign Policy (3 Jan 2017)
Tsygankov AP (2013) Moscow's soft power strategy. Current History 112(756):259–264
Tsygankov A (2015) Vladimir Putin's last stand: the sources of Russia's Ukraine policy. Post-Soviet Affairs 31(4):279–303
Tucker JA (2007) Enough! Electoral Fraud, Collective Action Problems, and Post-Communist Colored Revolutions. Perspectives on Politics 5(3):535–551
Way L (2011) The Lessons of 1989. Journal of Democracy 22(4):13–23
Wilhelmsen L (2006) A Democratic Peace Revisited. Measuring Democracy in International Relations. Cand. polit. thesis, Oslo University. http://www.duo.uio.no/bitstream/handle/10852/14081/39136.pdf?sequence=3 Accessed 8 Jan 2017
Wilson JL (2015) Russia and China Respond to Soft Power: Interpretation and Readaptation of a Western Construct. Politics 35(3–4):287–300
Zakaria F (1997) The rise of illiberal democracy. Foreign Affairs 76(6):22–43
Zakaria F (2003) The future of freedom: Illiberal democracy at home and abroad. W.W. Norton & Co, New York

Author Biography

Dr. M.G.D. Rothman is Associate Professor of International Security Studies at the Department of War Studies, Netherlands Defence Academy. He defended his Ph.D. in 2004 at Purdue University (IN, USA) on globalisation as a context for justice.

Chapter 4
'NATO Needs More Than Planes and Tanks and Guns'—Western Strategic Communication in the 1970s and 1980s and Its Implications for Today

Floribert Baudet

Abstract This chapter discusses a prime example of Western strategic communication in the final decades of the Cold War, i.e., the promotion and protection of human rights. Western human rights promotion was part of the ideological struggle between the two blocs and contributed to weakening Moscow's hold over its empire, but on the domestic scene it also created unrealistic expectations of what Western states' foreign policy should be about and ought to be able to achieve.

Keywords Human rights · Strategic communication · CSCE

Contents

F. Baudet (✉)
Faculty of Military Sciences, Netherlands Defence Academy, Breda, The Netherlands
e-mail: FH.Baudet@mindef.nl

P.A.L. Ducheine and F.P.B. Osinga (eds.), *Netherlands Annual Review of Military Studies 2017*, NL ARMS, DOI 10.1007/978-94-6265-189-0_4

4.1 Introduction

The focus in this chapter will be on Western human rights protection and promotion during the Cold War, which was a key contribution to winning this global conflict.[1] A discussion of these efforts serves two purposes. Firstly, it will show that the "novel" concepts in the field of strategic communication and non-kinetic influencing have clear historical counterparts, even within the military, and that "new" therefore is not really new. Western policy makers understood the necessity of a coherent narrative that should perform two functions: (a) convince the population in each NATO member state that the alliance would be successful in defending core Western values because its member states formed a community of values rather than a mere temporary coalition, and (b) convince the Soviet Union and its allies that it was committed to defend these values by military means if necessary. In the early 1970s a third element was added: partly for domestic reasons, Soviet human rights practices were put centre stage; they became a litmus test of Soviet good will, and improvement in this field was made a condition for a stable relationship between the two blocs.

Secondly, the study of historical counterparts of contemporary strategic communication will offer the opportunity to identify a number of mid to long-term effects of such activities. Whereas contemporary strategic thinkers and the military organizations that draw upon their ideas lavishly use acronyms to suggest that strategic communication and non-kinetic operations are miracle devices that enable victories without fighting, a study of past counterparts will reveal the complexities and inherent contradictions involved. These may serve as a warning against over-optimistic claims about the possibilities to predict and control their effects and outcome. They also give additional credence to critical analyses of more contemporary efforts such as those to "sell" the mission in Afghanistan to the wider public.[2]

4.2 Origins

There is considerable debate as to when human rights actually became a topic in international relations. Some, like Samuel Moyn, have argued that our current understanding and embrace of human rights stem from the 1970s; these should be attributed to the failure of other ideologies such as socialism and liberalism to deliver.[3] In Moyn's words, 'Our idealism is one born of disappointment, not of

[1] Baudet 2001.

[2] De Graaf and Dimitriu 2015.

[3] Moyn 2010, pp. 5–7, 47.

horror or of hope'.[4] The dominant view however is that human rights owe their attraction to the horrors of the Second World War, and the Holocaust in particular.[5]

By contrast, while acknowledging the impact of the Second World War, I propose to connect the rise of the human rights paradigm to the early Cold War, when in a deliberate act of framing, a number of European countries founded the Council of Europe as a "statement against the totalitarian countries of Europe".[6] In spite of overwhelming evidence to the contrary—after all it was only four years after the end of the Second World War, and a number of European states were involved in savage wars of decolonization—the Council defined Europeanness as respect for human rights. Its key document, the European Convention on Human Rights (signed in 1950), was meant to serve as a bulwark against the lure of totalitarianism. By the time it entered into force in 1953, this bulwark had been furnished with additional and credible defences. After the communist take-over in Czechoslovakia in February 1948, Western European states joined forces in the Western Union. A year later, on April 4, 1949, its members joined with Canada and the United States to found NATO.

It was obvious that individually the European members would be unable to withstand a Soviet onslaught and that forming an alliance was a potentially sound solution to this problem. Still, several if not all of its member states found it necessary to provide additional arguments as to why it would be advisable to commit to closer cooperation. One reason may have been that in several countries there was evidence of strong anti-American sentiments, which of course posed a risk for the cohesion of the alliance.[7] Be that as it may, NATO itself also was very much aware of the necessity to broaden popular support and its Paris-based information office NATIS tried to lead efforts to promote the "Atlantic community". This was a community of like-minded states founded on the principles of democracy, individual liberty and the rule of law. This community, of which NATO was said to be an expression,[8] was always somewhat problematic. Not only was Portugal a corporatist dictatorship when it joined, but NATO member states included Italy, and from 1952 even Greece and Turkey, that could hardly be said to border the Atlantic Ocean. To complicate matters, NATO was both said to protect this community (intimating that such a thing was actually in existence) and help create it (which of course implied that it didn't exist as yet).

In NATO's early days this uncertainty may explain why the Dutch *Defensie Studie Centrum* (more or less equivalent to the advanced staff course and consisting of excellent mid-career officers of all branches, and a number of civil servants) paid disproportionate attention to the question of the nature and substance of the Atlantic

[4] Moyn 2014, p. 60.

[5] Cf. Judt 2005; Forsythe 2012; Mazower 2004, pp. 386–388; Borgwaldt 2005; Lehners 2014, p. 31.

[6] See Baudet 2016.

[7] Frey 2009; Kroes 2000.

[8] Treaty of Washington, 4 April 1949, Preamble. For a discussion, see Smith 2000.

community and how to sell it to the Dutch population.[9] Regardless, it was only one amongst many institutions, media outlets, and military journals and training programs in Western countries to promote human rights.[10] Central coordination by NATIS was rejected though because of each member state's national sensitivities and cultural peculiarities.

4.3 Setting the Stage

Over time, societal change and, perhaps, also the effect of this steady campaign led to groups of people asking that their government finally practiced what it preached.[11] Thus, in the United States the Civil Rights Movement wanted the government to enforce the rights granted to blacks under the US Constitution. Much less acute, but an expression of similar sentiments, were the calls for equal pay and for an end to the automatic firing after wedlock of female civil servants in the Netherlands, the decriminalization of homosexuality, and "democratization" of the armed forces.[12] The early 1960s also saw the foundation of *Amnesty International* and a number of human rights NGOs focusing on specific countries and areas, whereas the mid-1960s witnessed the first state complaint for human rights violations under the European Convention of Human Rights when a number of states targeted Greece.

Meanwhile, after the near-catastrophe of the Cuba Crisis in October 1962, the Cold War entered into a period of relaxation of tensions with each side (mostly tacitly) acknowledging that an all-out nuclear confrontation could not be won. By 1969, at the political level, the Soviet bloc came up with a proposal to discuss issues relating to security in Europe. NATO indicated it was willing to discuss this, but at the behest of a number of European countries insisted on the inclusion of humanitarian issues in these talks. For obvious reasons, the Federal Republic of Germany wanted to facilitate family reunification, while others hope to achieve a more lasting improvement of inter-bloc relations.

When, from 1969, it seemed that a pan-European security conference would indeed be held in the near future, NATO member states started to formulate proposals that would promote what they termed "the free flow" (of information, individuals, and ideas, that is, a concept that encompassed a number of human rights). Research based on archival sources shows that a number of countries, such as Canada, the Netherlands, Britain, and France, felt this "free flow" could

---

[9] Baudet 2011, pp. 273–275; Baudet 2013.

[10] Baudet 2016, p. 137.

[11] Sargent 2015, pp. 6–7.

[12] Schuit and Taverne 2004, pp. 210, 258; Boot 2014.

transform East-West relations to such an extent that the division of the continent could be overcome.[13] Other Western states may have felt the same way.

By this time, the Soviet Union had also expressed a desire to draft a declaration on principles of good inter-state behaviour. It did so from a desire to establish its doctrine of "peaceful coexistence" as a principle of international relations. This doctrine held that the two blocs could coexist, while ideological competition continued. This implied that each had its own set of intra-bloc norms, and would thus legitimize the doctrine of limited sovereignty of socialist states. This doctrine, better known under its Western name of "Brezhnev Doctrine", held that socialist states had the right to militarily intervene when the fruits of socialism were under threat in another socialist country. Of course, Moscow would be the judge in this.

Whereas the Nixon-Kissinger team was willing to accept the Soviet domination of Eastern Europe as a fact of life, the Western European states refused to legitimize the Brezhnev Doctrine. Instead, they hoped to bind the Soviet Union by international law, human rights law, and the UN Charter, thus outlawing military intervention Czechoslovakia-style. So when in early 1973, the Soviet bloc as part of talks on the agenda of the proposed conference formally stated its desire to include a declaration on principles, Denmark, on behalf on the nine member states of the European Community, insisted that full respect for human rights be included as one of these.

Several months later, when NATO members had to decide on whether to formally participate in the upcoming security conference, Dutch foreign minister Max van der Stoel reminded his colleagues that NATO could hardly

> appear sincere in the defense of our "freedom, common heritage and civilization of our peoples, founded on the principles of democracy, individual liberty and the rule of law" – and therefore we can hardly expect to convince – if we ourselves, the members of this Alliance, do not, one for one, practice what we preach. Mr Chairman, freedom and democracy are more than lofty ideals; they are, as we know, exacting principles for conducting the business of govern-ment, requiring a constant struggle. And it is here that these principles, which I just quoted straight from the preamble of the treaty, do not remain a dead letter. (…) NATO needs more than planes and tanks and guns – it needs widespread popular support. In order to maintain the credibility and vitality of the alliance, it is essential that the basic requirements of individual liberty and political democracy are being respected everywhere.[14]

These words were directed at Portugal, Greece and possibly Turkey. Portugal and Greece were dictatorships, and after a coup in 1971, in Turkey the military severely limited civilian politicians' leeway. Van der Stoel's speech, though unprecedented, reflected a growing awareness, a sense of urgency even that fact and narrative should "match". In order to maintain public support and remain credible in the eyes of their opponent, NATO member states had to live up to their solemn

[13] Baudet 2001; Cotey Morgan (forthcoming); Cotey Morgan 2008.

[14] Archives of the Ministry of Foreign Affairs (The Hague) (AMFA), archive of the Directorate of Atlantic Cooperation and Security (DACS), 1965–1974, inventory number 703. Address by M. van der Stoel, Copenhagen, 14 June 1973.

promises; furthermore, Western policy makers increasingly believed that there could be no lasting and stable relationship with the communist states unless these states stopped paying lip-services to human rights and the rule of law, but actually started to respect them.

Not surprisingly, in the fall of 1973, the Dutch, also acting on behalf of the Nine, filed a proposal to have the right to self-determination included in any list of solemn principles that had to guide East West relations. The original text explicitly allowed a regime change. This of course was unacceptable to Moscow, but the Dutch, with the support of Greece and the remaining eight states of the European Community held firm. In the end, a toned-down version was in fact included in the Final Act that concluded the conference on 1 August 1975, and so was the principle of respect for human rights—each individual had inalienable rights that he had the right to know and act upon. The "Helsinki Final Act" also implicitly banned the Brezhnev Doctrine and additionally, it devoted an entire chapter to commitments with regard to the free flow of information, individuals and ideas, which had now been rebranded "Co-operation in Humanitarian and Other Fields".

While often presented in realist school IR literature as a major political defeat, since the Act also accepted the post-1945 borders, in recent years there has been a re-evaluation of its contents. The view that the Final Act actually represented a Western diplomatic victory is now widely accepted.[15]

4.3.1 Contradictions

From 1975, when the Final Act was signed, NATO members publicly vowed to treat progress in the field of human rights as the litmus test for détente and of Soviet good will. Implementation of the Final Act was crucial. Both the alliance and individual Western states did indeed try to forge a linkage between détente, disarmament and human rights. They held that détente could not be proclaimed, it should be the result of renewed trust, and thus be based on actions rather than mere words. As security was indivisible, Western trust of Soviet intentions could be reduced if the Soviet Union and its allies would live up to its human rights commitments. Likewise, if there was trust between the two sides, then there would be no need for an arms race, or the deployment of new nuclear weapons.[16] So far so good, but policy makers in the West soon found that this linkage was problematic. While rhetorically logical, and no doubt to a large extent sincere, this insistence was highly problematic.

---

15 Cf. the analysis in Romano 2009. See also Wenger, Mastny and Nuenlist 2008. For the Final Act itself, see www.osce.org/helsinki-final-act.

16 Witness speeches of key NATO leaders in Helsinki, such as US President Gerald M Ford and West German Chancellor Helmut Schmidt, see OSCE 1975a, b.

Part of the explanation for this was domestic in nature. Détente had been unpopular with many on the political right as it was seen as selling out on the West's founding principles. The Nixon-Kissinger tandem, and their German counterparts of Brandt and Bahr, were resented to a certain extent for accepting the status quo. Large sections of the public were not prepared to do this. Likewise, on the left there was a sense that the arms race was immoral and support for NATO was waning. It has been suggested that politicians stressed the importance of human rights as they would help re-establish a common ground in the domestic arena.[17] After all, both left and right subscribed to the idea of human rights itself. But as soon as they were elevated to a foreign policy goal, and put centre stage in a formal linkage with disarmament, disagreements arose as to which human rights mattered most, and whether they should really take precedence over strategic considerations—or economic ones for that matter. When confronted with the reality of nuclear annihilation, wasn't the only human right that truly mattered the right to life? And should one complain about violations that took place in allied states?

Thus, rather than uniting the home front, human rights split it, and accusations of hypocrisy abounded. The left accused the right of supporting right-wing dictatorships and racist regimes such as *Apartheid*-South Africa, whereas the right accused the left of turning a blind eye to human rights violations in the Soviet bloc and in newly-independent left-leaning former colonies. Not only did this lead to parliamentary strife, it also spurred discontent on the part of the growing number of human rights NGOs that felt the Western governments' actions fell short of their promises. This ruined a number of reputations. Finally, the public at large may have concluded from the quibbling that politics is nothing but organized hypocrisy.

As the declarations about linkage were not primarily meant for domestic consumption, but were instead intended for the Soviet leadership, I now turn to reactions in the Kremlin. Unsurprisingly, Brezhnev and his fellow leaders had no intention of submitting to the Western interpretation of the document right away. While some, like KGB chief Yuri Andropov, even warned against signing the document for it legitimized Western interference, foreign minister Andrei Gromyko famously remarked, "we are still the masters in our own house".[18] The Soviets widely published the Final Act—it was even projected on the wall of a building in Tashkent in the Uzbek SSR[19]—and presented it as a stunning victory of the Soviet bloc and as proof of the West's recognition of Moscow's hold over it. This contributed to a sense of defeat in Western public opinion.

Behind closed doors however, sentiments in the Kremlin were more subdued. The Soviet leadership resented the linkage that had been put in place by Congress in the 1974 Jackson-Vanik amendment to a US/USSR trade agreement. An increase in the emigration of Soviet Jews would be rewarded by a raise in the number of tons of

[17] This view contrasts with Suri's analysis that détente, not human rights promotion, was the answer to these changes: Suri 2003.

[18] Zubok 2007, p. 237.

[19] Interview J.L.R. Huydecoper van Nigtevecht, Dutch Ambassador to Moscow 1974–1977.

American wheat the Soviet Union could obtain. After a number of bad harvests the Soviets needed these deliveries and felt the United States attempted to blackmail the Soviet leadership. They reacted by severely reducing the number of exit visa for Jews.[20] The Western intransigence with regard to self-determination and its insistence on individual human rights during the negotiations that led to the Final Act was interpreted as a concerted attempt at subversion, especially so when Western states rejected Moscow's concerns over for instance the treatment of Moluccans and British mineworkers. Furthermore, the Soviets held that while respect for human rights was important, it was only one among ten Principles in the Final Act and the document itself stipulated that these were equally valid. They claimed that the West violated the principles of non-intervention and of sovereign equality when they insisted the Soviet Union and its allies should honour their commitments in the field of human rights. They felt that the West was not interested in peaceful coexistence, but continued the Cold War by other means. This accurate analysis hardened the resolve of the communist leaderships and contributed to the deterioration of East-West relations from 1975 onward.

This development was reinforced by the fact that the Final Act stimulated the rise of dissident groups throughout the Soviet sphere of influence. While the agreement did not produce the dissident movement, it became a focal point in their activities because the leaders of their countries had voluntarily signed the document and reaffirmed a number of commitments that were also in their constitutions. After 1975, new dissident groups arose and demanded that the regimes made good on these commitments. This was cause for serious concern in Moscow, because it tested its sway over its satellites and Moscow repeatedly instructed its allies on how to deal with these dissidents.

This rise of the dissident movement caused concern in several Western capitals as well, because their existence complicated matters. The West had reached tacit agreement with a number of regimes on closer economic relations and access to Western loans in return for gradual internal liberalization, an independent foreign policy, or both. The rise of the dissident movement in Poland in 1976–1977 and then of the independent trade union Solidarity, from 1980, however put this liberalization at risk, and until 1981 the West largely refrained from criticizing the Polish authorities. Repressive Romania, ruled by the megalomaniac Nicolae Ceauşescu, pursued a rather independent foreign policy which the West hoped to stimulate. The Final Act's stipulations on sovereign equality were to assist Romania in this, and the West deliberately ignored that country's internal repression. In other countries, such as Czechoslovakia, there was no liberalization or independent foreign policy that could be put at risk. Accordingly, the Western reaction to the crack-down of Charter `77 was vehement, which of course led to the accusation of selectivity.[21]

---

[20] On this: Buwalda 1997.

[21] Baudet 2001, pp. 99–252.

The West was also alarmed at the pace of events. While the European Nine, Canada and a number of Neutral states had hoped to promote more room for individual aspirations, and, in the end pave the way for a peaceful transition in Eastern Europe, they had done so without a time table. They had never expected groups that demanded that governments respected their international and constitutional obligations in the field of human rights to emerge within two years of the Act's signing. As the communist regimes were based on force rather than on popular support, fear of destabilization now came into the mix. In Poland, the regime seemed on the verge of collapse after its legalization of Solidarity in August 1980. The trade union attracted millions of members, until thinly veiled Soviet threats to intervene caused the Polish communist leader, General Jaruzelski, to declare Martial Law in December 1981, and imprison Solidarity's leadership. Paradoxically, in the West this crack-down was greeted with some relief —behind closed doors, of course. Public manifestations were much different and the West put pressure on Jaruzelski to release the Solidarity leadership and reopen a dialogue with the trade union. Poland faced a diplomatic (and financial) boycott. However, Western estimates as to Jaruzelski's room for manoeuver and intentions differed considerably and within three years the West's unified front had started to crumble. By 1987, only the United States and the Netherlands clung to the hard-line approach that NATO's members had agreed upon in December 1981.

This was only one out of many examples that show the complexity of harmonizing strategic communication and strategic interests. By the mid-1980s Western politicians warned that their domestic audiences were becoming disillusioned because "the only thing that the CSCE process had been producing was words, and concrete and tangible results had not been achieved."[22] While no doubt this was somewhat exaggerated, by this time the West had become deeply divided over the wisdom of the linkage they established in Helsinki ten years earlier. In the end, though, the unity of effort was secured and the CSCE follow-up meeting in Vienna (1986–1989) saw the Soviet bloc accept human rights as a legitimate topic of international relations. In January 1989 the concluding document of the Vienna Meeting established a political mechanism by which states could be summoned by other states to explain their behaviour with regard to human rights.[23] In accepting this, the communists gave up their monopoly of power as this mechanism legitimized the activities of human rights activists, dissidents and others who now finally had some breathing space, and, indeed, acted upon this. By year's end, the regimes in Eastern Europe had collapsed.

[22] AMFA, DACS, 1985–1989, file '10 jaar slotakte 16 juli 1985–18 januari 1986'. Address by H. van den Broek, Helsinki, 1 August 1985.

[23] Leatherman 2003, p. 187.

4.4 Epilogue

The case of Western Cold War human rights promotion illustrates the complexities involved in strategic communication and non-kinetic influencing. Western governments made respect for human rights the litmus test for détente, raising expectations with their domestic audiences and with various audiences in their target countries. But when things came to a head, time and again the fear for the stability of the regimes proved more important. In many cases, this meant that dissidents found that support of the West was in words only, and only as long as dissidents did not tilt the fragile balance between the two blocs. This is hardly surprising as instability could well lead to a military conflict with the Soviet Union, but it meant that Western governments often found themselves facing domestic criticism for not doing enough to promote and protect human rights, and indignation from communist regimes for meddling in their internal affairs, which complicated arms limitation talks and contributed to the so-called "Second Cold War" of the late 1970s and early 1980s.

The analysis presented above should however not be construed as to vindicate the realist critique of human rights advocacy, for eventually this advocacy paid off. And Henry Kissinger, who had been a staunch opponent of this advocacy, later admitted he was wrong.[24] What transpires, though, is that in influencing non-kinetically, an alliance encounters numerous challenges. Firstly, it will need to devise a common theme, which will need to derive from a common strategy. At the same time, given that the alliance's adversary is also listening, it cannot detail its strategy in full, leaving room for ambiguity. This may serve to obscure internal divisions, as it did for NATO during the early 1970s, but it may also complicate both internal relations and the relation to the adversary since acts and words do not seem to match. This, of course, may also be intentional as part of a strategic deception, but then it must be clear to all alliance members how the words relate to the acts.

Secondly, it will have to take into account not only the target audience but also other audiences that happen to be listening in. Expectations of domestic audiences (and the interests of various information services, branches of the armed forces and the like) may vary and thus complicate or even negate the formulation and then propagation of a single narrative.

Thirdly, as part of strategy, communication is a part of a clash of wills, and the target audience no doubt will come up with a narrative of its own. Thus one sees a number of competing narratives. In open societies such narratives will most likely be scrutinized, but, to give an example, the Soviet Union's Cold War posture as peaceful was rather successful in that it won the sympathy of thousands that had no communist sympathies. In closed societies it seems far more difficult to achieve such a success, either because intended audiences have no access to our message or because they have no means to scrutinize their authorities' narrative. This suggests

[24] Kissinger 1999, pp. 635–663.

that formulating and executing a coherent strategic communications policy is easier for our opponents than it is for Western governments. While no doubt far more attractive than thermonuclear influencing, in the 1970s and 1980s non-kinetic influencing was no recipe for instant success. There is no reason to think it is now.

References

Baudet FH (2001) "Het heeft onze aandacht" - Nederland en de rechten van de mens in Oost-Europa en Joegoslavië, 1972–1989. Boom, Amsterdam

Baudet FH (2011) The ideological equivalent of the Atomic Bomb. Journal of Transatlantic Studies 9(4):269–281

Baudet FH (2013) Het Vierde Wapen - Voorlichting, propaganda en volksweerbaarheid, 1944–1953. Boom, Amsterdam

Baudet FH (2016) A statement against the totalitarian countries of Europe. Cold War History 16 (2):125–140

Boot CYE (2014) Het leger onder vuur - De Koninklijke Landmacht en haar critici, 1945–1989. Ph.D. thesis Leiden University

Borgwaldt E (2005) A New Deal for the World: America's Vision for Human Rights. Harvard University Press, Cambridge

Buwalda P (1997) They did not dwell alone. Praeger, New York

Cotey Morgan M (2008) North America, Atlanticism, and the Helsinki Process. In: Wenger A, Mastny V, Nuenlist C (eds) At the Roots of European Security: The Early Helsinki Process Revisited, 1965–1975. Routledge, London, pp. 25–45

Cotey Morgan M (forthcoming) Helsinki 1975: Détente, Human Rights, and the Transformation of the Cold War. Princeton University Press, Princeton

de Graaf BA, Dimitriu G (2015) Strategic Narratives, Public Opinion and War: Winning domestic support for the Afghan War. Routledge, London

Forsythe D (2012) Human Rights in International Relations. Cambridge University Press, Cambridge

Frey M (2009) Decolonization and Dutch-American relations. In: Krabbendam H, van der Minnen C, Scott Smith G (eds) NL/USA 400 years of Dutch-American relations. Boom, Amsterdam, pp. 609–620

Helsinki Final Act, http://www.ocse.org/helsinki-final-act Accessed 1 October 2016

Judt T (2005) Postwar - A History of Europe Since 1945. Penguin, London

Kroes R (2000) Anti-Americanism and Anti-Modernism in Europe. In: Stephan A (ed) Americanization and Anti-Americanism: the German encounter with American culture after 1945. Berghahn Books, New York

Kissinger H (1999) Years of Renewal. Simon and Schuster, New York

Leatherman J (2003) From Cold War to Democratic Peace: Third Parties, Peaceful Change, and the OSCE. Syracuse University Press, Syracuse, NY

Lehners JP (2014) Pleading for a new history of human rights. In: Mihr A, Gibney M (eds) The SAGE Handbook of Human Rights, Volume I. Sage, London, pp. 22–38

Mazower M (2004) The strange triumph of human rights, 1933–1950. The Historical Journal 47 (2):379–398

Moyn S (2010) The Last Utopia. Human Rights in History. Harvard University Press, Cambridge

Moyn S (2014) The Future of Human Rights. SUR - International Journal on Human Rights 20 (1):57–64

OSCE (1975a) Speech by US President Gerald Ford, Helsinki (1 August 1975) http://www.osce.org/documents/16090?download=true Accessed 1 October 2016

OSCE (1975b) Speech by Helmut Schmidt, Chancellor of the Federal Republic of Germany, Helsinki (1 August 1975) http://www.osce.org/documents/16088?download=true Accessed 1 October 2016
Romano A (2009) From Détente in Europe to European Détente: How the West Shaped the Helsinki CSCE. Peter Lang, Brussels
Sargent D (2015) A Superpower Transformed: The Remaking of American Foreign Relations in the 1970s. Oxford University Press, Oxford
Schuit K, Taverne E (2004) Dutch Culture in a European Perspective: 1950, Prosperity and Welfare. Palgrave MacMillan, London
Smith M (2000) NATO Enlargement during the Cold War: Strategy and system in the Western Alliance. Palgrave, London
Suri J (2003) Power and protest. Global revolution and the rise of détente. Harvard University Press, Cambridge
Wenger A, Mastny V, Nuenlist C (eds) (2008) At the Roots of European Security: The Early Helsinki Process Revisited, 1965–1975. Routledge, London
Zubok V (2007) A failed empire: the Soviet Union in the Cold War from Stalin to Gorbachev. University of North Carolina Press, Chapel Hill

Author Biography

Dr. F.H. Baudet is Associate Professor of Strategy and Military History at the Department of War Studies, Netherlands Defence Academy. He defended his Ph.D. in 2001 at the University of Utrecht on the Netherlands' (human rights) policy for Eastern Europe and Yugoslavia (1972–1989).

Chapter 5
The Snake Oil of Stabilisation? Explaining the Rise and Demise of the Comprehensive Approach

Thijs Brocades Zaalberg

Abstract Following the withdrawal of large scale military contingents from Afghanistan and Iraq, enthusiasm in the Western world for what has been called 'armed state-building' seems to have sunk to an all-time low. This is not entirely surprising given the high stakes and the meagre results of the massive efforts by American, European and other military personnel and civilians in the wake of forced regime change in these countries in 2001 and 2003. As a result, the so-called comprehensive approach—the NATO-concept for integrating military and civilian efforts in post-intervention stabilization efforts—has also lost much of its shine. This concept emerged in recognition of the fact that such operations required concerted efforts of a variety of state and non-state organizations. In this sense, the comprehensive approach was the organizational vehicle to engage with and influence the existing post-conflict economic, political and social structures and processes in failing and fragile states. Treating the concept in a broader historical context, this chapter sketches and explains its roots, rise and the early signs of demise. It argues that it is justified to question many of the assumptions underlying the recent idealistic interpretations of comprehensive approaches and to criticize the actual levels of civil-military integration in Afghanistan. It also acknowledges the inevitability, given current threats and a loss of appetite for counter-insurgency and state-building, of a shift of focus to conventional warfare and waging ground wars by proxy. However, like US institutions in the wake of the Vietnam War—we run the risk of throwing out the baby with the bathwater if we do not continue to build on past mistakes, thereby shaking of the naivety of the post-9/11 era and anchoring the many lessons that were learned. After all, the type of conflict we get embroiled in is mostly not of our own choosing.

Keywords Comprehensive approach · State-building · Counter-insurgency · Peace operations · Civil-military cooperation

T.W. Brocades Zaalberg (✉)
Faculty of Military Sciences, Netherlands Defence Academy, Breda, The Netherlands
e-mail: tbrocades@gmail.com

T.W. Brocades Zaalberg
Institute for History, University of Leiden, Leiden, The Netherlands

P.A.L. Ducheine and F.P.B. Osinga (eds.), *Netherlands Annual Review of Military Studies 2017*, NL ARMS, DOI 10.1007/978-94-6265-189-0_5

Contents

5.1 Introduction

Only a decade ago,[1] the idea of an integrated approach that combined military, political and socio-economic means in stabilization operations, counter-insurgency and state-building held great promise. As NATO gradually expanded its area of operations of the International Security and Assistance Force (ISAF) in Afghanistan after 2004 and the 2007 US 'Surge' of troops and civilian capacity in Iraq started to bear fruit, the so-called comprehensive approach seemed a unifying concept that one could hardly object to. Who but a number of orthodox humanitarians in the NGO community defending the 'humanitarian space' or some war mongers propagating singular reliance on punitive military strikes could possibly disagree with better cooperation—and preferably integration—of security, reconstruction and diplomatic efforts when building peace and (re)constructing war torn states?[2] The comprehensive approach seemed to offer the best chance of winning the peace and —in more violent missions—it would ideally contribute to the ideal of 'winning without killing' by relying on other than military means.

For NATO and many of its more active member-states, Afghanistan emerged as a laboratory for the comprehensive approach that emerged after 2004 as the Alliance-concept for integrating military and civilian efforts in theatre. On the national level, various governments embraced related concepts such as the '3D-approach' (Defence, Diplomacy and Development), the 'Whole of Government Approach' or the 'integrated approach' for their own interdepartmental processes. In the meantime the NATO military also temporarily embraced the related Effects-Based Operations-concept (EBO). Recently, however, enthusiasm in the Western world for what has been labelled 'armed state-building' or 'nation-building under fire' has sunk to an all-time low.[3] This is not surprising given the high stakes and the meagre long-term results of the sacrifices and the massive efforts by American, European and other military personnel and civilians in

[1] The author would like to thank Andrew Gawthorpe and Frans Osinga for their comments. He is indebted to the former for the quote by Sherard Cowper-Coles that was used in the title.

[2] For the seemingly self-evident character of the benefits of the CA, see De Coning and Friis 2011, pp. 257–258.

[3] Gates 2009, pp. 29–30; Gentile 2013.

Afghanistan as well as Iraq in the wake of forced regime change in 2001 and 2003. As a result, the comprehensive approach has also lost much of its shine.

This chapter focuses on the Afghan case as it defined NATO's approach. However, it argues that the explanation for the concept's impending demise is more complex than the disillusionment with fifteen years of civil and military efforts in Central Asia, or ten years of 'armed state-building' in Iraq. A second cause has been the tendency to treat the comprehensive approach as a relatively new phenomenon, which made it vulnerable to critique on the basis of the Afghan and Iraqi cases. The third factor undermining the credibility of the concept has been the tendency by policymakers to abuse the rosy image of cooperation and integration as a substitute for strategic guidance, thus inflating the concept into something it was not: strategy. Fourth, increased scepticism and the impending neglect seems to be caused by the tendency in academic circles to question the very assumptions underlying the current interpretations of the comprehensive approach: liberal state-building and the liberal peace paradigm. And finally, as a result of emergent threats, there is a recent shift of focus from counter-insurgency type operations to conventional warfare and waging war by proxy.

5.2 The Roots of the Comprehensive Approach

In 2010, NATO Secretary General Anders Fogh Rasmussen called the comprehensive approach the primary lesson from Afghanistan. In his effort to amplify the need for better civil-military integration, he stated that "[t]he days when the military could defeat the enemy, then hand the baton off to the civilians and go home, are past us."[4] By embracing this common notion that in history warfare was a rather straightforward affair, with clear end-states and a neat division between military and political and civilian responsibilities, Rasmussen built on what has been called the new complexity myth.[5] Although the increased complexity of integrating civil and military efforts in theatres such as Afghanistan is hard to deny—particularly with the information dimension moving centre-stage—such approaches in irregular warfare, post-conflict operations, domestic counter-terrorism as well as conventional warfare have been an integral part of the historical landscape since ancient times. Mostly, the requirement to integrate civil and military efforts were met by handing civil powers to military commanders, something modern democracies are uncomfortable doing. History is rife with examples of soldiers acting as governors, from the Roman proconsuls governing newly conquered and potentially rebellious territory, to Allied commanders exercising civil powers during and after the Second World War.

---

[4] Rasmussen 2010.

[5] Mountcastle 2016, pp. 47–53.

Up to the capitulation of Nazi Germany, General Eisenhower and his subordinate commanders did so primarily in support of military objectives: the defeat of the enemy. They delegated military government in liberated and occupied territory to a vast Civil Affairs apparatus made up of administrators in uniform. Directly after the surrender, occupation through military government and constabulary operations became the primary strategic objective. Even though Allied and German civilian officials gradually took over from Civil Affairs—12,000 military personnel in the US occupation zone alone—they continued to do so under military command until 1948. On the relation between military operations, governance and political objectives F.S.V. Donnison wrote in 1966 in the official British history of military government in the Second World War: "Modern war consumes governments and administrations in its path, leaving anarchy and chaos behind. If authority and the necessary minimum order and administration are not at once re-established, disorder and subversion can all too quickly erode the victory that has been won in the field."[6] Unlike during World War II, US policymaker consciously ignored this vital aspect of war in 2002–2003—with devastating consequences up to the present day.

Also French, British, Dutch and other imperial officers fighting 19th and early 20th Century 'small wars' often acted as proconsuls. Nowhere was Von Clausewitz's dictum that war is "the continuation of policy with the addition of other means" more tangible down to the lower levels of command than in the imperial arena. In support of a grand colonial designs such as the *mission civilatrice*, French generals like Joseph Gallieni and his disciple Hubert Lyautey created a civil-military blueprint for pacification called the *tache d'huile* or 'oil spot' method for Indochina, Madagascar and Northern Africa around 1900. What they euphemistically labelled 'peaceful penetration' was geared towards gaining the loyalty of the population through political and economic incentives, such as the creation of marketplaces near French military outposts to draw the local population to safe and prosperous areas. While simultaneously targeting the insurgents outside the oil spots, the 'pacified' areas were gradually to merge.[7] However, in practise colonial 'pacification' remained an extremely violent affair. French control over 'future subjects', like that of the expanding British, Dutch and other expires under guise of their versions of the 'civilizing mission', depended on carrots and the co-optation of local elites as well as on sticks in the form of (collective) punishment and exemplary force—or the credible threat thereof.[8] All adhered to the colonial wisdom that sticks made carrots more effective.

During the era of decolonisation that followed the Second World Wars, the colonial powers experimented in various ways with integrated approaches. The Dutch were least successful when Indonesian nationalists rebelled immediately and massively upon their return in late 1945. Lacking either civilian capacity for

[6] Donnison 1966, p. 361.

[7] Porch 2002, pp. 132, 211.

[8] The British and the Dutch launched similar notions around 1900 in the form of 'White man's burden' and 'Ethische Politiek'.

administration and policing or a specialized military branch for civil affairs during the crucial counter-insurgency phase from 1947 to 1949, the Dutch failed to create a successful and balanced mechanism for civil-military co-operation. Despotic but ineffective military rule increasingly became the norm in a period when—under the eyes of an increasingly critical global audience—wielding sticks freely became harder.[9] Fighting the much smaller communist revolt in Malaya in the 1950s, the British successfully introduced what became known as the 'war by committee system', the foundations of which was formed by the District War Executive Committees. Chaired by the colonial district administrative officer, the committee included the battalion or company commander and the police superintendent. They worked together to conduct day-to-day operations supported by an integrated intelligence committee of soldiers and police. For each of the nine Malayan states a similar triumvirate, a State War Executive Committee, was created. In Kuala Lumpur, the high commissioner chaired the Federation Executive Council that issued policy directives and controlled finance. It took some time for the system to function smoothly, but it proved to be the most effective implementation tool for a comprehensive strategy aimed at segregating the insurgent from his popular base and a viable policy that accepted decolonisation as inevitable. [However, its ultimate success hinged on a viable overall policy that accepted that true decolonization as inevitable—something the Dutch and the French we not ready to do in Indonesia and Algeria.]

In 2007, at a time when the tendency to abuse the sanitised lessons of the Malayan Emergency for Afghanistan and Iraq was at its height, Hew Strachan reminded his audience at a counter-insurgency conference that—in the context of Malaya and other wars of decolonization—"[w]hen we speak about 'hearts and minds', we are not talking about being nice to the natives, but about giving them a firm smack of government."[10] The British approach was about authority, even though political and social reform did accompany firm government. In Malaya it also hinged on massive forced resettlement. The British colonial apparatus has rightfully been called a police state—yet "a police state with a conscience".[11] In recent years, however, particularly in relation to the Mau Mau rebellion in Kenya in the 1950s, historians have called this latter qualifier into question. Nevertheless, it is undeniable that during several campaigns, the British reached a reasonable degree of success by prioritizing the political dimension of counter-insurgency and by the creation of a system for civil-military cooperation within a predominantly civilian context.

The British committee system was the alternative to the 'single commander system' mostly preferred by the French. This is not to say that French did not reach any level of success. Having learned from experience in Vietnam in the early 1950s, the French campaign in Algeria saw a reasonable degree of co-ordination between

[9] Brocades Zaalberg 2015, pp. 67–83.
[10] Strachan 2007, p. 8.
[11] Beckett 2001, p. 92.

the civil and military efforts and scored many tactical victories in pacifying the country. Peter Paret argued in his powerful analysis of French counterinsurgency doctrine: "[T]he integration of political and military measures that is a part of every war must not only determine the grand lines of strategy but also effectively inform the tactical decisions. The French met this requirement for all-encompassing low-level integration by extending the responsibility of the armed forces …" Under Governor General Jacques Soustelle, the French addressed the acute problem of under-administration by creating the *Sections Administratives Specialisées* (SAS), manned by military personnel also known as the *kepis blues* because of their distinctive headgear. The French created some 400 S.A.S. detachments, "each under an army lieutenant or captain who was an expert in Arabic and Arab affairs and could deal with every conceivable aspects of administration; from agronomy, teaching and health, to building houses and administrating justice."[12]

In order to "breach the front of misery", Soustelle initiated an impressive program of agrarian reform while he increasing school-building. He also made Arabic an official, obligatory language in schools.[13] The SAS played an important role in executing these programs. For their impressive socio-administrative work under dangerous circumstances in remote areas the SAS lacked sufficiently qualified personnel and despite the praise for much of their work, some teams tarnished the reputation of the SAS by transforming their offices into centres for intelligence gathering "where torture was not unknown." The French reached high levels of tactical and even operational success, but only after committing 450.000 troops and at the cost of moral bankruptcy due to institutionalized torture that structurally undermined popular support in the theatre and in metropolitan France. Massive forced migration without sufficient adjoining socio-economic measures further undermined the effort. Paret concluded in 1964 that civil-military integration was better achieved by leaving to civilian personnel the areas of internal war that "despite their military implication" were essentially social, economic and political. "The role of civilians in revolutionary warfare remained, however, largely unexplored in Indochina and Algeria." French counter-insurgency theory and practise was "almost wholly military in character."[14]

The same has been said of the US dominated phase of the Vietnam War from 1965. The American approach is often presented as the epitome of a militarized answer to a complex insurgency that effectively combined all instruments of communist power. However, the Americans eventually reached high levels of civil-military integration when compared to recent operations. Like in recent years, there was no lack of awareness at the top, as shown by general Westmoreland's September 1965 statement that "[this] is a political as well as a military war … the

[12] Horne 2006, p. 109.

[13] Horne 2006, pp. 108–109.

[14] Paret 1964, p. 125. The fundamental source of failure in Algeria, however, was on the political level. Like the Dutch in the East-Indies, the French declined to offer independence, and thus refused to address Arab political and socio-economic aspirations.

ultimate goal is to regain the loyalty and cooperation of the people. It is abundantly clear that all political, military, economic and security (police) programs must be integrated in order to attain any kind of success."[15] Theory and practice took long to merge due to slow learning and conventional military reflexes is the US Army. It took two years from the arrival of the first large-scale US combat units to the creation of the hybrid Civil Operations and Revolutionary Development Support (CORDS).[16] From 1967, the history of civil-military integration in Vietnam is very much the story of CORDS, whose four regional branches attached to the four US corps headquarters mirrored it civilian-led central organisation in Saigon. A three-star general headed each corps with a deputy commander for CORDS, usually a civilian. CORDS rapidly increased to 8,327 personnel who were eventually operational in all 44 provinces and 250 districts. Fifteen percent of its personnel were civilians focussing on community development and supporting the Vietnamese province chiefs establishing their authority. The rest was military personnel, most of whom handled security issues such as civil defence and police matters.[17]

Despite these impressive numbers and many accomplishments, Andrew Gawthorpe concludes in his recent book on US state building in Vietnam, that "CORDS suffered from tardiness and slow lessons learned". According to its first leader, Robert Komer, it was simply "too little, too late."[18] Treating the program in the context of recent operations, Gawthorpe's final verdict is even harsher:

> CORDS boasted thousands of personnel, unity of command, a sophisticated operating concept based on an understanding of its main antagonist, theoretically close integration with the host government, and even a training institute dedicated to producing personnel skilled in understanding the country in which it operated. Still it failed. No doubt in the future there will be those who say they can succeed in its stead, but they ought to have to reckon with its failure all the same.[19]

The key problem was that South-Vietnam not only malfunctioned as a state, but that it hardly existed as a nation. Chasing after "grand designs of nation-building", many of the Americans working to prop up the Saigon regime while fending of the Vietnamese Communist movement were guilty of hubris.[20]

Due to the massive disillusionment with the grandiose US scheme in Vietnam, as well as failed attempts by some European powers to retain their colonies, virtually all the lessons learned in relation to comprehensive approaches went overboard on both sides of the Atlantic. Over-ambitious and unrealistic policy

---

[15] Westmoreland 1965.

[16] Military and civilian agencies had tried to buttress the South-Vietnamese governments for nearly a decade by the time of the Americanization of the war in 1965. See Carter 2006.

[17] For a resurgence of interest in CORDS, see Andrade and Willbanks 2006; Coffey 2008; U.S. Army, *FM 3-24*, 2006.

[18] Gawthorpe 2017. For "too little, too late", see Coffey 2006, p. 32.

[19] Gawthorpe 2017.

[20] Gawthorpe 2017.

objectives and over-reliance on fire-power may have been the main culprits, but operational civil-military designs such as CORDS and counter-insurgency lessons in general were discarded in the process. Particularly in the US, as a result of the 'no more Vietnams' mantra, a distaste for anything resembling counter-insurgency and state-building resulted in an all but singular focus of conventional warfare as well as a strong tendency to divorce military operations from politics. During the remainder of the Cold War, corps commanders could return to preparing for high intensity operations in "what was to all intents and purposes a policy-free zone."[21] This belief was reinforced by swift operational success serving limited policy objectives in the Iraqi desert during the 1991 liberation of Kuwait.

Given the leading American role in NATO, this tendency had serious repercussions in the 1990s when Alliance militaries became heavily involved in engagements of lower intensity, particularly peace operations. As in colonial days, the proliferation of peacekeeping operations and the simultaneous expansion of UN mandates into the domain of state-building required civil-military integration down to the lowest levels of command. In this sense, Cold-War military preparations proved an interlude rather than the gold standard. The UN-system, with a Special Representative of the UN Secretary General (SRSG) heading both the civilian and military efforts, facilitated a fairly appropriate model for comprehensive mission design in for instance Namibia, Mozambique, El Salvador and Cambodia. Although some of them were slow to deploy, subsidiary UN braches for development and refugees and attached ad hoc organisations for elections and policing were relatively easy to embed within the civilian component operating parallel to the military component. The quality of part of the civilian personnel and the military capacity of some peacekeeping contingents was not always up to standard and the integration of civil and military efforts on the tactical level, for instance in Cambodia, was often the result of improvisation under pressure. Acceptable results were nevertheless gained at fairly low cost and key lessons were learned in those operations in the early 1990s. However, peacekeeping experienced a major crisis of confidence due to dilemmas primarily related to the use of force in Somalia, Rwanda and Bosnia in the years 1993–1995. As a result of these dramatic failures, many of the lessons—including those related to civil-military cooperation—went overboard when NATO stepped into the peacekeeping arena in the mid-1990s.[22]

Under US leadership, NATO-style 'green' peacekeeping by the enforcement-capable Implementation Force (IFOR) in Bosnia re-established confidence in the military side of peacekeeping from 1996. However, as a result of US-inspired fears of 'mission creep' and 'military nation-building' it did so at the cost of creating a rift between military and civilian components and implementation. Rapid military success was obtained at the expense of longer-term implementation of the 1995 Dayton Peace Accord, which already fell halfway between the limited aim of ending the Bosnian war and building a self-sustaining peace. But

[21] Strachan 2007, pp. 206–207.

[22] Brocades Zaalberg 2006, pp. 243–250.

most of these structural malfunctions were redressed in the late 1990s. In Bosnia from 1997, NATO military forces gradually stepped up their support for civilian agencies. Military support to civil implementation of the Dayton Accords became essential for any level of progress in Bosnia. Despite institutional civil-military segregation in Kosovo and the non-NATO intervention in East-Timor in 1999 military peacekeepers were even more generous in their support and even temporarily took over principal functions of the host government in the realm of civil administration and policing.

These trusteeships-like mission were labelled 'Empire Lite' by some and discarded as neo-imperialism by others.[23] Despite certain levels of success, they created much resentment in the United States, particularly in Republican circles. Condoleezza Rice, who was to become George W. Bush's Security Advisor, called them "naïve and wasteful" during the 2000 electoral campaign.[24] She famously added in relation to the role of the military in Kosovo: "Carrying out civil administration and police functions is simply going to degrade the American capability to do the things America has to do. We don't need to have the 82nd Airborne escorting kids to kindergarten."[25] But while the Bush-administration declared it would stop using US troops for low-intensity operations like peacekeeping and or to facilitate state-building, it would soon learn the hard way that the type of military engagement we engage in is mostly not of our own choosing. It tends to find us. Driven by the terrorist attacks of September 2001 and the unintended effects of its own predominantly military response, the US would set out on the most ambitious state-building efforts under fire since the Vietnam War—but with even less preparation. By 2008, Condoleezza Rice admitted: "We now recognize that democratic state building is now an urgent component of our national interest."[26]

5.3 The Rise of a Concept

The neglect of the political dimension of war had been clearly present in the phase directly following 9/11, a period when—particularly in the US—targeting enemy organizations was believed to in itself produce political outcomes consistent with national vital interests. The enthusiasm with which the comprehensive approach was embraced in its various reincarnations after 2004 was clearly a reaction to the limited results of earlier segregated approaches and the resurgent tendency into divorce military operations from policy. The broader emphasis on the need for increased civil-military integration was also the result of what Robert Egnell has

[23] Ignatieff 2003; Porch 2011, p. 247; Williams 2011.

[24] Rice 2000.

[25] Gordon 2000.

[26] Rice 2008.

described as "a theoretical inclination to increasingly connect security and development."[27]

NATO's Afghanistan experience was quintessential for the development of the concept. After the US forces toppled the Taliban regime, US forces in Operation Enduring Freedom (OEF) continued to conduct anti-terrorist operations. The internationally sponsored 2002 Bonn Agreement excluded a dialogue with the Taliban thus leaving a political peace settlement unexplored. It foresaw in the deployment of NATO's International Security and Assistance Force (ISAF), which started operating in Kabul with under five thousand military personnel in 2002 in support of the emerging Afghan government. The aim of linking up development and reconstruction with military effect in Afghanistan was complicated by the Alliance's dominant military character.[28] It lacked civilian institutions and implementing agencies, something the UN-system, the OSCE and the European Union (EU) did encompass—and with which Alliance forces had learned to cooperate reasonably well on the Balkans. Meanwhile, the civilian UN Assistance Mission in Afghanistan (UNAMA) had few staff and—compared to the Balkans and East-Timor—a weak mandate to advise and monitor the Afghan government in the capital.

Without the ambition to duplicate such civilian capacity, the Alliance started to develop its own concept for linking up its military capacity to civilian efforts on the operational and tactical level, resulting in the Comprehensive Approach. It evolved out of the Concerted Planning and Action initiative that Denmark had pushed into Alliance discussions from 2004 in response to the various interventions of the previous decade.[29] The ensuing process also brought home to the member states the need for better interdepartmental cooperation on a national level in order to integrate the three key elements of state power: political, military and economic. Around the time of the CPA-initiative, that the Canadians first coined the term '3D-approach'.[30]

On the tactical level in Afghanistan, NATO had latched on to the Provincial Reconstruction Team (PRT)-model. The US military in OEF had gradually started to expand its role by developing these small hybrid civil-military units in support of its anti-terrorist operations by liaising with local administrators and performing

[27] Egnell 2013, p. 239.

[28] As a tactical implementation tool in the early years of NATO's engagement in and around Kabul, the Alliance had little more than an outdated NATO concept for Civil-Military Co-operation (CIMIC), which had its foundations in Cold War preparations for conventional war-fighting. Conceptually it was unsuccessful in adapting to the requirements of complex peace operations in the late 1990s. It remained tactical in nature and its emphasis had been on supporting military objectives rather than enabling the military to make a coherent contribution to broader political objectives, thereby neglecting a true integration of civilian and military efforts. This made CIMIC unsuitable at the time for meeting the Alliance's main operational challenges as they expanded from peace operations on the Balkans to counter-insurgency in Afghanistan. See Brocades Zaalberg 2006 and 2008.

[29] Williams 2011, p. 65.

[30] Martin 2004.

small-scale 'hearts and minds'-projects. Various European NATO-partners willing to contribute to ISAF's expansion in the relatively safe Northern and Eastern provinces after 2003 emulated and built upon the US model. The PRT-model was embraced for a number of reasons. First of all, for NATO it provided a stop-gap for the void left by the lack of an integrated civil-military mission design. Second, the so-called 'reconstruction' teams were easy to sell politically. They were cheap but highly visible national contributions to ISAF and their mission had a positive ring to it.[31] And finally, they became a way of mustering the member states' national civilian resources in the absence of sufficient international governmental organizations for state-building. Such organizations often refused to operate in the increasingly insecure Afghan provinces, partly out of fear of becoming embroiled in the US-led 'War on Terror' that ISAF was obviously closely associated with. For similar reasons, international NGO's would often keep even more distance from the military.

Obviously, the PRT-model did little to enhance uniformity and coherence in the Alliance's approach as it resulted in national experiments in the various Afghan provinces that they were assigned to. The overarching purpose of the teams was to assist the Afghan government in enhancing its legitimacy and authority in the provinces by developing local capacities to conduct reconstruction and provide security as well as good governance. Unsurprisingly, various interpretations of the mission led to enormous divergence in accents on administration, development or security sector reform. There were substantial differences in size (varying from 80 to 400 personnel), civilian contributions (from 30 civilians on 100 a strong PRT to 1 to 160 PRT personnel) and development funds available.[32] PRTs have been criticized for emphasizing projects in support of military goals and their short-term focus, while their development projects characterized as "costly, wasteful, lacking in quality and often not taking into account community needs".[33] Their actual influence over the Afghan provincial administration was very limited. Despite initiatives to create a coherent approach, ISAF Headquarters had little control over the operations conducted by what amounted to 26 PRTs by 2008.[34] By that time, it had become clear that NATO had taken a giant leap into the dark. From the summer of 2006, the British, Canadian and Dutch expanded what was presented as the PRT-model—but now with heavily armed battlegroups for protection—to Afghanistan's volatile South. Stirring up a hornets' nest of renascent Taliban and

[31] PRT reconstruction funds were limited and the teams did not require a large infantry capacity for actual stabilization operations. They only used several platoons for force protection for their bases and core elements, such as the mission teams that liaised with the Afghan authorities and other power-brokers.

[32] For an early and relatively positive review of PRT-activities that nevertheless emphasizes the vast differences in approach, see Jakobsen 2005, p. 6. For a summary of the critique on the PRT-concept, see Hazelbag 2016, p. 52.

[33] Cornish 2007, p. 38; see also Egnell 2010.

[34] Jakobson 2008.

other insurgent groups, NATO forces engaged in full-blown counter-insurgency operations for the first time in the Alliance's history.

Meanwhile, the contributing countries had all started developing increasingly sophisticated versions of an integrated civil-military approach on a conceptual level, with the UK at the forefront of institutionalizing a 'whole of government' approach.[35] The Dutch embraced the notion of the 3D-approach with particular fervour in 2006–2007.[36] By the time the Netherlands deployed its Task Force Uruzgan (TFU) in Southern Afghanistan's Uruzgan province, it was confidently presenting itself as a forerunner along with, or even ahead of its partners in the South. Around this time the US government threw its weight behind the idea that development was a critical part of winning the peace and bolstered the initiative within NATO. It was then that the concept received its label at the NATO Riga Summit in November 2006.[37] With its new etiquette, the Comprehensive Approach and its various national descendants, often still little more than a philosophy, were perceived by many in political, diplomatic and military circles as actually something new and fresh.[38]

5.4 Early Signs and Causes of Demise

The very concept that generated such enthusiasm from 2006 has started to cause scepticism in recent years. At the time, notions related to the comprehensive approach such as 'winning the hearts and minds', the 'ink-spot method', a 'population centric approach', 'the people are the prize' and 'clear-hold-build' suggested vision and fostered hope. Now, they seem rather hollow.[39] In this sense, the aftermath of the large scale military operations in Afghanistan and Iraq bears a strong resemblance to the post-Vietnam War era.[40] Apart from disillusionment and the ultimate failure to deliver in Afghanistan, which will be treated in the next paragraph, recent scepticism has various origins and forms.

First of all, 'comprehensiveness', 3D, EBO and related constructs have been plagued by their inherent vagueness and broadness. After several years as the

[35] For these developments, see Grandia 2015; Hazelbag 2016; UK Ministry of Defence 2006.

[36] Bot 2006.

[37] The NATO Riga Summit Declaration noted that "today's challenges require a comprehensive approach by the international community involving a wide spectrum of civil and military instruments". Thereto NATO leaders tasked members "to improve coherent application of NATO's own crisis management instruments" and strengthen practical coordination with relevant international organizations. NATO 2006; Friss and Binnendijk 2007.

[38] Hazelbag 2016, p. 52.

[39] The fact that most of these notions were derived from earlier counter-insurgency experience and theory was often unknown in NATO circles around 2006.

[40] In The Netherlands, despite its prominent place in policy documents, the comprehensive approach leads a "waning existence". See Hazelbag 2016, p. 364.

British ambassador in Kabul, former diplomat Sherard Cowper-Coles lamented of military planners and civilian stabilization 'experts' always came offering more and better "joined-up" plans with plenty of "cross-cutting" themes, which he called "the snake oil of stabilisation".[41] Weariness was also created by the broad scope and bizarre level of ambitions of the tasks listed in doctrine. The famous 2006 US Field Manual 3-24 Counterinsurgency Operations placed the integration of civilian and military efforts at the heart of successful counterinsurgency operations as all efforts focused on supporting the local populace and the host-nation government. Subsequently, it prescribed that "[c]ounterinsurgents take upon themselves responsibility for the people's well-being in all its manifestations", including security from insurgent and criminal violence, the provision of economic needs and "essential services, such as water, electricity, sanitation, and medical care" as well as sustainment of "key social and cultural institutions."[42] In short, the military field manual prescribed counter-insurgents to take upon them the responsibilities of a state like Denmark in an environment like Afghanistan.[43]

Effect-Based Operations (EBO in NATO parlance Effects Based Approach to Operations), the other concept that revolved around improving the implementation of an interagency process, also contributed in a negative way. As US Army Major-General David Fastabend stated during a 2007 RUSI conference on Managing Complex Crisis:

> I went to Iraq believing EBO was merely useless, an attempt to build a doctrinal theology around the notion that *actions* have *consequences*. But in its more radial interpretation EBO advocates a strict planning focus on outcomes isolated from actor and method and this in turn leads to operational planning that rapidly devolves into a ridiculous essay, a listing of aspirations: 'let us eliminate corruption, isolate the border, prevent sectarian tension.' Such aspirations, with no consideration of who must do what by when, are worse than useless; they are damaging because they conceal the need to make hard choices.[44]

EBO was useful in making the military think and plan in terms of direct and indirect effects rather than on the basis of capacity, but as a result of its presumptuous methodological rigor for implementing over-ambitious versions of the comprehensive approach, the 'EBO'-gospel disappeared far quicker than it had emerged.[45]

This brings us to the second factor undermining enthusiasm for the comprehensive approach. The tendency to avoid hard choices was not just present on the tactical and operational level, it was actually the consequence of the political abuse

---

[41] Cowper-Coles 2012, p. 144.

[42] U.S. Army, *FM-3-24*, 2006, p. 55.

[43] U.S. Army *FM- 3-24* resonated what Michael Shafer in his critical analysis of the Vietnam war presented as the Hearts and Minds theory, which was based on what he called the three big 'oughts': the government countering an insurgency ought to secure the population from insurgent coercion, they ought to provide competent, legal, responsive administration and they ought to meet rising expectations with higher living standards. See Grandia Mantas 2013.

[44] Fastabend 2007.

[45] Mattis 2008, pp. 18–25. For a critical analysis of EBO, see Jobbagy 2009.

of comprehensive approach as it facilitated decision-makers avoiding these strategic choices. The idea that the key to success was simply to increase coordination between the military, international civilian agencies and their equivalents among the host government became a sort of gospel, suggesting expertise and the existence of an actual operational plan. This was exactly what made it so attractive to political leaders during the hard sell of ISAF expansion to the violent South and the later prolongation of what was undeniably a counter-insurgency operation from 2006 onward. However, amongst most NATO partners a taboo attached itself to the term counter-insurgency until 2008, as the preeminent operational approach in Afghanistan because of its negative, violent and colonial connotation.[46] The comprehensive approach and its national equivalents thus also became a substitute for counter-insurgency.[47]

Third, it has also been argued in academic circles that the comprehensive approach is flawed on a more fundamental level as the assumptions underlying it are wrong. Its close association with the hubris surrounding liberal state-building in the wake of 9/11 interventionism are mainly to blame.[48] "The basic shortcoming of the comprehensive approach", Isabelle Duyvesteyn has argued recently, "is that it is based on the Western model of state formation." The Bonn Agreement propagating a centralized democratic state in Afghanistan is one of the obvious examples. "The democratic state with a capitalist market economy", Duyvesteyn continued, "is the product of a specific historical process that cannot be transplanted one on one to other parts of the world. It ignores the fact that in many cases civil wars are related to ongoing processes of state formation."[49] Related critique results from the danger of the intervention force trying to support the government in fragile state actually undermining the legitimacy of local authorities, particularly by robbing them of their core function of establishing a monopoly on the use of force.[50]

A similar point has been made by scholars attacking the neo-classical counter-insurgency (COIN)-doctrine, who strongly question the premise of liberal peace-building on which much of it seems founded.[51] For instance, in his crusade against COIN-theory Douglas Porch criticizes the use of 'hearts and minds' tactics

[46] Jockel 2014, pp. 62–64; De Bont 2011, p. 217; Siccama 2011, p. 265.

[47] This was clearly the case in The Netherlands, where policymakers tried to avoid entanglement in the debate about whether their Task Force Uruzgan was engaged in a 'fighting mission' or a 'reconstruction mission'. Missions were sold under the guise of 3D-mantras such as "no reconstruction without security, no security without reconstruction", thus giving equal importance to all lines of operations and doing a bit of everything. In the Dutch case, military commanders facilitated political indecisiveness in setting priorities and creating hierarchy between 'ends' and 'ways' by claiming—at least in public—that the debate of whether the Dutch had come to Uruzgan to fight an insurgency or reconstruct was irrelevant as their units were doing both. Dimitriu and De Graaf 2015; Boom 2007; Katsman 2012. For a similar statement by the NATO Secretary General, see Rasmussen 2010.

[48] For a summary of this argument, see Grandia Mantas 2013.

[49] Duyvesteyn 2013.

[50] See Angstrom 2008; Lake 2006; Duyvesteyn 2013.

[51] Porch 2013; Gentile 2013.

and stabilization and reconstruction efforts. He argues that the semi-privatized nation building of today undermines, bypasses and diminishes the state instead of modernizing it. "Billions of dollars go missing, the arbitrary power of warlords is enhanced, and politicians appeal to sectarian divisions. Meanwhile the actions and attitudes of policemen and soldiers often decrease the legitimacy of the state, which in turn is unable to provide basic services or improve security and economic conditions."[52] These critics are correct in arguing that projecting Western values onto foreign populations in the form of "freedom, democracy, and free enterprise" has emerged as the single, but certainly not always viable recipe for dealing with fragile states. The question remains, however, whether either an operational approach such as counter-insurgency or the related comprehensive approach—a way of combining civil and military means to whatever political end is formulated by the responsible politicians—are to blame for recent failures. It is also debatable whether the comprehensive approach is solely dependent on this idealistic and value-driven model.

Finally, enthusiasm for the comprehensive approach has been tempered by a clear shift in priority amongst Western powers to conventional warfare and waging war by proxy. The former is driven by a mounting conventional military threat with the East-West tensions on the rise and has caused some to "feverishly advocate" the military to "keep close to the traditional war fighting business."[53] The latter is clearly the result of war wariness—particularly with 'boots on the ground'-type counter-insurgency efforts like in Iraq and Afghanistan. The radically decreased level of Western ambition has been felt in conflict like those in Libya and Syria. It has resulted in what US Defence Secretary Robert Gates in 2009 called "the indirect approach" to counter-insurgency, by which he meant the building the capacities of partner governments through Security Forces Assistance.[54] A different form of this indirect approach has been used to support rebel organizations fighting their governments in for instance Libya and Syria.

5.5 Anything but Comprehensive: The Afghan Legacy

Before we discard the concept out of disillusionment on the basis of the Afghan experience, we need to take a closer look at whether it really is the benchmark against which the concept should be judged. Have NATO and its partners been anywhere near implementing a comprehensive approach? Analysing it in its full complexity is clearly outside the scope of this chapter, but a quick scan of recent studies and first-hand accounts shows that the Western record is poor. Cedric de Coning and Karsten Friis concluded there is "a considerable gap between the degree

[52] Porch 2011, p. 253.

[53] Rietjens et al. 2013, p. 273.

[54] Gates 2009, pp. 29–30.

to which the benefits of coherence are held to be self-evident at the policy level, and a persistent inability to achieve coherence at the operational level".[55] It can even be argued that, at least up to 2008, the main reason why NATO and its partners failed was that from the international policy level down to the tactical level the overall effort was anything but comprehensive.[56]

The naivety with which the initial intervention in Afghanistan was launched resulted in extremely poor interagency planning and coordination for the phase following regime-change. For this, the Bush-administration's knee-jerk responses to 9/11 are to blame rather than NATO. But subsequently, as the 'post-conflict phase' was internationalized, the campaign's overall objective became extremely diffuse. It has been amply summarized as "shifting between state-building, counter-terrorism, saving NATO's reputation and keeping the United States happy (a prime concern among European troops contributors)"[57] Counter-insurgency, the sole operational concept that—despite some of its flaws—had the potential of combining some of the objectives while incorporating many aspects of the comprehensive approach, was ignored until quite late in the mission. Nevertheless, we should avoid what Emile Simpson called "investing in operational ideas a policy-like quality" as was done with counter-insurgency in both Iraq in 2007–2008 and Afghanistan 2009–2010."[58]

Cowper-Coles may have derided the elaborate comprehensive plans briefed to him, but he did not fundamentally question the need for a better integrated civil-military approach. Instead, he argued that these optimistic plans would always collide with Afghan realities. For them to work, at least three key components were missing: sufficient Western military and civilian resources, the capacity and even the will of the Afghans to take over where NATO had stepped in, and the ability to cut the insurgents off from their sanctuary across the Pakistani border.[59] Even though many commanders eventually adopted a mind-set that was receptive to the comprehensive approach, the resources and structures did not sustain such an approach until 2009. It was only after U.S. General Stanley McCrystal's reforms and the injection of vast amounts of extra U.S. troops and "surge" of civilian personnel, that the civil-military effort started to gain momentum.[60] Still many flaws remained, particularly on the international level. In his plea for a comprehensive approach in 2010, Rasmussen was rather harsh in his verdict:

> [A]t the international level, this lesson has simply not yet been learnt. Let me illustrate it by a concrete example. The European Union does both development and police training in Afghanistan. Nevertheless, the EU and NATO do not plan or coordinate together. For

[55] De Coning and Friis 2011, pp. 257–258.

[56] Lindley-French 2007; Simpson 2013, p. 147.

[57] Ucko 2012, p. 70. See also Hazelbag 2016, p. 274.

[58] Simpson 2013, p. 244.

[59] Cowper-Coles 2012, pp. 142, 144.

[60] Simpson 2013, p. 147. On the complex process to draw on civilian specialists, see Gates 2014, pp. 207, 343.

> political reasons totally separate from Afghanistan. The same is basically true of NATO and the United Nations. I consider this to be an unacceptable waste of resources and effectiveness.[61]

He also deplored the extremely poor cooperation with international NGO's, who wanted to stay neutral. What he failed to mention was that even the military side continued to suffer from extremely confusing military command and control arrangements, starting with the tension between NATO's ISAF and the separate American-led Operation Enduring Freedom.

Interestingly, while Rasmussen criticized civil-military cooperation at an international level, he lauded integration on the national level, where "NATO governments have generally moved towards a 'whole of Government' approach to Afghanistan. Diplomats, defence ministries and development experts sit together, plan together and operate together, including in Provincial Reconstruction Teams all over Afghanistan."[62] The Secretary General was referring to the situation after four years of gradual national improvements in the provinces, adaptations facilitated by McCrystal's operational level innovations and increased capacity. Still, Rasmussen's optimism is contentious. Even the tactical case of the Dutch in Southern Afghanistan's Uruzgan province, which has been hailed as a model for the integration of military and civilian efforts,[63] shows how painfully slow the learning process was and how true integration of efforts was rare. British, Canadian, Danish and American contingents eventually experienced similar learning curves while often meeting with higher levels of insurgent resistance.

During the period 2006–2010 it proved hard for the TFU to reach any level of true comprehensiveness, let alone effectiveness in a complex and violent environment. Civilian capacity was scant. Starting off with just one political advisor and one development advisor on a 1500-strong military organisation, it took extremely long for the Dutch governmental civilian capacity to reach twelve civilians and be anywhere near its pretentious claim of effectuating a 3D-approach. The situation further improved in the course of 2009 as the Dutch introduced a civilian-led PRT and dual headed civil-military leadership for all the Task Force Uruzgan. But on other levels, comprehensiveness was still nowhere to be found. As late as 2010, police training and mentoring was handled by no less than five different civilian and military organisations within Uruzgan province alone.[64]

As Lenny Hazelbag concluded in a recent detailed study on interdepartmental cooperation in relation to Uruzgan and Helmand province, the level of coherence in The Hague was mostly limited to coordination and information exchange. A surprising number of officials even regarded coordination on many topics related

---

[61] Rasmussen 2010.

[62] Rasmussen 2010.

[63] See, for instance, The Liaison Office 2010, p. vi; Solomon 2009; Jockel 2014, p. 10. Also Jockel sees the Dutch as having less difficulty conceptualizing and implementing a '3D' approach than the Canadians, at least in their capitals.

[64] As witnessed by author with the TFU in Uruzgan in 2008 and 2010.

to Uruzgan unnecessary. They merely kept their colleagues at the other departments informed. Despite frequent and pleasant contacts and cooperation that was overall judged as satisfying, complex problems were simply reduced to understandable and manageable parts. This seemingly allowed defence, diplomatic and development officials to tackle them within their own departmental boundaries. True integration of various efforts was only reached in a limited amount of dossiers.[65]

Sometimes, civil-military rifts were more serious. Progress was not only hampered by the complex Afghan tribal structures and incomprehensible power-relations, but also by interdepartmental competition in The Hague trickling down to Uruzgan and the diverse interests of military and civilian partners operating in the province. The Ministry of Foreign Affairs would set restrictions on cooperating with certain local warlords that held key positions close to the central government mainly because they failed to meet the Dutch government's moral standards. In another recent elaborate analysis, Martijn Kitzen criticizes the rather inconsistent manner in which the TFU operated, particularly in the way it engaged local key leaders until late 2008. This was mainly the result of faltering strategic guidance, that hardly went beyond the earlier mentioned 3D-type guidelines, and the late arrival and poor use of an important civil assessment on local power relations. In the end, Kitzen is fairly positive about tactical level improvisation and the eventual innovations. A truly comprehensive approach was nevertheless often hampered by on the one hand 'enemy centric' priorities in military circles[66] and on the other hand civilian tendencies to go it alone on the development side. Governmental civilians and military personnel operating within the same Task Force Uruzgan would pursue their own agenda's. Ministry of Foreign Affairs' political advisors and development advisors embedded within the PRT initiating a separate 'Track 2' program for covert development assistance outside the Afghan Development Zones (ADZ's), the modern version of the French colonial 'oil spots'. These projects, which lasted until the end of the Dutch mission, were far better funded than the overt quick-impact and development campaign by the PRT. However, TFU-staff was hardly informed of these projects. Therefore, they would impede unity of effort and did not contribute to a concerted plan. One TFU commander angrily called them "a war within a war" and a potential hazard.[67]

In the end, the relatively successful Dutch mission in Uruzgan was hampered by two fundamental strategic mismatches. First, on top of the separate Track 2 program the departments of Foreign Affairs and Development heavily sponsored a top-down approach to state-building by channelling the vast majority of reconstruction funds through the highly corrupt Afghan government in Kabul. This left very little to spend on truly integrated security, governance and development programs that would have contributed to bottom-up state-building. This may have suited

[65] Hazelbag 2016, pp. 210, 361.

[66] Consecutive PRT-commanders complained about the enemy-centric priorities within the TFU and the Battlegroup. Grandia Mantas 2009, p. 19.

[67] Kitzen 2016, p. 436.

Afghanistan better than its top-down variant. Secondly, the Dutch approach encompassed a long term vision to redress the tribal power balance that involved high levels of social engineering. The underlying idea may have been right, but the Dutch tried to do so in the context of a short-term commitment of first two, and after prolongation, four years. They also tried to do so by going against the grain of the local American approach that was primarily defined by U.S. Army Special Forces operating in the province.[68] The well intentioned Dutch attempt at re-balancing power-relations made the local partners that the TFU had been propping up extremely vulnerable after its withdrawal in 2010. The Dutch in Uruzgan, like most other modern counter-insurgents, clearly suffered from lack of means and time. By the time they improved their performance, political consideration in The Hague forced them to withdraw from their costly project. ISAF as a whole suffered the same fate when most of the NATO member states started to draw down their contingents not long after the important innovations of 2009–2010.

5.6 Conclusion

By treating the comprehensive approach in a broader historical context, this chapter has sketched and explained the continuities, the rise and the early signs of demise of the comprehensive approach. It has emphasized the historical inevitability of combining the key elements of state power to resolve conflict, to win wars and to consolidate political outcomes. The main difference between the comprehensive approach in conventional warfare and counter-insurgency is that in the latter type of conflict—like in complex peace operations—civil-military balance and integration is required down to the lower tactical level. David Ucko has argued quite convincingly that over the past decade counter-insurgency has functioned as "a powerful, corrective to the unhelpful tendency in the US military, but also elsewhere in the West, to divorce military affairs from political considerations, and to treat war as an elaborate targeting drill."[69] As we have seen, it was actually the comprehensive approach centrally embedded in classic counter-insurgency theory that formed the core ingredient of this antidote.

It may be correct to question the liberal-state building assumptions underlying recent interpretations of the comprehensive approach and it is certainly possible to criticize the actual levels of civil-military integration in Afghanistan. However, it is

---

[68] Dutch-American tension was strong within Uruzgan, but also U.S. policy towards Uruzgan lacked coherence. U.S. Special Forces cooperated intensively with the two main local power brokers that they saw as key allies in the struggle against the Taliban. Meanwhile the TFU was trying to curtail and even sideline these related warlords, despite their strong ties to the presidency in Kabul. American State Department officials, including those in Uruzgan, seemed to share the Dutch view. Kitzen 2016, pp. 382–383, 456.

[69] Ucko 2012, p. 78; for a similar note: Brigadier-general T.R. McMaster keynote lecture during the annual Society of Military History Conference, 10–13 May 2012, Washington D.C.

incorrect to conflate the comprehensive approach with strategy. In strategic terms it will never be more than a key element of the way in which the means—military and civilian—are combined towards certain ends at the various levels of command. Comprehensive approaches will have to serve whatever goal is set on the political level, including—but not necessarily—liberal state-building. Often, and certainly in any post-conflict phase, such organizational constructs will be required to create the necessary level of influence.

The impending neglect of the comprehensive approach is risky. Lenny Hazelbag did most of his research on interdepartmental civil-military cooperation at a time when the belief in the concept seemed unquestioned. However, he has signalled that even in The Netherlands the concept leads a waning existence after 'Uruzgan'—despite the lip service being paid in policy documents. He warns that the lessons of intense interdepartmental cooperation in The Hague during the Uruzgan mission are rapidly fading.[70] In military circles, in light of the emerging conventional military threat posed by Russia, there is a real risk of educating a generation of officers that turn away from comprehensive thinking. Dutch PRT-commander lieutenant-colonel Gerard Koot claimed he had "3D programmed in my genes" prior to and during his 2007 deployment. Many key commanders during the deployments in Afghanistan and Iraq were increasingly thinking of war in terms of "armed politics".[71] The question is how long this will last. Given the current threats and a loss of appetite for counter-insurgency and state-building, a shift of focus to conventional warfare and waging ground wars by proxy is inevitable. However, like US military and civilian institutions in the wake of the Vietnam War and during the early years of the George W. Bush-administration—European militaries run the risk of throwing out the baby with the bathwater. Trying to avoid another Afghanistan or Iraq altogether may be understandable and wise, but the experience of the past decades suggests that Western militaries do not dictate the type of wars they will become engaged in. The experience in Afghanistan and Iraq has demonstrated once again the perils of neglecting hard-won lessons learned, discarding valuable doctrines and operational concepts and focusing on just one type of armed conflict instead of preparing for the varieties of wars.

References

Andrade D, Willbanks JH (2006) CORDS/Phoenix - Counterinsurgency Lessons from Vietnam for the Future. Military Review 86(1):77–91

Angstrom J (2008) Inviting the Leviathan. External Forces, War and State-Building in Afghanistan. Small Wars and Insurgencies 19(3):374–396

Beckett IFW (2001) Modern Insurgencies and Counter-insurgencies: Guerrillas and their Opponents since 1750. Routledge, London

[70] Hazelbag 2016, p. 363.

[71] Grandia Mantas 2009. For "armed politics" see: Simpson 2013, p. 147.

Bont E de (2011) Onder Taliban en Krijgsheren: Nederland en de Oorlog in Afghanistan. Nieuw Amsterdam, Amsterdam

Boom J (2007) Rustig aan in Uruzgan. De Groene Amsterdammer (5 January 2007)

Bot B (2006) The Dutch Approach: Preserving the Trinity of Politics, Security and Development. Minister of Foreign Affairs' speech presented at the SID and NCDO Conference on Security and Development (7 April 2006), The Hague

Brocades Zaalberg T (2006) Soldiers and Civil Power - Supporting or Substituting Civil Authorities in Modern Peace Operation. Amsterdam University Press, Amsterdam

Brocades Zaalberg T (2008) The Historical Origins of Civil-Military Cooperation. In: Rietjens SJH, Bollen MTIB (eds) Managing Civil-Military Cooperation - A 24/7 Joint Effort for Stability. Ashgate Publishers, Aldershot

Brocades Zaalberg T (2015) Civil-Military Cooperating during the faltering Pacification of Java, 1947–1949. British Journal of Military History 1(2):67–84

Carter JM (2008) Inventing Vietnam: The United States and State Building, 1954–1968. Cambridge University Press, Cambridge

Coffey R (2006) Revisiting Cords: The Need for Unity of Effort to Secure Victory in Iraq. Military Review 86(2):24–34

Cornish S (2007) No Room for Humanitarianism in 3D Policies: Have Forcible Humanitarian Interventions and Integrated Approaches Lost their Way? Journal of Military and Strategic Studies 10(1):1–48

Cowper-Coles S (2012) Cables from Kabul: The Inside Story of the West's Afghanistan Campaign. Harper Press, London

de Coning C, Friis K (2011) Coherence and Coordination: The Limits of the Comprehensive Approach. Journal of International Peacekeeping 15(1–2):243–277

Dimitriu G, De Graaf B (2015) Fighting the War at Home: Strategic (Counter-) Narratives and the Case of the Dutch Deployment in Uruzgan, 2006–2010. In: De Graaf B, Dimitriu G, Ringmose J (eds) Strategic Narratives, Public Opinion and War: Winning Domestic Support for the Afghan War. Routledge, New York

Donnison FSV (1966) Civil Affairs and Military Government: Central Organisation and Planning. Her Majesty's Stationary Office, London

Duyvesteyn I (2013) Strategic Illiteracy - The Art of Strategic Thinking in Modern Military Operations. Inaugural Lecture (10 June 2013), Leiden University

Egnell R (2010) Winning "Hearts and Minds"? A Critical Analysis of Counter-Insurgency Operations in Afghanistan. Civil Wars 12(3):283–303

Egnell R (2013) Civil–military coordination for operational effectiveness: Towards a measured approach. Small Wars & Insurgencies 24(2):237–256

Fastabend DA (2007) RUSI Conference 2007 Managing Complex Crisis. http://www.rusi.org/downloads/assets/Managing_Complex_Crisis.pdf. Accessed 1 May 2008

Friis AP, Binnendijk H (2007) The Comprehensive Approach Initiative: Future Options for NATO. Defence Horizons 58:1–6

Gates RM (2009) A Balanced Strategy. Foreign Affairs 88(1):28–40

Gates RM (2014) Duty: Memoires of a Secretary at War. WH Allen, London

Gawthorpe AJ (2017) To Build as well as Destroy: The American Experience of Nation-Building in South Vietnam, 1965–73. Cornell University Press, Ithaca (forthcoming; citations are from the conclusion)

Gentile G (2013) The Wrong Turn. America's Deadly Embrace of Counterinsurgency. The New Press, New York/London

Gordon MR (2000) The 2000 Campaign: The Military. Bush would stop US Peacekeeping in the Balkan Fights. New York Times (21 October 2000)

Grandia Mantas M (2009) The 3D Approach and Counterinsurgency - A Mix of Defence, Diplomacy and Development. The Case of Uruzgan. Master thesis, Leiden University

Grandia Mantas M (2013) Shafer Revisited – The Three Great Oughts of Winning the Hearts and Minds: Analysing the Assumptions Underpinning the British and Dutch COIN Approach in Helmand and Uruzgan. Small Wars & Insurgencies 24(4):731–750

Grandia Mantas M (2015) Deadly Embrace: The Decision Paths to Uruzgan and Helmand. Unpublished PhD thesis, Leiden University

Hazelbag L (2016) De geïntegreerde benadering in Afghanistan: tussen ambitie en praktijk. De interdepartementale samenwerking in Nederland en het Verenigd Koninkrijk. Unpublished PhD thesis, Erasmus University, Rotterdam

Horne A (2006) A Savage War of Peace: Algeria, 1954–1962. New York Review Books Classics, New York

Ignatieff M (2003) Empire Lite. Nation-Building in Bosnia, Kosovo and Afghanistan. Penguin, Toronto

Jakobsen PV (2005) PRTs in Afghanistan: Successful but not sufficient. DIIS Report 2005:6. Danish Institute for International Studies, Copenhagen

Jakobson PV (2008) NATO's Comprehensive Approach to Crisis Response Operations: A Work in Slow Progress. DISS Report 2008:15. Institute for International Studies, Copenhagen

Jobbagy Z (2009) From Effects-based Operations to Effects-based Force: On Causality, Complex Adaptive System and the Biology of War. Unpublished PhD thesis, Budapest

Jockel JT (2014) Canada and The Netherlands in Afghanistan. Martello Papers Series, no. 38. Centre for International and Defence Policy, Kingston

Katsman M (2012) Militairen hebben geen nationale grenzen meer - Een interview met C-LAS Mart de Kruif. Atlantisch Perspectief 1

Kitzen M (2016) The Course of Co-Option: Co-option of Local Power-Holders as a Tool for Obtaining Control over the Population in Counterinsurgency Campaigns in Weblike Societies. With Case studies in Dutch Experiences During the Aceh War (1873-c.1912) and the Uruzgan Campaign (2006–2010). Unpublished PhD thesis, University of Amsterdam

Lake D (2006) Building Legitimate States after Civil Wars: Order, Authority and Institutions. Paper prepared for the Workshop on Building Peace in Fragile States, University of California (1–2 December)

Lindley-French J (2007) Presentation at COIN Conference, The Hague (14 December)

Martin P (2004) Address as Prime Minister on the occasion of his visit to Washington DC (29 April 2004)

Mattis JN (2008) USJFCOM Commander's Guidance for Effects-based Operations. Parameters 38 (3):18–25

Mountcastle C (2016) The Myth of the New Complexity. Military Review (March–April):47–53

NATO (2006) Riga Summit Declaration (29 November 2006) http://www.nato.int/docu/pr/2006/p06–150e.htm. Accessed 1 March 2017

Paret P (1964) French Revolutionary Warfare from Indochina to Algeria: The Analysis of a Political and Military Doctrine. Praeger, London

Porch D (2002) Wars of Empire. Cassell & Co., London

Porch D (2011) The Dangerous Myth and Dubious Promise of COIN. Small Wars & Insurgencies 22(2):239–257

Porch D (2013) Counterinsurgency. Exposing the Myths of the New Way of War. Cambridge University Press, Cambridge

Rasmussen AF (2010) Afghanistan and the Future of Peace Operations. Speech by the NATO Secretary General at the University of Chicago (8 April 2010) http://www.nato.int/cps/en/natolive/opinions_62510.htm. Accessed 1 March 2017

Rice C (2000) Campaign 2000. Promoting the National Interest. Foreign Aff 79(2):45–62

Rice C (2008) Rethinking the National Interest. Foreign Affairs 87(4):2–26

Rietjens S, Soeters J, van Fenema PC (2013) Learning from Afghanistan: Towards a compass for civil–military coordination. Small Wars & Insurgencies 24(2):257–277

Siccama JG (2011) New and Innovative Counterinsurgency-related Books. In: Brocades Zaalberg T, Hoffenaar J, Lemmers A (eds) Insurgency and Counterinsurgency: Irregular Warfare from 1800 to the Present. Netherlands Institute of Military History, The Hague, pp 265–270

Simpson E (2013) War from the Ground Up. Twenty-First Century Combat as Politics. Scribe, Brunswick

Solomon J (2009) U.S. Takes Dutch Military as Role Model in Afghanistan. The Wall Street Journal (30 April 2009)
Strachan H (2007) British Counter-Insurgency from Malaya to Iraq. RUSI Journal 152(6)
The Liaison Office (2010) The Dutch Engagement in Uruzgan: 2006–2010. TLO Report
Ucko DH (2012) Wither Counterinsurgency. The Rise and Fall of a Divisive Concept. In: Rich P, Duyvesteyn I (eds) The Routledge Handbook of Insurgency and Counterinsurgency. Routledge, London
UK Ministry of Defence (2006) Joint Discussion Note 4–05: The Comprehensive Approach. Development, Concepts and Doctrine Centre, Shrivenham
US Army/Marine Corps (2006) Field Manual 3-24/MCWP 3-33 Counterinsurgency Operations
Westmoreland WC (1965) COMUSMACV, MACV Directive 525–4 (17 September)
Williams MJ (2011) Empire Lite Revisited: NATO, the Comprehensive Approach and State-Building in Afghanistan. International Peacekeeping 18(1):64–78

Author Biography

Dr. T.W. Brocades Zaalberg is Associate Professor of Intelligence and Security at the Department of War Studies, Netherlands Defence Academy and a University Lecturer in Contemporary Military History at Leiden University. He holds a Master of Arts degree from the University of Groningen. He is the author of *Soldiers and Civil Power: Supporting or Substituting Civil Authorities in Modern Peace Operations*, which earned him his Ph.D. in history from the University of Amsterdam in 2005.

Chapter 6
Winning Without Killing: The Case for Targeted Sanctions

Francesco Giumelli

Abstract Sanctions are a frequently employed diplomatic instrument to exert influence. However, it is also a relatively poorly understood instrument that, while seductive because it seems easy to employ and relatively risk-free, is also criticized for lack of effectiveness. In particular in the past decade UN sanctions have undergone a significant development. The chapter presents how sanctions changed from their classical and comprehensive form to a more recent targeted version. Whereas comprehensive sanctions aimed at whole states, today the targets of sanctions are mostly individuals, entities and specific economic sectors. The situations in which sanctions have been used and adapted to achieve foreign policy objectives with the simultaneous objective to reduce their humanitarian impact has grown overtime, for instance by including the utilization of sanctions in post-conflict contexts. Especially, the objective of this chapter is to review the opportunities and challenges that characterize targeted sanctions in order to provide recommendations on how to enhance their effectiveness. The argument of the chapter is that targeted sanctions are more demanding in terms of knowledge and maintenance compared to comprehensive sanctions. If the international community, or individual states, wish to rely effectively on this foreign policy instrument, then ways to acquire more information about targeted societies and threats that are to be countered as well as the creation of institutional capacity to monitor and enforce sanctions is key to making sanctions useful.

Keywords Economic sanctions · Targeted sanctions · Coercion

F. Giumelli (✉)
Faculty of Arts, University of Groningen, Groningen, The Netherlands
e-mail: f.giumelli@rug.nl

P.A.L. Ducheine and F.P.B. Osinga (eds.), *Netherlands Annual Review of Military Studies 2017*, NL ARMS, DOI 10.1007/978-94-6265-189-0_6

Contents

6.1 Introduction

The revolution in military affairs with targeted bombing and the advent of drone warfare tactics have changed the nature of conflicts and reduced the human costs for using force. However, the lower casualty rate on the offensive side has not been equalled with the same decrease on the civilian side of the defensive party. Since the war in Vietnam, the 'collateral damage' has not been overlooked as a variable for determining the success of foreign policy actions. Victory was not enough to justify the use of force, and the humanitarian consequences of interventions became a central concern for foreign policy actions. Sanctions could be placed in a very similar context since the United Nations embargo imposed on Iraq after its invasion of Kuwait was made illegitimate by the 500.000 children allegedly killed by the decision of the Security Council. In such case, the use of sanctions was criticized regardless from the reason that led to the imposition and the objective that could be achieved because the humanitarian consequences are deemed disproportionate compared to the harm they were supposed to fight.

This chapter deals with the evolution of sanctions according to the principle of winning without killing. Especially, the objective of this chapter is to review the opportunities and challenges that characterize targeted sanctions in order to provide recommendations on how to enhance their effectiveness. The argument of the chapter is that targeted sanctions are more demanding in terms of knowledge and maintenance compared to comprehensive sanctions. If the international community, or individual states, wish to rely effectively on this foreign policy instrument, then ways to acquire more information about targeted societies and threats that are to be countered as well as the creation of institutional capacity to monitor and enforce sanctions is key to making sanctions useful. The chapter presents how sanctions changed from their classical and comprehensive form to a more recent targeted version. Whereas comprehensive sanctions aimed at whole states, today the targets of sanctions are mostly individuals, entities and specific economic sectors. The situations in which sanctions have been used and adapted to achieve foreign policy objectives with the simultaneous objective to reduce their humanitarian impact has grown overtime, for instance by including the utilization of sanctions in post-conflict contexts. Thus, the utilization of targeted sanctions to achieve policy objectives is in the spirit of *winning without killing*.

The chapter is divided in four sections. The first section presents the evolution of sanctions from comprehensive to targeted. The second section illustrates the different ways in which sanctions can be used and how sanctions can affect the behaviour of actors in the international system. The third section discusses the opportunities and challenges that targeted sanctions present if compared with their comprehensive ancestors. Finally, the effectiveness of targeted sanctions is dependent upon specific measures that could be undertaken and this last section analyses ways in which targeted sanctions can be improved.

6.2 The Evolution of Sanctions: From Comprehensive to Targeted

Sanctions are an old practice in international affairs. The wide spread understanding of sanctions is an economic embargo on a state, defined as the prohibition of exporting and importing goods to and from other states. This view is historically rooted in the practice of sieges, when fortresses or fortified cities were surrounded by aggressors with the objective to force them to surrender. From the Megarian decree, with which Athens banned Megara from trading with the Delian League 4 centuries B.C., till the decision to impose sanctions on Iraq after the invasion of Kuwait in 1990, the understanding of sanctions had not changed.[1] In the 1990s, the United Nations has changed the nature of sanctions through its practice of targeting individuals and entities as well as economic sectors.

The Security Council of the United Nations imposes sanctions under Chap. 7 of the Charter, which deals with countries' 'Action(s) with respect to threats to the peace, breaches of the peace, and acts of aggression.' As established under Art. 41 of the Charter, the Security Council is authorized to impose 'measures not involving the use of armed force' that 'may include complete or partial interruption of economic relations and of rail, sea, air, postal, telegraphic, radio, and other means of communication, and the severance of diplomatic relations'.[2] As such, given that the UN is an association of *states,* sanctions were interpreted as being directed towards states, as the cases of, for instance, Southern Rhodesia, South Africa and the former Yugoslavia seem to suggest.[3] For instance, Milosevic was included in the sanctions list only once he was out of office, and Saddam Hussein was not sanctioned by the United Nations until 2003. Why did then the UN start to impose sanctions on individuals?

The evolution of sanctions from their comprehensive nature to a targeted form that took place in the 1990s is due mainly to three factors. First, comprehensive sanctions were not deemed to be effective. The intense debate that took place since

[1] Tsebelis 1990.

[2] United Nations 1945.

[3] Cortright and Lopez 1995.

the late '60s led Baldwin to write that one of the things in common among all the scholars of sanctions was their 'universal tendency to denigrate the utility of such tools of foreign policy'.[4] Drezner reports a number of statements by practitioners and academics that would confirm this: Kennan, Nixon, Schultz, Galtung, Wallensteen, Pape, Doxey, Friedman, and others all say that economic sanctions are not effective.[5] The perspective of sanctions in the 1990s is best described by looking at the debate that took place between Pape, Elliot and Baldwin in the journal *International Security.* Pape triggered the discussion by analysing the findings of a study that set the success rate of sanction at 34% of the times,[6] and countering this by offering his assessment that that figure should in fact be lowered substantially to as low as 5%.[7] This contribution sparked the responses from Elliot, suggesting a more optimistic view of sanctions, and by Baldwin, who invited to consider the comparative utility of sanctions vis-à-vis other foreign policy instruments. However, the general perception remained negative.[8]

The second factor that favoured the evolution towards targeted sanctions was the need to avoid the human sufferance on civilians caused by comprehensive sanctions. Possible backlashes from the imposition of sanctions that could have the adverse effect of strengthening targets—the so-called rally-around-the-flag-effect—were already known,[9] but growing attention was paid to the humanitarian consequences of imposing comprehensive and, therefore, non-discriminatory measures.[10] The hardship that was imposed on the population of Iraq[11] and Haiti[12] proved to be more harmful for innocent civilians than for the elites who they were intended to punish. The Iraqi case became unfortunately known for the widely reported study that blamed UN sanctions for the death of 500,000 children[13] and sanctions against Iraq and Haiti 'have paid only minimal political dividends at a very high price in human terms'.[14] Sanctions were ineffective and they were harming the people that were supposed to protect. A comment made by the Economic and Social Council suggested that states should be responsible for the humanitarian consequences that

---

[4] Baldwin 1985.

[5] Drezner 1999.

[6] Hufbauer et al. 1990.

[7] Pape 1997.

[8] Naturally, the academic debate was richer, but the practitioners' view was dominated by the slight effectiveness of sanctions. For different ways to assess the success of sanctions, see Lindsay 1986; Barber 1979; Nincic and Wallensteen 1983.

[9] Galtung 1967.

[10] Mueller and Mueller 1999; Naylor 2001; Weiss and All 1997.

[11] Clawson 1993; Hoskin 1997.

[12] Gibbon 1999.

[13] Ali and Iqbal 1999; Alnasrawi 2001.

[14] The Financial Times 1995.

their policies had in other countries,[15] and the Secretary General Boutros-Ghali called on states to avoid comprehensive measures in the future.[16]

Third, the individualization of responsibility principle allowed for targeting individuals and entities. Affected by the tragedies of the disintegration of Yugoslavia, the genocide in Rwanda and the protracted conflicts in Sierra Leone, the international community was under pressure to provide a solution to situations that were intolerable in the *new world order* after the end of the Cold War. In the attempt to deal with the reconstruction of communities after conflicts, the growing demand for justice was dealt with the attempt by international courts to identify individual responsibilities for wrongdoings done in the name of states. The unique and novel character of such transitional justice efforts was represented by the fact that these courts were after *individual responsibilities* rather than those of states. Parallel to these creations, the Rome Statute establishing the International Criminal Court (ICC) represents the institutionalization of a norm that was observed only since the crises in Yugoslavia, Rwanda and Sierra Leone,[17] thus creating a landmark decision to formalize the existence of an international individual responsibility.[18] The world of sanctions was also affected by this trend. In August 1998, the US embassies in Kenya and Tanzania were targeted by terrorist attacks in which over 200 people were killed. The Security Council reacted promptly with Resolution 1189 condemning the action and calling on states to adopt effective and practical measures to respond. The Security Council passed Resolutions 1193 (1998) and 1214 (1998) linking the situation in Afghanistan and international terrorism, but the main step was made on 15 October 1999. The Security Council, taking note that Usama bin Laden and his associates had been indicted by the United States of American for the bombings in Kenya and Tanzania and determining that the Taliban authorities had failed to comply with Resolution 1214, passed Resolution 1267 imposing financial restrictions and a travel ban on a list of individuals identified by the sanctions committee. There is still some ambiguity in Resolution 1267. If on the one hand, the Taliban are targeted as they are acknowledged as the leaders of Afghanistan, on the other, they are defined as 'the Afghan faction known as the Taliban' and, therefore, they are not recognized as state authorities and banned as outlaws.[19]

The Pandora's Box was open. On the one hand, the UNITA sanctions committee led by the Canadian Ambassador Robert Fowler published a report with names of individuals responsible for supporting or being involved in sanctions busting activities, and, on the other hand, the Security Council passed Resolution 1333 (2000) adding Usama bin Laden, individuals and entities associated with him and with Al Qaida organization to the list of targets of the 1267 committee. The

[15] United Nations 1997.

[16] Von Braunmühl and Kulessa 1995.

[17] Shraga and Zacklin 1994; Schabas 2006.

[18] Sunga 1992.

[19] United Nations 1999.

Security Council sanctioned individuals for actions that could have not been justified and/or motivated by the pursuant of national duties. This transition has been accompanied by the discussion taking place in three different international conferences organized by like-minded states known as the Interlaken, Bonn-Berlin and Stockholm processes. The first conference on the feasibility of financial sanctions and how to make targeted financial sanctions more effective was held in 1998 and 1999 in Interlaken, Switzerland. The Interlaken process, which was initiated by Switzerland, triggered further efforts from Germany and Sweden, which organized and sponsored two similar initiatives. The Bonn-Berlin aimed at improving the effectiveness and the Stockholm process focused on the implementation process of sanctions, the challenges posed by the legislation of nation-states, and on the opportunity to evade the sanctions by the targets.[20]

The Security Council legitimized the practice for an international organization to restrict the freedom of individuals and non-state actors when international peace and security were threatened. The 9/11 terrorist attacks strengthened this trend and the 1267 list grew exponentially in the following months to the thousands since states felt compelled, in solidarity with the US, to share information on terrorist activities and individuals known to their services.[21] Targeting individuals and entities became the norm of sanctions practices and the onset of such trend was welcomed by the international community since sanctions had become 'smart'.[22] Given the little guarantees about the 'smartness' of sanctions, the label soon changed into the less problematic 'targeted sanctions'. Biersteker wrote in 2009 that 'All UN sanctions today are targeted sanctions'.[23]

A targeted sanction policy has been defined as 'one that imposes coercive pressures on specific individuals and entities and that restricts selective products or activities, while minimizing unintended economic and social consequences for vulnerable populations and innocent bystanders'.[24] In fact, sanctions targeting specific sectors of state economies had been practiced before, such as in Libya and Haiti in the 1990s and on South Africa in the 1980s. Instead, sanctions targeting individuals and non-state actors gained importance with the aforementioned 1267 regime in 1999.[25]

Currently, there are four types of targeted sanctions that are imposed: arms embargoes, travel bans, commodity boycotts and financial restrictions. Arms embargoes regard the prohibitions to trade with either dual-use or military related equipment and services with individuals, entities and actors located in specific regions. Travel ban relates to the prohibition for people to travel and access the territory of specific states. Commodity boycotts refer to the limitation to trade

[20] Wallensteen and Staibano 2005.

[21] Hudson 2007.

[22] Cortright and Lopez 2002.

[23] Biersteker 2009.

[24] Cortright et al. 2002: 2.

[25] Biersteker 2009.

specific goods and services. This could be the case for natural resources, such as diamond and timber that can be linked to the enrichment of specific groups, but also strategic sectors of economies, such as the exploration of natural resources. Other forms of diplomatic measures could be also listed here. There have been numerous efforts of creativity to design sanctions in the recent years, but all of them would fall under these categories. The next section presents why sanctions are used.

6.3 The Purpose of Sanctions: Coercing, Constraining and Signalling

Sanctions have been used in support of or as alternative to the use of force in the past. Military expeditions are normally accompanied by the denial of resources and technologies to direct opponents, therefore sanctions are imposed as a prelude to military operations. However, sanctions have also been described as a 'silent and deadly remedy' by President Wilson in the aftermath of WWI.[26] The Great War was a shocking event in Europe and the League of Nations, warmly advocated by the United States, was supposed to bring collective security mechanisms according to which threats posed by one state would have been countered by all the others. This was the logic of sanctions, whose imposition would have cut off the trouble maker state from resources and relations with the others so to annul the threat without having to wage war for that. The evolution of targeted sanctions changed the nature of the instrument in two ways. First, there is a weaker link with the use of force as sanctions are frequently used independently from the military instrument, but the substituting element is still central. Second, sanctioning sub-national actors highlight even further that sanctions are not only imposed to change the behaviour of targets, as wrongly assumed by many, but they can also constrain and signal actors in the international system.[27]

Targeted sanctions are currently used in scenarios that may have been dealt with force in the past. For instance, sanctions are currently used in counter-terrorism and non-proliferation regimes. Despite the risk of threats originating from states and non-state actors, the utilization of sanctions with the objective of regulating the market of import/export of dual-use goods is more an instrument of governance rather than a classical foreign policy instruments. For instance, the decision to deal with threats originating from Iraq in 1990 and Afghanistan in 2001 was dealt with through the application of military force. Victory was sought mainly through the employment of the military instruments. In both cases, sanctions preceded military operations. However, targeted sanctions are often imposed without referral or expectation that force will be used.

---

[26] Fouley 1923, p. 71.

[27] Giumelli 2011.

This is true for terrorism and proliferation threats, but targeted sanctions have also become an essential component for comprehensive strategies towards the consolidation of democracies in post-conflict phases. For instance, the utilization of sanctions to support the stabilization process in Liberia, Sierra Leone, Tunisia and Egypt, only to mention a few, have shown how sanctions can be used in ways that substitute for the use of more forceful instruments. Here targeted sanctions aims at undermining the role of conflict spoilers in the stabilization of countries, therefore force is often left to the local enforcement agencies.[28]

Finally, targeted sanctions can still be used with a more classical understanding of foreign policy instruments. For instance, the case of the US and EU sanctions on Russia can be seen in this light. Allegedly threatening the independence of Ukraine, Russia was hit by targeted sanctions on individuals, entities and specific sectors of the economy related to the exploration industry of natural resources. The objective is to ensure that Moscow halts its support to the rebel forces in the Donbass region and it collaborates with the international community about the annexation of Crimea in 2014. Once again, targeted sanctions have been imposed with the intent to deal with the challenge posed by Russia without using military force, which is often connected to casualties and humanitarian consequences.

Thus, sanctions are used as an alternative to the use of force, but how can they lead to *victory*? The debate on effectiveness has been dominated by the so called 'naïve logic of sanctions'. The naïve logic assumes that senders[29] intend to change the behaviour of targets by inflicting economic pain that, consequentially, would convince targets to modify their policies. In fact, sanctions operate along the lines of at least three different logics of power. First, sanctions do indeed attempt to change the behaviour of targets. Here, the logic is that sanctions would have to add a cost to specific behaviours that senders dislike, such as the Russian interference with Ukrainian domestic issues, and specific actions are required by senders. Second, there are sanctions that do not ask any active action, but rather a passive one from the targets. Whereas targets need to act to comply with senders' requests, targets need to halt their action in order to see the sanctions lifted. This latter can occur either because targets change their mind about their objective or because they are unable to achieve them due to the impact of sanctions. Finally, sanctions send strong signals in international politics. Leaving aside the domestic groups of sender societies, which alone can sometimes explain the imposition of sanctions in specific cases, there are two other signals that can be sent. The first set of signals is intended for the explicit target of sanctions. The imposition of sanctions is a widely understood as an exceptional measure taken in exceptional times, therefore targets can alter their expectations about the future if they persist with certain actions. Secondly, sanctions are always imposed in front of a larger audience, therefore the

[28] There are examples, for instance the use of force employed to arrest Laurent Gbagbo in Cote d'Ivoire, see Simonen 2012.

[29] Coherently with the literature on sanctions, senders are the actors that impose sanctions and targets are the actors that are sanctioned.

signal is meant for the audience rather than for the explicit target. Norms and practices in the international system also contribute to forming expectations regarding future decisions. Regardless of the implications that certain sanctions may have on the immediate threat, by communicating a norm they also have implications for other similar actors that contribute to the dynamics of international politics.

Coercing, constraining and signalling are logics of power that, if connected to the practice of sanctions, interpret well the twenty-first century shift towards achieving objectives in international politics while minimizing the human sufferance in doing so.

6.4 Opportunities and Challenges

The evolution to targeted sanctions was also caused by the need to increase the effectiveness of comprehensive ones. The Targeted Sanctions Consortium (TSC) on UN targeted sanctions found that the effectiveness rate has not changed much compared to the past, but the debate is still ongoing on whether sanctions today are more or less effective than in the past. Instead, given the fundamental differences between targeted and comprehensive sanctions,[30] the attention should focus on the opportunities and challenges that targeted sanctions present rather than focusing exclusively on the issue of effectiveness.

Targeted sanctions have fewer unintended effects when compared to comprehensive sanctions, therefore similar rates of effectiveness would still be accompanied by fewer 'collateral damages'. At the same time, wrongdoers are explicitly named, which helps in identifying more specifically the set of undesired behaviours at the international level. This represents also a constraint for senders as they need to justify the imposition of sanctions to a much greater extent compared to the past.

From a foreign policy perspective, targeted sanctions are more flexible and offer more diplomatic options compared to comprehensive sanctions. Whereas a sender could only impose or remove sanctions, targeting offers the possibility to change the listing and to modify it with the listing criteria as well. The suspension of sanctions was available even in the past, but the possibility to offer individual deals to targeted people is unique to targeted sanctions. Exemptions were also available in the past, but the degree to which exemptions are utilized today is immensely more complicated than in the past. Targeted sanctions can be linked to goods, but also to services and to specific financial transactions. Individuals can be listed, but directly and indirectly associate individuals can also be touched by sanctions. The room for manoeuvring for a sender is substantially larger than in the past.

Targeted sanctions can be used not only against societies, but also within them. This extends the scenarios under which sanctions can be used. For instance, targeted sanctions can be used to support democracy and to promote human rights,

[30] Giumelli 2015.

they can sustain the consolidation of democratic institutions and they can be used to counter the proliferation of weapons of mass destructions and the financing of terrorism. The innovative nature of targeted sanctions lies in their nature to discriminate between sub-national actors, which was not a feature of comprehensive sanctions. Today, senders often impose sanctions against one of the parties in the domestic context of the targeted society. Sanctions have caught up with other foreign policy instruments, such as economic aid and diplomacy, which have been used in that way also in the past.

Finally, targeted sanctions are not only likely to have fewer humanitarian consequences, but they are also likely to be cheaper than comprehensive sanctions. For instance, one can take the example of the UN sanctions on Iraq in 1990 and consider the costs that oil dependent countries had to sustain after the imposition of sanctions. The sanctions on Russia today presented the EU and the US with a certainly manageable cost compared to what it would have been if they had decided to impose a full fledge embargo on Russia. Costs for targeted sanctions are more limited and they are easier to estimate in advance.

At the same time, targeting present challenges that are new to the world of sanctions. First, the implementation of targeted sanctions is far more complex than for comprehensive ones. The identification of individuals and entities requires very detailed knowledge about targeted societies that is very difficult to attain. Designing targeted sanctions is extremely complicated and more detailed information is required.

Similarly, monitoring and enforcing targeted sanctions require institutional capacities that need to be built at the international, state and local levels. In essence, targeted sanctions borrow practices from domestic systems and apply them at the international level, where no institutional mechanism has been constructed for ensuring a coherent and consistent implementation across states. For instance, who carries out investigations about possible busting activities? Who can provide the correct evidence for appropriate listing? And, quite relevant, who needs to pay for these functions? These questions are in need for answers if we are to enhance the adoption of targeted sanctions in the future.

The lack of knowledge on the state side has led to a reliance on private actors for the implementation of sanctions. For instance, the freezing of assets is decided by public authorities, but it is carried out by financial institutions that have listed individuals among their clients. When it comes to import/export of goods, technologies and services, it is up to companies to verify not to be in violations of sanctions. The collaboration between private and public actors will be critical to the correct implementation of sanctions and, therefore, to the achievement of the foreign policy goals that motivated the imposition of sanctions. On the contrary, the behaviour of private actors can change the nature of sanctions as designed by governments. For instance, this is the case for sanctions overcompliance. While public regulators are designing sanctions to reduce the scope of their impact on specific individuals, the lack of guidelines provided spread fears among financial institutions so that all transactions towards a society are blocked, de facto transforming sanctions from targeted to comprehensive.

The second set of problems that regard targeted sanctions is made of legal challenges. Surprisingly, when the Security Council decided to opt for sanctions on individuals and entities, few anticipated that the same individuals and entities may have claimed individual rights. Therefore, the problem consists in establishing criminal proceedings-like institutions at the international system without having the guarantees that exist at the national level. For instance, the initial listings after 09/11 were done with no communication with listed individuals. Additionally, mistakes could not be rectified as procedures were not available to individuals to react to decisions of the Security Council, if not by using the byzantine path of complaining with the authorities of their state of citizenship or residence. The problem became evident after the well-known Kadi case. In 2003, a Saudi citizen named Mr. Yassin Abdullah Kadi appealed against the EU Regulation (EC) No 881/2002 implementing UNSCR 1390 (2002) at the Court of First Instance (now the General Court) of the European Union.[31] According to Kadi, the UN procedure violated due process principles that were recognized by EU treaties. The 2008 judgement by the European Court of Justice determined that the procedure to target individuals with sanctions, even when it is by the decision of the Security Council, should follow due process and fair trial tenets. Individuals have the right to know the reasons that motivate such restriction, they have the right to be heard and they should have the instruments to remedy to mistakes made by the authority. Consequently, further court cases and discussions have established a number of practices that make more difficult imposing sanctions on individuals and non-state actors. For instance, if freedom to property and movement is to be restricted, evidence need to be shown to justify such decision. However, evidence gathered in one country does not necessarily respect the principle for evidence gathering in force in other countries. Targeted sanctions become, then, foreign policy decisions that are strictly limited by pre-existing principles and legislations. Furthermore, sanctions on entities can also open disputes about the capacity of complying with commercial contracts signed before the imposition of sanctions. Legal challenges risk to either nullify the impact of sanctions or to undermine the utility of targeted sanctions.

6.5 Improving Targeting Strategies

Targeting has become the norm in the world of sanctions. The objective is to reduce the humanitarian impact while maximizing the pressure on individuals that are responsible for undesired behaviours. If *winning* is intended as achieving policy objectives, then targeted sanctions is an instrument that depicts well the theme of this edited volume of *winning without killing* since the evolution of targeting was mainly inspired by the need to reduce the humanitarian impact of comprehensive embargoes. While targeted sanctions present policy makers with a number of

[31] *Kadi v. Council of the European Union* 2005.

opportunities, as illustrated above, there are also several challenges that need to be dealt with if targeted sanctions are to be useful in foreign policy.

First, senders need to know more about the societies/threats that they intend to deal with when sanctions are imposed. Effective design of targeted sanctions presupposes good knowledge about targets and the societies wherein targets operate. While targeting government officials appears rather simple, targeting others is a more complex issue. For instance, knowledge is necessary to identify the network of power that sustain governments and constitute powerful groups. However, this is only half of the effort, as there is need to know both who they are and how they operate. The level of knowledge required is, often times, also not achieved by the enforcement authorities of states, but gathering intelligence is central to a successful strategy of winning without killing.

Second, institutional capacities and international collaboration are necessary to ensure that the rationale of decision-makers coincides with the political outcome of sanctions. For instance, whereas police forces in cooperation with judicial authorities investigate and prosecute actions aiming at circumventing the laws, little is done in the area of sanctions. The most active institution is the Office for Financial Asset Control (OFAC) at the US Treasury, which has carried out extensive investigations and imposed fines on several US and non-US based banks for circumvention of sanctions imposed by US authorities.[32] This example is unmatched by other institutions. The Office for the Implementation of Financial Sanctions (OFSI) established at the UK treasure is an attempt to go into this direction and shows a trend that is consolidating overtime, but it is too soon to assess their effectiveness. In any case, both OFAC and OFSI do respond to different governments and implement different sanctions regimes. The UN has established institutions to support the judicial review of the listing, such as the Focal point for delisting and the Office of the Ombudsperson, and to collect information to assess the impact of sanctions, such as the UN Panels of experts. However, more is to be done if sanctions are to be taken seriously and are to achieve the expected foreign policy objectives.

Third, targeted sanctions need to be used strategically. This means that the opportunity to use sanctions should be assessed according to the context and to the other tools that are used at the same time. Indeed, targeted sanctions should be designed to coerce, constraint or signal specific actors in the targeted society. The same targeted sanctions regime can be used with different intents according to the list of individuals, so certain individuals may be coerced, others constrained and others signalled. Strategic reasoning should lead the design of targeted sanctions and it should regard all the foreign policy instruments adopted at the same time that either strengthen or weaken the utility of sanctions in *winning* in international politics.

---

[32] Giumelli and Levi 2016.

6.6 Conclusions

The evolution of sanctions from comprehensive to targeted has been determined by the need to enhance effectiveness while reducing the humanitarian impact that sanctions had on innocent civilians. The assessment of sanctions in Iraq, former Yugoslavia and Haiti in the '90s was extremely negative and something had to change. Thanks to the emergence of the international individual responsibility principle, sanctions started to target individuals and entities as well as economic sectors. The hope was to increase effectiveness while decreasing unintended humanitarian consequences. The contemporary form of targeted sanctions fits well under the theme of this edited volume that focuses with efforts of *winning without killing* in international politics.

This chapter has shown that even if it is difficult to reach a definitive conclusion about the effectiveness of targeted sanctions, they represent a departure from the conventional understanding of sanctions that has both opportunities and challenges for policy-makers. They present opportunities because targeted sanctions can operate at a subnational level, like foreign aid and diplomacy, and therefore are more ductile when it comes to foreign policy. They present challenges because they target individuals and entities, therefore more accurate knowledge is needed to carefully design sanctions and legal guarantees need to be considered. Consequently, sanctions can be effectively designed, monitored and enforced if senders have enough knowledge and institutional capacities to investigate and prosecute sanctions busters. The acquisition of more information and the establishment of institutions to monitor and enforce sanctions are necessary steps towards more effective utilization of sanctions to achieve policy objectives (*winning*) with limited humanitarian consequences (*without killing*).

References

Ali MM, Iqbal SH (1999) Sanctions and Childhood Mortality in Iraq. The Lancet 355 (9218):1851–1857

Alnasrawi A (2001) Iraq: economic sanctions and consequences, 1990–2000. Third World Quarterly 22(2):205–218

Baldwin DA (1985) Economic statecraft. Princeton University Press, Princeton NJ

Barber J (1979) Economic Sanctions as a Policy Instrument. International Affairs 55(3):367–384

Biersteker TJ (2009) Targeted sanctions and individual human rights. International Journal 65 (1):99–117

Clawson P (1993) How has Saddam Hussein Survived? Economic Sanctions, 1990–93. INSS McNair Papers, Institute for National Strategic Studies, Washington DC

Cortright D, Lopez GA (1995) Economic Sanctions: Panacea or Peace building in a Post-Cold War World? Westview Press, Boulder CO

Cortright D, Lopez GA (eds) (2002) Smart Sanctions: Toward Effective and Humane Sanctions Reform. Rowman and Littlefield, Lanham MD

Cortright D, Lopez GA, Gerber L (2002) Sanctions and the search for security: challenges to UN action. L. Rienner Publishers, Boulder CO

Drezner DW (1999) The sanctions paradox: economic statecraft and international relations. Cambridge University Press, Cambridge/New York

Foley H (ed) (1923) Woodrow Wilson's Case for the League of Nations. Princeton University Press, Princeton NJ

Galtung J (1967) On the Effects of International Economic Sanctions: With Examples from the Case of Rhodesia. World Politics 19(3):378–416

Gibbons ED (1999) Sanctions in Haiti: human rights and democracy under assault. Praeger, Westport CON

Giumelli F (2011) Coercing, Constraining and Signalling. Explaining and Understanding UN and EU Sanctions After the Cold War. ECPR Press, Colchester

Giumelli F (2015) Understanding United Nations targeted sanctions: an empirical analysis. International Affairs 91(6):1351–1368

Giumelli F, Levi G (2016) anzioni: alle imprese europee la multa arriva dagli USA. LaVoce.info (3 June 2016) http://www.lavoce.info/archives/41389/sanzioni-alle-imprese-europee-la-multa-arriva-dagli-usa/ Accessed 14 March 2017

Hoskin E (1997) The Humanitarian Impacts of Economic Sanctions and War on Iraq. In: Weiss TG (ed) Political gain and civilian pain: humanitarian impacts of economic sanctions. Rowman & Littlefield Publishers, Oxford/Lanham MD

Hudson A (2007) Not a Great Asset: The UN Security Council's Counter-Terrorism Regime: Violating Human Rights. Berkeley Journal of International Law 25(2):203–227

Hufbauer GC, Schott JJ, Elliot KA (1990) Economic Sanctions Reconsidered. Institute for International Economics, Washington, DC

Kadi v. Council of the European Union (2005) 'Kadi, Yassin Abdullah and Al Barakaat International Foundation v. Council of the European Union and Commission of the European Communities,' Joined Cases C-402/05 P and C-415/05 P. http://eur-lex.europa.eu/legal-content/EN/TXT/HTML/?uri=CELEX:62005CJ0402&from=EN. Accessed on 14 March 2017

Lindsay J (1986) Trade Sanctions as Policy Instruments: A Re-Examination. International Studies Quarterly 30(2):153–173

Mueller J, Mueller K (1999) Sanctions of Mass Destruction. Foreign Affairs 78(May/June):43–53

Naylor RT (2001) Economic warfare: sanctions, embargo busting, and their human cost. Northeastern University Press, Boston MA

Nincic M, Wallensteen P (eds) (1983) Dilemmas of Economic Coercion. Sanctions in World Politics. Praeger, New York

Pape RA (1997) Why Economic Sanctions Do Not Work. International Security 22(2):90–136

Schabas W (2006) The UN international criminal tribunals: the former Yugoslavia, Rwanda and Sierra Leone. Cambridge University Press, Cambridge

Shraga D, Zacklin R (1994) The International Criminal Tribunal for the Former Yugoslavia. European Journal of International Law 5(1):360–380

Simonen K (2012) Qui s'excuse s'accuse… An Analysis of French Justifications for Intervening in Côte d'Ivoire. International Peacekeeping 19(3):363–376

Sunga LS (1992) Individual responsibility in international law for serious human rights violations. Martinus Nijhoff, Dordrecht/London

The Financial Times (1995) Personal View: How to Keep Sanctions in Proportion (18 May 1995)

Tsebelis G (1990) Are Sanctions Effective? A Game-Theoretic Analysis. Journal of Conflict Resolution 34(1):3–28

United Nations (1945) Charter of the United Nations

United Nations (1997) The Relationship Between Economic Sanctions and Respect for Economic, Social and Cultural Rights, General Comment 8. Committee on Economic, Social and Cultural Rights. UN Doc E/1998/22 (4 December 1997)

United Nations (1999) Resolution 1267 (1999). Adopted by the Security Council at its 4051st meeting on 15 October 1999. UN Doc S/RES/1267 (1999)

Von Braunmühl C, Kulessa M (1995) The Impact of UN Sanctions on Humanitarian Assistance Activities. A Report on a Study Commissioned by the United Nations Department of Humanitarian Affairs. Gesellschaft für Communication Management Interkultur Training, Berlin

Wallensteen P, Staibano C (eds) (2005) International Sanctions: Between Words and Wars in the Global System. Frank Cass, London/New York

Weiss TG, All E (1997) Political gain and civilian pain: humanitarian impacts of economic sanctions. Rowman & Littlefield Publishers, Oxford/Lanham MD

Chapter 7
International Mediation in Syria's Complex War: Strategic Implications

Berma Klein Goldewijk

Abstract This chapter offers a critical inquiry of the current process of United Nations (UN)-Arab League mediation in the Syrian war, in order to explain the effectiveness and implications of international mediation as a non-kinetic (non-coercive) asset in strategies that address this war. The Syrian war is perceived as a civil war with features of an evolving proxy war, which has in many ways come to define twenty-first century wars. Whereas it is often argued that multilateral mediation in Syria has largely failed at the new fault lines of major-power rivalries, it will be demonstrated here that mediation's effectiveness as a non-kinetic strategic means needs to be interpreted in a much broader perspective than the usual dichotomous notions of success or failure may account for. Moreover, the limitations will be shown of traditional concepts in mediation theory that have largely defined the conditions for mediation onset, such as a mutually hurting stalemate and the ripe moment for conflict resolution. The chapter outlines the implications in view of critical scholarship on war and non-linear strategic approaches, involving notions such as uncertainty and unpredictability. By addressing these topics, the study brings various bodies of literature together, problematizes some core concepts and assumptions, identifies consequences, and suggests ways forward in research.

Keywords Multilateral mediation · International negotiation · Conflict resolution · Institution-building · Non-kinetic strategic means · Proxy warfare · Critical war studies · Non-linearity · Uncertainty · Unpredictability · Clausewitz

---

B. Klein Goldewijk (✉)
Faculty of Military Sciences, Netherlands Defence Academy, Breda, The Netherlands
e-mail: bmjb.kleingoldewijk@mindef.nl

P.A.L. Ducheine and F.P.B. Osinga (eds.), *Netherlands Annual Review of Military Studies 2017*, NL ARMS, DOI 10.1007/978-94-6265-189-0_7

Contents

7.1 Introduction

The critical question 'what is war' is always asked under the duress imposed by "this war," as Etienne Balibar aptly remarks.[1] This war, today, is the Syrian war: the most decisive war we are witnessing with its (il)legitimacy, causes and origins, and its immediate or future consequences. The gruesome carnage in Syria has brought about continuous international mediation efforts. The overarching question addressed here is to what extent international mediation should be considered an effective means of non-kinetic strategies in today's proxy warfare, and what the implications would be for future policy and strategic approaches.

The use of strategies that involve non-kinetic instruments has recently gained particular relevance in reinvigorated forms of proxy warfare. The Syrian conflict, but also Russia's annexation of the Crimea and its intervening operations in Ukraine,[2] have called wide attention to the primacy given to non-kinetic means of warfare. More specifically, the war dynamics in Syria are making proxy war a renewed and significant feature of contemporary conflict.[3] However, international mediation scholarship has not yet studied mediation in this perspective as a non-kinetic strategic means. Likewise, the challenges of warfare by proxy have not specifically been examined in the mediation literature as yet. The assumption here is that both aspects may have considerable implications by setting the need for transforming conventional approaches and theories of international mediation.

Definitions of international mediation[4] are fairly uncontroversial and fit very well in the setting of non-kinetic approaches. Mediation, as a diplomatic channel of power and influence, is conducted by a single third-party or in the form of

---

[1] Balibar 2008, p. 366.

[2] See also Sauer 2017.

[3] Mumford 2013a, b.

[4] Mediation is "a process of conflict management where disputants seek the assistance of, or accept an offer of help from, an individual, group, state, or organization to settle their conflict or resolve their differences *without resorting to physical force* or invoking the authority of the law [italics mine]." Bercovitch et al. 1991, p. 8. See also Hellman 2012, p. 591.

multiparty activities to involve antagonistic actors. Mediation involves *third actors supporting* the conflicting actors to work out their own agendas and solutions towards a settlement. Negotiation is different and rather involves efforts to resolve disputes and reach agreements *between* the warring actors themselves. Negotiation can be brought in as part of a mediation process, as will be seen below in the Syrian mediation process. Both mediation and negotiation integrally fit as means into constructive non-kinetic strategies that aim at cooperation between adversaries to reach settlements or stabilisation, conflict resolution or state- and institution-building. In Syria, mediation is undertaken jointly by a coalition of international and regional actors, the UN and the Arab League, in addition to wider distinct attempts by states (such as Russia, Turkey, or the United States) or by non-state organisations.

The common spectrum of failure of multilateral mediation in Syria will be assessed here in connection to the complex character of this war, as both a civil war and a proxy war. Joint mediation by the UN-Arab League has been employed since the start of the Syrian war and continues to date, but has turned out to offer limited results: it is commonly held that it overtly failed to bring about conflict resolution in the sense of stability and peace. This mediation process will be studied through the prism of critical war studies. As explained by Tarak Barkawi and Shane Brighton, critical war studies primarily focuses on what war is, on war's excess and its generative aspects, encompassing the new political order and society that war generates.[5] This view will be explored as relevant for international mediation and as a key support of multilateral stabilisation and reconstruction efforts.

Although studies on mediation and negotiation may credibly have developed critical approaches, *systematic critical scholarship* has not emerged in these specific fields.[6] Making critical knowledge on war relevant to present challenges may be of help for understanding and interpreting mediation as a non-kinetic strategic means to address proxy warfare. Similarly, a non-linear approach to strategy, as developed by Alan Beyerchen[7], with key notions such as uncertainty and unpredictability, will be assimilated here in regard of its potential contributions for identifying the implications of failed international mediation for future strategy. In line with these approaches, some conventional core concepts and assumptions of mediation will be questioned.

Drawing on the case of the Syrian war does not imply that this chapter integrates a full designed case study in the strict methodical sense. The paper mainly uses an interpretivist methodological approach: as a theoretically distinct contribution in social sciences, the focus of this approach is contextual and the knowledge generated appreciates the non-linearities of understanding the real world - to be

[5] Barkawi and Brighton 2011.

[6] A few exceptions should be noted, amongst them Jones 2000 and Hellman 2012, who in fact developed critical scholarship on mediation.

[7] Beyerchen 1992-1993; also Osinga 2007.

assumed in the role of theory.[8] The chapter proceeds in two main parts, the first focuses on the specifics of strategy and proxy war, particularly in Syria, the second on the process of joint UN-Arab League mediation in the light of thinking about war and strategy. The chapter ends with conclusions and implications.

By studying international mediation as a non-kinetic tool in proxy war, and by taking the perspective of critical war studies and non-linear strategy, this article breaks some new ground—advancing beyond previous achievements in mediation studies while exploring how new knowledge can be created. By taking this approach, the chapter advances a constructive strand of non-kinetic strategic interventions in war, foregrounding mediation—even if it fails. This may support an interpretive understanding of how non-kinetic tools may fit into future international policies and strategies that make use of all available (both kinetic and non-kinetic) options.

7.2 Strategy and Proxy War in the Syrian Context

States are increasingly pursuing their foreign policy goals through local non-state proxies along with international and regional forces, thus changing the character of warfare to involve organised proxy violence. Along these lines, Russia's annexation of the Crimea and its intervention in Ukraine (2014)[9] have been perceived as the scene of a proxy war between the West and Russia, just like Yemen is viewed as a war by proxy for Saudi–Iranian coalitions. The scale of the integrated use of cross-border infiltration by non-state armed proxy actors such as militias or private military organisations, with formal command systems and sustained lucrative financing architectures, has created concerns in the West about a 'new way of war.' As a matter of fact, proxy warfare today is increasingly coupled with strategies that use both kinetic and non-kinetic means, such as cyber information operations, whereby the use of cyberspace functions as a platform from which to (in)directly, or from a distance, wage war, combined with more traditional capabilities.[10]

Roughly defined, in conventional approaches strategy is seen as the direction and use made of means and ways for the purposes of policy. In military operational settings, strategic performance can broadly be categorised as kinetic and non-kinetic. The most prevalent kinetic strategies use coercive means, involving offensive measures to eliminate, capture or neutralise opposing forces and objects.[11] Kinetic effects, then, are (primarily) lethal effects in the physical domain.[12] Non-kinetic strategies use more subtle forms of power, with restraint and

[8] See Schwartz-Shea and Yanow 2011; Mearsheimer and Walt 2013; Zweibelson 2015, 2016.

[9] Sauer 2017.

[10] Proxy warfare, then, may have some overlaps with but is also different from hybrid warfare, conceived as a strategy that links conventional means with insurgency tactics. See Lanoszka 2016.

[11] Roberts and Everton 2016, p. 30.

[12] Ducheine 2016, p. 203.

non-coercive instruments.[13] They may appear in basically two distinct forms: either involving an effort to *undermine* opposing forces by, for example, economic sanctions, lawfare, and cyber operations, or by attempting *cooperation* with adversaries via diplomacy, mediation or negotiation. Non-kinetic effects may occur both in the non-physical and physical domain.[14] Jointly, kinetic and non-kinetic approaches represent a 'full spectrum' approach, involving a broad array of means and methods to generate effects throughout the physical and non-physical domains.[15] A focus on non-kinetic approaches, therefore, does not negate that military strategy and operations may require "the use of all capabilities, military and otherwise, kinetic, lethal, non-kinetic, and non-lethal," with effects ranging from "constructive to disruptive".[16]

Obviously, this way of warfare, with proxy features and the use of both kinetic and non-kinetic means, is not an entirely qualitative novelty since many of these tactics have since long been used in both Western operations and pre-Soviet Russian practice. Proxy wars, indeed, are have a long past: the use of 'proxy' actors in the context of irregular conflict, as well as in wars between states, spans nearly the full military history. During the Cold War, they came largely to be seen as international superpower-induced wars fought on the territory of a third party. States developed the capacity to stay out of the immediate conflict dynamics, by shaping the desired strategic outcome of a conflict through indirect engagement, and by choosing and resourcing proxies that would offer fewer risks than the direct commitment of the own armed forces.[17] Today, however, as Mark Galeotti remarks, such proxy operations are "distinctive in terms of the degree"[18] to which predominance is given to non-kinetic means of warfare, and how these fit in closely with political and military command structures. What is new, according to this view, is the *configuration* of contemporary political, military, technological, informational and social components of the contexts in which new wars are being fought.

Major changes in the character of modern and asymmetric warfare point at a potential further upsurge in proxy strategies by states, as Andrew Mumford has clarified.[19] Actually, proxy warfare does not incur the financial and human costs of direct military intervention and seems to offer a convenient means of achieving foreign policy goals. Moreover, there is a notably decreased public and political appetite in the West for large-scale military interventions, in particular after the military and strategic failures in Iraq and Afghanistan.[20] War by proxy, then, has

[13] Roberts and Everton 2016, p. 30.

[14] Ducheine 2016, p. 203.

[15] Ducheine 2016, pp. 201, 210.

[16] Ducheine 2016, p. 206.

[17] Hughes 2014, p. 523.

[18] Galeotti 2016, p. 282.

[19] Mumford 2013a, b.

[20] Hughes 2014.

been called by various names: "war on the cheap",[21] or in a more elegant formulation by Philip Towle, "vicarious belligerency".[22] In sum, proxy warfare has in many ways again come to define twenty-first century wars, involving armed conflicts that share some or all of the above-mentioned features in varying degrees.[23]

The character of the Syrian war shifted and has over time drawn in multiple regional and international forces, including Russia and Iran, Saudi Arabia and Qatar, Turkey, the United States (US) and other Western powers. The political crisis in Syria turned into a civil war in the wake of the Arab Spring in 2011. Whereas the principal accountability lies with the regime of president Bashar al-Assad in Syria, which sparked a brutal and violent response to initially peaceful demonstrators, it was the opposition's armed response that transformed the existent crisis into a civil war in 2012. It soon appeared that there are, within and between the main Syrian blocks, multiple fault lines of contention, including ideological, ethnic, religious, and sub-state ties, making the Syrian civil war a fluid war with a constantly changing combination of such affiliations.[24]

Steadily, this complex civil war evolved into a regional and international proxy war where the Baathist regime, as well as the non-regime forces, have become almost fully reliant on support from major external powers aiming at preventing defeat at all costs while giving this war a proxy reach with global geostrategic implications.[25] First, the Syrian Assad regime endures by virtue of direct political and military support from Iran, the involvement of Hezbollah (based in Lebanon) as well as Iraqi and Afghan militias, and finally, as of 2015, military and diplomatic supportive interventions by Russia. Second, the manifold Syrian non-regime opposition forces are mainly supported by Saudi Arabia and Qatar, Turkey and Western coalitions. Relatively independent Kurdish fighters, in some areas joining and at other places combatting oppositional rebel factions, are opposed by Turkey but backed by a few other states in the region and the US. Third, a vast network of Islamist militias, including Jabhat Fateh al-Sham (formerly known as Jabhat al-Nusra), Al-Qaida and the Islamic State of Iraq and al-Sham (IS), rely on supportive channels from regional and international beneficiaries. Moreover, longstanding conflicts have reappeared in Syria, with Turkey being distrustful of the Kurdish involvement in the Syrian opposition, and Iran and Saudi Arabia fighting their war in Syria over hegemony in the region. On both the policy and operational levels, Western major powers, as well as Russia, Iran and Turkey, are involved in sophisticated proxy strategies in Syria.

---

[21] Mumford 2013a, p. 40.

[22] Towle 1981, pp. 21, 23, 26.

[23] The field of proxy wars is widely recognised to be underdeveloped in studies on war, security and the military. See Mumford 2013a, b.

[24] Phillips 2015, p. 358; Hughes 2014, p. 523; Hill 2015; Lundgren 2016.

[25] Hughes 2014.

Overall, it can be stated that the agendas of local, regional and international actors in Syria are disjointedly incoherent and collectively virtually incompatible.[26] In fact, the use of proxies is not without risks and such risks are multiplied when the strategic priorities of the patron states and their proxies come to diverge. As Brown observes, "[t]he motivations of proxies and the interests of the countries employing them are hardly ever sufficiently close, and the command-and-control arrangements sufficiently tight, to ensure that the battlefield behaviour of proxies will not distort the military strategies and political objectives of their patrons".[27]

From a strategic perspective, in the sense Hew Strachan and Clausewitz use the notion of strategy, any interpretation of the Syrian war needs to be linked to views on the war itself. Interpreting what made the political crisis in Syria a war directly links to such a reflection on the nature of war. What promptly stands out here is the apparently counterintuitive insight from Clausewitz about war itself: *war starts with defence*, not attack. In Strachan's words: "it is the defender who resorts to war to oppose the invader's intentions".[28] In fact, as Beyerchen confirms, when an opponent uses one-sided military force, without defensive response, the outcome is not a war but a massacre. Likewise, total non-resistance would be no war at all.[29] It is precisely this element of mutuality that is to be seen as constitutive to what war is: the dynamics of war is inherently interaction. If war, as Clausewitz argues in one of his three definitions of war, "is nothing but a duel [Zweikampf] on a larger scale … an act of force to compel our enemy to do our will",[30] then the nature of war is intrinsically linked to the counterforce and counterweight of an opponent. Thus war emerged, in line with war's antagonistic and reciprocal nature, when the opposition forces in Syria decided that the unilateral atrocious force applied by the Assad regime became intolerable.

Proxy war not only involves another way of conducting war but may have significant implications for interpreting the role of multilateral mediation as a non-kinetic strategic asset, as will be demonstrated below. Though mediation is widely thought to influence the outcome of conflict resolution in terms of mutual agreements, this is, however, not a straightforward matter. This assumption has led to a dichotomous notion of conflict resolution outcomes in terms of *success or failure*, negating a fuller interpretation of mediation that may appear relevant in a strategic perspective.[31]

---

[26] Lucas et al. 2016, p. 26.

[27] Brown 2016, p. 243.

[28] Strachan 2007, p. 11.

[29] Beyerchen 1992–1993, p. 67.

[30] Von Clausewitz [1832] 1976, p. 75. For the German edition, see Von Clausewitz [1832] 1991.

[31] Bercovitch and Gartner 2006, p. 329.

7.3 Joint UN-Arab League Mediation Process: Limitations and Failure

Syria's long-stalled international diplomacy represents the prototype minefield of mediation. The Syrian case illustrates how multilateral mediation, particularly the UN-Arab League joint effort, has been limited. In a wider perspective, this complicated case may shed light on the effectiveness and implications of mediation as a non-kinetic strategic asset.

Kofi Annan, appointed as UN-Arab League Joint Special Envoy to Syria, from 23 February 2012 until he resigned on 2 Aug 2012, appeared to be the best-suited candidate to find a way out of the Syrian civil war because of his personal authority and legitimacy as former UN Secretary General. Within a few weeks after his appointment, Annan presented his six-point peace plan to end the Syrian war, involving political negotiations in the framework of a UN-supervised ceasefire.[32] At the time Annan and his mediation team started, the Syrian civil war had created serious tensions between the major external powers. A Western military intervention, though unlikely, remained highly uncertain and could not be ruled out.[33] Facing widespread reluctance of the international community to armed intervention in Syria, Annan's initial priority was to reduce the level of uncertainty inside and outside Syria, intending to create a framework for political talks between the main actors.

However, as Richard Gowan appropriately remarks, by lowering the level of uncertainty, Annan reduced his own leverage: the Syrian government concluded that it would not face international military intervention for failing to cooperate with Annan's mediating team and assumed that international efforts would be limited to arming parts of the opposition.[34] In addition, it has been argued that Kofi Annan lacked a strategy at that time to address the manifold opposition's obstinacy and its escalating military campaigns: convinced by the Libyan precedent that the West would intervene on its behalf, the opposition held on to unrealistic demands in the mediation process.[35] In fact, until mid-2014, external military support to the opposition by Western and Gulf countries remained mostly limited to providing training and equipment. Russian and Iranian support for the Assad regime was complemented by military advisers, the involvement of the Iranian Revolutionary Guard, and fighters provided by Hezbollah, Iraqi and Afghan militias.[36] This restrained military intervention only changed in the second half of 2014.

[32] Gowan 2013, p. 1.

[33] Gowan 2013, p. 1.

[34] Gowan 2013, pp. 3–4.

[35] Crocker et al. 2015; Hill 2015; Hinnebusch and Zartman 2016.

[36] Lucas et al. 2016, p. 24.

Annan's mediation efforts evolved in three steps.[37] First, he focussed on the Syrian government and the opposition, due to mounting (inter)national pressure to oust the Syrian government, only to discover that the regime of President Bashar al-Assad was unwilling to accept any political compromise, whereas the opposition forces, still at that time, were too divided to agree on such. Both sides were not welcoming mediation based on a common motive: both held out the conviction they could *win* the war by escalating the level of violence. All sides lost interest in mutual negotiations precisely at those moments the balance of power seemed to shift in their favour. As a second step, Annan turned to regional powers (the Arab League, Turkey) but was unable to pressure them to stop financing and arming the opposition, after which, thirdly, he turned mediation efforts towards Russia and US-intricacies. Annan's six-point plan for ending the Syrian civil war ultimately failed and he resigned.[38] His plan expected the Syrian government to make major concessions, which was the reason why he openly blamed the Syrian government's refusal to implement his plan. He further explained his mediation failure for reasons of mounting oppositional military actions and a lack of consensus in the UN Security Council that was sharply divided on how to deal with the crisis:[39] the United States, Great Britain, and France calling for tougher international action against the Assad regime, and Russia and China repelling.

Lakhdar Brahimi, an Algerian diplomat and negotiator, served from 17 August 2012 until his resignation on 14 May 2014, succeeding Kofi Annan as Joint Special Envoy for mediation in Syria. He is widely known from the innovative Brahimi Report (2000) in his earlier capacity as Chair of the Panel on UN Peace Operations. Brahimi faced an even more terrifying and obstinate mediation environment after US intelligence in April 2013 brought out that president Assad had twice used chemical weapons.[40] The situation escalated into a full-scale civil war with a larger range of actors with incompatible interests, with the Syrian turmoil spilling over into sectarian clashes in Lebanon and Iraq—affecting, in turn, Syrian forces, and with an increasing Syrian presence at that time of Hezbollah but also IS. During Brahimi's time, the Assad regime, in August 2013, allegedly used Sarin gas in Ghouta (a Damascus suburb), killing 1,400 people.[41] In this climate of an internationalising war, where the conflict began to be exposed by militant Islamic groups as a zero-sum religious war, Brahimi swiftly decided to skip Annan's first step, spending little time mediating between the regime and opposition. He rather pursued a strategy that focused on the US and Russia but made little headway in the face of mutual distrust and competing interests. Brahimi's efforts climaxed in the Geneva II Conference, with the first face-to-face talks among Syrian parties, which

[37] Lundgren 2016; Hill 2015.

[38] Gowan 2013, p. 1.

[39] Crocker et al. 2015, p. 144.

[40] Crocker et al. 2015, p. 144.

[41] Crocker et al. 2015, p. 144.

failed. Brahimi explained this failure with the argument that the Syrian conflict was not '*ripe*' for resolution and that he had no leverage in this regard.[42]

Staffan de Mistura, an Italian-Swedish diplomat with a decades-long career with the UN, since 10 July 2014 to date is serving as third Joint Special Envoy to Syria. The mediation environment again changed considerably. In late 2014, the US-led coalition with proxy forces in Syria began military operations against IS, in response to IS expansion in Iraq and Northern Syria threatening Kurdish areas. Over the course of 2015, Assad's position became ever more perilous in a setting of increased local rebel victories and simultaneous offensives of militant Islamic groups, notably IS. Russia then decided to intervene in Syria with a two-pronged strategy of both military interventions in support of Assad and diplomatic initiatives without requiring Assad's departure.[43] In this new context of *overtly* military involvement of major powers, de Mistura has succeeded in an international mediation approach of temporary truces, small steps, with a series of short cease-fires brokered by Russia and/or the US across Syria, mainly intended to create access for humanitarian aid to besieged areas.[44] Though such truces are short-term and often violated, de Mistura appears to follow through this mediation approach.

In sum, the three successive UN-Arab League mediation missions harvested partial results and failed in producing a definitive agreement or outcome at the time of writing. So far, no international institution has been able to replace these international mediation efforts. Various separate mediation initiatives by the US and Western allies, as well as by Russia, Iran and Turkey, have not yet cast their role as trusted and sustained efforts in the Syrian war. UN mediation may have been well worth the attempts in view of harsher alternatives, such as fully unchecked violence. However, attempts to synchronise the leverages of multiple international mediators, including various efforts by Russia, failed thus far.

7.4 Key Issues and Challenges: Multilateral Mediation in Syria's War

An interpretation of the process of multilateral mediation in the strategic context of Syrian warfare may raise various key issues and challenges that tend to go beyond the common explanations from the mediation literature. Generally, the main factors used to explain mediation processes in this body of literature are either linked to the key conflicting actors, involving questions of representation, legitimacy, and lack of commitment to the mediation process, or linked to the mediators, in terms of

[42] Hinnebusch and Zartman 2016, p. 1.

[43] Lucas 2016, p. 24.

[44] Akpinar 2016.

insufficient coordination, third party interests, lack of neutrality and inclusiveness.[45] In the extant theory of mediation, where timing or ripeness of a conflict hold a central position as a prerequisite for mediation, such factors are often interpretively channelled in the perspective of a hurting stalemate.

7.4.1 Uncertainty and Unpredictability

The concept of non-linear strategy helps to interpret the difficulties encountered in the context of Syria. Kofi Annan could have used and exploited the extant uncertainties and unpredictability from the wider context (whether international military intervention by major powers would occur in Syria) as concrete elements of non-kinetic international mediation. It might be recommendable for future policies and strategies to explicitly acknowledge the degree of uncertainty and unpredictability about how the conflict might develop, make such contingencies integral part of the international mediation process, and link the dynamics involved to the leverage the mediators may offer. Richard Gowan similarly argues that "Annan could have opted for an alternative strategy that prioritised extending and using uncertainty to gain additional leverage".[46] However, when Annan resigned, he attributed the failure of the mediation process to factors such as the increasing militarization in Syria and the lack of unity in the Security Council,[47] interpreted as main obstacles, whereas he could have adapted his meditation strategy to changing dynamics and interactions in the context of proxy actors at the time.

7.4.2 Limitations of Ripeness and Hurting Stalemate

The concept of *ripeness*, in the sense that Brahimi used it to explain the failed mediation process under his mandate, is part of what widely has been called 'ripeness theory', which is mostly linked to the notion of a 'hurting stalemate' in the fields of mediation and negotiation.[48] With a strategy of repeatedly arguing that there is no military solution to the conflict, Brahimi intended to advance and ripen the moment and the perceptions by the actors of a hurting stalemate in Syria.[49] Some analysts use the same interpretation to argue that when Annan took up his mandate in the initial stages, the conflict "did not appear ripe for a negotiated

[45] Akpinar 2016; Hill 2015; Lundgren 2016. Also Bercovitch et al. 1991; Bercovitch and Gartner 2006.

[46] Gowan 2013, p. 5.

[47] Akpinar 2016, p. 2294.

[48] Zartman 2001, 2015; Pruitt 2015.

[49] Hinnebusch and Zartman 2016, p. 13.

settlement".[50] In these much-established views of mediation and negotiation, only a mutually hurting stalemate[51] would make a conflict ripe for settlement: this occurs when *both* parties become convinced that unilateral victory is unattainable in view of the increasing costs of persistent further violence—thus beginning to share a common interest in settlement.[52]

Ripeness theory, in fact, holds that as the costs of conflict rise towards a hurting stalemate, mediation becomes more feasible. This mediation approach, however, though it has its merits, is too limited to guide mediation in wars by proxy where regional and international geopolitical polarisation is involved. Ripeness theory seems to be designed for unipolar antagonisms, linked to an estimation of the antecedents and preconditions of mediation,[53] not for civil and proxy wars at the fault lines of great power interests that increasingly appear as a systematic threat to the international order.[54] In addition, ripeness theory does not explain how such a presumed 'ripe' moment is sustained after the start of a mediation process, or why parties would stay at the negotiation table until an agreement has been reached. If mediation should only take place under the most favourable circumstances of a hurting stalemate, the lead-up to the Geneva II process by Brahimi would never have occurred: he benefited at the time from US-Russian coordination that opened up a range of mutually acceptable solutions to address Syria's chemical weapons, resulting in a US-Russia tripartite brokered deal with the Syrian government.[55]

7.4.3 Coercion and Mediation: Military Intervention and the Use of Force

Another key issue regards the role of coercion, which links international mediation as a strategic means to the nexus between kinetic and non-kinetic effects-based policies. It is a widely held position that mediation and negotiation tend to be more effective when backed by the threat of military force or sanctions. Coercion, for example, can be involved as an offstage threat to pressure certain outcomes of conflict resolution efforts, as in the case of sanctions imposed on Iran during the multilateral negotiations rounds on Iran's nuclear programme.[56] In the Syrian

---

[50] Hinnebusch and Zartman 2016, p. 6.

[51] Touval and Zartman 1985 have claimed that a stalemate is a *sine qua non* for mediation to occur. See also Hellman 2012.

[52] Hellman 2012, p. 593.

[53] Pruitt 2015, p. 134.

[54] Hill 2015, p. 446.

[55] The deal was that the Syrian government would accede to the Chemical Weapons Convention and destroy its chemical weapons arsenal under the supervision of the Organization for the Prohibition of Chemical Weapons. See Crocker et al. 2015, p. 150.

[56] Simonen 2017; see also Orakhelashvili 2015.

context, thus far, UN-Arab League mediation has been accompanied by various means from the non-kinetic arsenal, including a wide array of sanctions (such as reducing diplomatic ties, travel bans, ceasing oil purchases) imposed on the Syrian government by the Arab League (suspending Syria's membership), the US, Western-European countries and Turkey. Coercion can also be used as a follow-up to failed mediation. By the time of writing, mediation in Syria has not been supported by traditional coercive military force from the Security Council under Chapter VII—even though the stakes became higher under Annan's mediation mandate in July 2012 when the Syrian regime first threatened with chemical and biological weapons if the country was invaded.[57] However, the Syrian war has rather reinvigorated the use of veto power by some of the permanent five members of the UN Security Council, which continues to trouble the mediation process. Recurrent vetoes of draft resolutions or the mere threat of a veto have persistently stalled negotiations within the Security Council and rendered it mostly passive in regard of Syria's mass atrocities.[58]

7.5 Conclusions and Implications for Non-Kinetic Strategy

This paper has studied mediation as a non-kinetic strategic means in Syria's evolving proxy war, by questioning what its effectiveness is, and what the implications would be for the development of future international policy and strategy

In view of understanding the effectiveness of international mediation as a non-kinetic strategic instrument, contemporary proxy warfare was taken as one of the most significant features of the present and future character of conflict—with the Syrian war in many ways coming to define such twenty-first-century war by proxy. Mediation as a non-kinetic strategic asset was thus framed as an integral part of profoundly changing realities of warfare at the fault lines of regional and global geopolitical rivalries, and the complexities and uncertainties involved. It is in this wider perspective that the context of proxy warfare was interpreted here as the main feature of current and future conflict. It can be concluded that multilateral mediation, at least in the Syrian case, has to face this context of reinvigorated proxy wars with manifold geopolitical polarisations involved. In response to the guiding question, to what extent mediation should be considered an effective asset in non-kinetic strategies to address proxy war, several other conclusions and implications emerge.

Specific insights may emerge from interpreting UN-Arab League mediation as a non-kinetic strategy in view of emergent trends of critical scholarship on war. Looking through this prism sheds light not only on the antagonistic and destructive side of war but also on the generative aspects of war's nature, the making and

[57] Crocker et al. 2015, p. 144.

[58] Webb 2014.

unmaking of the orders and structures of human life. The complex proxy war in Syria not only destroys political orders but contributes to shaping and producing them, encompassing the new political order and society that war generates—depending on the power balances of the manifold proxy actors involved. This generative aspect is of central importance for understanding mediation as a key pillar of multilateral stabilisation and reconstruction efforts. It links the non-kinetic power influences of mediation to a complex interplay between wars and state formation, as well as to how legitimate institutions are emerging from and are shaped by civil wars. Future policies and strategies that adopt non-kinetic elements have to move dynamically at this interface. In history, state formation has always been realised through armed conflict, civil war and interstate war. In effect, in Syria, civil war and proxy war are part of state-building and connect to what the post-war political order will look like. This stands in contrast to the most common and established perceptions of contemporary wars as the *outcome* of the disintegration, fragility, decay and collapse of political institutions. The generative features of the nature of war shift such a common reductionist understanding: the interpretation of the interactions and dynamics of proxies, as well as their positions in international mediation activities in Syria, will have deep impacts on the post-war political order.

If wars do not only destroy political orders but contribute to shaping and producing them, this will have serious implications for multilateral mediation. It shifts mediation's conventional focus (overcoming the antagonisms, working with the conflicting parties in order to reach agreements towards conflict resolution and bring the violence to an end) towards an explicit and principal emphasis on issues of legitimacy and institutionalisation of extant power relations in war. Statehood and legitimate institutions, when seen as part of historical and constant changes, cannot only be negotiated *after* but must be an integral part of mediation as a non-kinetic strategic asset *during* civil war. This is a relevant aspect of the effectiveness of international mediation in the present Syrian war and has implications. It first requires a shift in the interpretation of complex proxy battles over contending legitimacy claims: in future international policies and strategies, they have to be perceived as an integral and inherent part of mediation dynamics—not as an obstacle to mediation or a reason for resigning, as occurred in the UN-Arab League mediation process in Syria. Obviously, this regards the ways and means by which Syrian opposition forces, the Syrian Assad regime, regional and international actors negotiate their claims at political power, manage to institutionalise their military forces, and transform them towards the political domain of new state formation.[59]

The implications of these findings are different from what has been done in mediation studies before and need to be advanced beyond previous achievements in order to break some new grounds. Some forms of *critical scholarship* in the fields of mediation and negotiation would much be needed in order to ground the emergence of fresh theoretical frameworks that can critically interpret the complexities of the nature of war and strategy, as well as the benefits and burdens, and

[59] This point is also highlighted by Crocker et al. 2015, p. 151.

the gains and limitations of international mediation in the proxy war in Syria. First, critical scholarship can support international mediation in engaging with the complex and predominantly paradoxical aspects of war so as to avoid that the contingencies of a very fluid war, with manifold proxy actors, are eliminated from the mediation process. Uncertainty and unpredictability have to be built into international mediation instead of perceiving them as an obstacle, or removing and keeping them at a distance from the dynamics of the process. Second, critical scholarship also has the potential to overcome the limitations of established mediation theory that invariably tends to explain such complexities in reductionist terms of ripeness theory and hurting stalemates. This paper intended to set a first step in elaborating this perspective.

Third, military strategists and operational commanders, at all levels, have substantial and longstanding experience with mediation and negotiation. This is rarely systematically studied,[60] the less so in terms of non-linear strategies and critical studies of war. The military conduct mediation and negotiation on an everyday basis as part of stabilisation and peacekeeping, transition and reconstruction operations. Mediation and negotiation are among the most common instruments in achieving strategic mission objectives and pursuing operational and tactical goals. Such non-kinetic assets may be seen as an alternative or as a complement to military solutions that involve the use of force. Yet meditation and negotiation as part of international policies and military strategies are largely understudied, the more so when employed as non-kinetic strategic means. Positioning mediation as a non-kinetic strategic asset of stabilisation and conflict resolution, therefore, contributes unexplored viewpoints.

By foregrounding mediation as a constructive means, non-kinetic strategies were conceptualised here away from the more common non-kinetic instruments linked to disruptive, undermining, and deceptive stratagems. By engaging with this debate, an initial response is given to calls for more systematic critical scholarship to underpin the shift towards greater reliance on non-kinetic powers. In such a way, this chapter has contributed some new avenues. All in all, it has been revealed that a focus on non-kinetic strategic assets in contemporary war may trigger various pioneering research perspectives in need to be further elaborated towards a new comprehensive and critical research agenda on international mediation and negotiation.

References

Akpinar P (2016) The Limits of Mediation in the Arab Spring: The Case of Syria. Third World Quarterly 37(12):2288–2303

Balibar E (2008) What's in a War? (Politics as War, War as Politics). Ratio Juris 21(3):365–386

Barkawi T, Brighton S (2011) Powers of War: Fighting, Knowledge, and Critique. International Political Sociology 5(2):126–143

[60] Among the few exceptions are Goodwin 2005 and Wheeler 2013.

Bercovitch J, Anagnoson JT, Wille DL (1991) Some Conceptual Issues and Empirical Trends in the Study of Successful Mediation in International Relations. Journal of Peace Research 28 (1):7–17

Bercovitch J, Gartner SS (2006) Is there Method in the Madness of Mediation? Some Lessons for Mediators from Quantitative Studies of Mediation. International Interactions 32(4):329–354

Beyerchen AD (1992–1993) Clausewitz, Nonlinearity and the Unpredictability of War. International Security 17(3):59–90

Brown S (2016) Purposes and Pitfalls of War by Proxy: A Systemic Analysis. Small Wars & Insurgencies 27(2):243–257

Crocker CA, Hampson FO, Aall P, Palamar S (2015) Why Is Mediation So Hard? The Case of Syria. In: Galluccio M (ed) Handbook of International Negotiation: Interpersonal, Intercultural, and Diplomatic Perspectives. Springer International Publishing, Cham, Switzerland, pp 139–155

Ducheine PAL (2016) Non-kinetic Capabilities: Complementing the Kinetic Prevalence to Targeting. In: Ducheine PAL, Schmitt MN, Osinga FPB (eds) Targeting: The Challenges of Modern Warfare. TMC Asser Press, The Hague, pp 201–230

Galeotti M (2016) Hybrid, Ambiguous, and Non-Linear? How New is Russia's "New Way of War"? Small Wars & Insurgencies 27(2):282–301

Goodwin D (2005) The Military and Negotiation: The Role of the Soldier-Diplomat. Frank Cass, Abingdon

Gowan R (2013) Kofi Annan, Syria and the Uses of Uncertainty in Mediation. Stability 2(1):1–6. DOI:http://dx.doi.org/10.5334/sta.ax

Hellman J (2012) The Occurrence of Mediation: A Critical Evaluation of the Current Debate. International Studies Review 14(4):591–603

Hill THJ (2015) Kofi Annan's Multilateral Strategy of Mediation and the Syrian Crisis: The Future of Peacemaking in a Multipolar World? International Negotiation 20(3):444–478

Hinnebusch R, Zartman IW (2016) UN Mediation in the Syrian Crisis: From Kofi Annan to Lakhdar Brahimi. International Peace Institute (IPI), New York

Hughes GA (2014) Syria and the Perils of Proxy Warfare. Small Wars & Insurgencies 25(3):522–538

Jones DJ (2000) Mediation, Conflict Resolution and Critical Theory. Rev Int Stud 26(4):647–662

Lanoszka A (2016) Russian Hybrid Warfare and Extended Deterrence in Eastern Europe. International Affairs 92(1):175–195

Lucas S, Yakinthou C, Wolff S (2016) Syria: Laying the Foundations for a Credible and Sustainable Transition. RUSI Journal 161(3):22–32

Lundgren M (2016) Mediation in Syria: Initiatives, Strategies, and Obstacles, 2011–2016. Contemporary Security Policy 37(2):273–288

Mearsheimer JJ, Walt SM (2013) Leaving Theory Behind: Why Simplistic Hypothesis Testing is Bad for International Relations. European Journal of International Relations 19(3):427–457

Mumford A (2013a) Proxy Warfare and the Future of Conflict. RUSI Journal 158(2):40–46

Mumford A (2013b) Proxy Warfare: War and Conflict in the Modern World. Polity Press, Cambridge/Malden

Orakhelashvili A (2015) The Impact of Unilateral EU Economic Sanctions on the UN Collective Security Framework: The Cases of Iran and Syria. In: Marossi AZ, Bassett MR (eds) Economic Sanctions under International Law: Unilateralism, Multilateralism, Legitimacy, and Consequences. TMC Asser Press, The Hague, pp 3–21

Osinga FPB (2007) Science, Strategy and War: The Strategic Theory of John Boyd. Routledge, Abingdon

Phillips C (2015) Sectarianism and Conflict in Syria. Third World Quarterly 36(2):357–376

Pruitt DG (2015) The Evolution of Readiness Theory. In: Galluccio M (ed) Handbook of International Negotiation: Interpersonal, Intercultural, and Diplomatic Perspectives. Springer International Publishing, Cham, Switzerland, pp 123–138

Roberts N, Everton S (2016) Monitoring and Disrupting Dark Networks: A Bias toward the Center and What It Costs Us. In: Dawoody AR (ed) Eradicating Terrorism from the Middle East:

Policy and Administrative Approaches. Springer International Publishing, Cham, Switzerland, pp 29–42

Sauer T (2017) The Origins of the Ukraine Crisis and the Need for Collective Security between Russia and the West. Global Policy 8(1):82–91

Schwartz-Shea P, Yanow D (2011) Interpretive Research Design: Concepts and Processes. Routledge, New York

Simonen K (2017) The Strong Do What They Can and the Weak Suffer What They Must - But Must They? Fairness as a Prerequisite for Successful Negotiation (Benchmarking the Iran Nuclear Negotiations). Journal of Conflict and Security Law

Strachan H (2007) The Changing Character of War. Europaeum Lecture, delivered at the Graduate Institute of International Relations, Geneva (9 Nov 2006) Europaeum Oxford-Geneva Link Programme, Oxford

Touval S, Zartman IW (1985) Introduction: Mediation in Theory In: Touval S, Zartman IW (eds) International Mediation in Theory and Practice. Westview Press, Boulder CO, pp 7–17

Towle P (1981) The Strategy of War by Proxy. RUSI Journal 126(1):21–26

Von Clausewitz C [1832] (1976) On War (ed. and transl. Howard M, Paret P). Princeton University Press, Princeton

Von Clausewitz C [1832] (1991) Vom Kriege. Hinterlassenes Werk des Generals Carl von Clausewitz (Hahlweg W (ed) Vollständige Ausgabe im Urtext, 19th edn). Fred. Dümmlers Verlag, Bonn

Webb P (2014) Deadlock or Restraint? The Security Council Veto and the Use of Force in Syria. Journal of Conflict and Security Law 19(3):471–488

Wheeler M (2013) The Fog of Negotiation: What Negotiators Can Learn from Military Doctrine. Negotiation Journal 29(1):23–38

Zartman IW (2001) The Timing of Peace Initiatives: Hurting Stalemates and Ripe Moments. Global Review of Ethnopolitics 1(1):8–18

Zartman IW (2015) Mediation: Ripeness and its Challenges in the Middle East. International Negotiation 20(3):479–493

Zweibelson B (2015) One Piece at a Time: Why Linear Planning and Institutionalisms Promote Military Campaign Failures. Defence Studies 15(4):360–374

Zweibelson B (2016) Rose-Tinted Lenses: How American Functionalist Strategy Inhibits our Appreciation of Complex Conflicts. Defence Studies 16(1):68–88

Author Biography

Dr. Berma Klein Goldewijk is Assistant Professor of International Security Studies at the Department of War Studies, Netherlands Defence Academy. She obtained a master degree (1984) and a Ph.D. degree (1991) on contextual methodology in international religious studies from Radboud University Nijmegen, both cum laude.

Chapter 8
Lo and Behold: Let the Truth Be Told—Russian Deception Warfare in Crimea and Ukraine and the Return of 'Maskirovka' and 'Reflexive Control Theory'

Han Bouwmeester

All the truth in the world adds up to one big lie!
Bob Dylan—Things Have Changed (1999).

Abstract Since Russia occupied Crimea in March 2014, the Western world has been searching for explanations how this operation was conducted and why it came as a surprise for many military and international relations subject matter experts. The Western media followed by NATO called the Russian activities a form of hybrid warfare, while the Russians spoke about non-linear warfare. One the most important elements in this way of warfare is 'informationally', meaning the content as well as the spreading of information. The majority of the Russian actions took place in the information domain. Deception warfare, being a substantial element of information warfare, generates surprise, one of the principles of war. Deception is most of the time a combination of hiding the truth (dissimulation) and showing falsehood (simulation). During deception warfare two variants are used: (1) ambiguity-increased or A-type deception, creating confusion so that the opponent is unsure what to believe, and (2) misleading or M-type deception, reducing ambiguity by offering only one attractive but wrong alternative. Russia developed its own methods of deception over time. Most Russian deception warfare is based on 'maskirovka', the reflexive control theory, agitprop and disinformation. These four elements were constantly used by the Russian government, secret services and armed forces during the conflicts in Crimea and Ukraine's Donbass area in 2014, although Russia still denies every involvement in the latter operation.

Keywords Hybrid warfare · Non-linear warfare · Gerasimov doctrine · Information warfare · Deception warfare · Surprise · Maskirovka · Reflexive control · Disinformation · Spetspropaganda · Agitprop · Kompromat · Infotainment · Russia · Crimea · Ukraine · NATO

H. Bouwmeester (✉)
Faculty of Military Sciences, Netherlands Defence Academy, Breda, The Netherlands
e-mail: AJH.Bouwmeester@mindef.nl

P.A.L. Ducheine and F.P.B. Osinga (eds.), *Netherlands Annual Review of Military Studies 2017*, NL ARMS, DOI 10.1007/978-94-6265-189-0_8

Contents

8.1 Introduction

> Then, a second later, there was a haphazard burst of half a magazine on automatic. 'That's a bog standard AK 47M', Morland muttered to himself. Now the shooting sounded as if it was the work of a lone lunatic and no longer a trained professional. Morland saw the crowd part [in the streets of Riga] and start to run and, unable to see exactly where the shots had come from, he switched his binoculars there [...]. A TV crew from Russia Today conveniently close by, rushed to get their close-up shots: three young, ethnic Russian girls murdered by a crazed, doubtless [Latvian] 'nationalist', gunman. But shot, Morland would swear, by one or more Spetsnaz snipers from the top of the university building in Riga. Those terrible images would be circulating the globe in moments. Any more victims? No. That extra burst had definitely been for deception and had been aimed to miss. But Moreland knew that the black-fleeced Russian with the mobile phone, the man in his telescope, was the man who had planned this atrocity [...]. 'One day I'll get you, you bastard', he vowed.[1]

Captain Tom Morland, a fictive leader of a five-man training squad from the reconnaissance platoon of 1st Battalion, the Mercian Regiment, is one of the main characters in Shirreff's book *2017 War with Russia*. Another protagonist is the man in black fleece, Major Anatoly Nikolayevich Vronsky of the Russian 45th Guards Spetsnaz Regiment.[2] Retired General Sir Richard Shirreff, a former NATO's Deputy Supreme Allied Commander in Europe, wrote the book with a reason. Shirreff is very concerned about Russian presence in Crimea and the Donbass region, although Putin has always denied any official Russian involvement in the latter operation. The book, which reads as a thrilling novel, dramatizes a potential Russian hybrid scenario for the Baltics in the spring of 2017. This chapter is not

---

[1] Shirreff 2016, pp. 102–103.

[2] Major Anatoly Nikolayevich Vronsky is probably a fictive descendant of Count Alexei Kirilovich Vronsky, a cavalry officer, just like General Sir Richard Shirreff, and a lover of Princess Anna Arkadyevna Karenina in Leo Tolstoy's novel 'Anna Karenina'. See Tolstoy 2016.

going to give away the plot of Shirreff's book, but in many ways the book does paint a realistic picture of Russia's renewed notion of modern warfare.

Many politicians and high-ranking officers in NATO, both serving and retired, currently consider Russia as the number one threat for NATO.[3] The North Atlantic Council (NAC) of NATO spoke initially about 'the challenges posed by Russia and their strategic complications'.[4] In their Warsaw Summit Communiqué, issued in July 2016, the NAC made it clear that the Russian threat had become a top priority: "Russia's aggressive actions, including provocative military activities in the periphery of NATO territory and its demonstrated willingness to attain political goals by the threat and use of force, are a source of regional instability, fundamentally challenge to the Alliance, have damaged Euro-Atlantic security, and threaten our long-standing goal of a Europe whole, free, and at peace".[5]

The main purpose of this NLARMS is to explore to what extend the use of non-kinetic capabilities at strategic and operational level might contribute to obtain political objectives. The aim of this chapter is to assess the seriousness of the current non-kinetic Russian threats. Especially the way Russia is operating in the information domain, is considered a real danger. The background of this concern is perhaps best explained by NATO's former Supreme Allied Commander, General Phillip Breedlove, when he called 'informationally'[6] as the most impressive part of Russia's hybrid approach.[7] Breedlove emphasized that the Russians are able to exploit a conflict situation and turn it into a false perception. "All they do," he said in an interview, "is to frame a false narrative, get it out rapidly, and then sustain it through all of the tools that are out there. And they are very good at it!"[8] Breedlove points out that the Russians not only prevail in the information domain, they are also masters in creating new perceptions to deceive their opponents as well as public opinion.

This chapter starts with highlighting some relevant developments in military thinking, especially those in the information domain. It resumes with Western experience in deception and the Russian view on information warfare, followed by the essence of deception warfare. Later on, the nexus between deception warfare and the Russian tenets '*maskirovka*' and the 'Reflexive Control Theory' shall be subsequently analysed, interlaced with examples of the Russian takeover of Crimea in 2014.

---

[3] Among these politicians and top officers are former British Foreign Secretary, Mr. Philip Hammond (Winning 2015), current U.S. Chairman Joint Chiefs of Staff, General Joe Dunford (McClearly 2015), current U.S. Chief of the Army Staff, General Mark Milley (Leipold 2015), and former NATO's Supreme Allied Commander in Europe, Retired Admiral Jim Stavridis (Stavridis 2016, p. ix).

[4] North Atlantic Council 2014.

[5] North Atlantic Council 2016.

[6] 'Informationally' means the content as well as dissemination of information.

[7] U.S. Department of Defense 2015.

[8] U.S. Department of Defense 2015.

8.2 Neglecting and Rediscovering the Information Domain

Let us first go back to Captain Moreland, who was impressed by what he saw in the streets of Riga as described at the beginning of this chapter. Moreland, a young captain, did not grow up with the notion of deception warfare.[9] Most officers of his generation do not have a profound understanding of what deception really means and how and why it should be an integral part of a military plan. Deception has not been taught or encouraged anymore at the Western military academies for decades. During World War II the Allies were known for their resourcefulness. Operations like *Operation Bodyguard* intended to trick the Germans during the invasion of June 1944.[10] Unfortunately, the Western world has lost interest in deception. One of the main reasons for this lack of interest had started already during the Cold War. Most NATO countries were focused on physical manoeuver operations, stopping the armoured columns of the Warsaw Pact, their supporting fleet of fighters and bombers, and on ensuring the unrestricted use of transatlantic sea lines of communication. By and large, the consensus opinion was that deception plans were not serious business nor an important part of military strategy.[11]

After the collapse of the Warsaw Pact, the armed forces of most NATO countries became involved in peace and stability operations, e.g. in the Balkans, Iraq and Afghanistan. A main idiosyncrasy of such operations, especially for those in a force contributing role like most NATO countries did, is to be transparent and not to betray the local population and public opinion.[12] With the proliferation of social media, and in reaction of Taliban's sophisticated information operations, Western militaries became increasingly aware that acting in the information domain is of vital importance. The Taliban, using so-called night letters, audio cassettes and later social media, constantly disseminated a consistent narrative grounded in the Islamic faith, and intimidated Afghan people not to support ISAF. In response NATO attempted to develop a coherent strategic communications strategy.[13]

These are not the only reasons why deception warfare is still not very *en vogue* in the Western world. Most Western democracies and their armed forces still embrace a chivalrous code when conducting wars. They represent state systems in which freedom, human rights and superior ethical standards are paramount, and they do not want to lose this moral high ground. Deception warfare as a form of

---

[9] Deception never became a principle of war. The collective opinion among most war studies researchers and authorities was that deception contributes to and strengthens other principles of war, being 'surprise' the most obvious one.

[10] Operation Bodyguard was the code name for an overall deception plan employed by Western Allied states prior to the invasion of north-west Europe in June 1944. The plan was intended to mislead the German High Command as to what time and where the invasion would take place. See Brown 2007.

[11] Kitzen 2012, pp. 1–8.

[12] DP 3.2 2014, pp. 7-3 and 7-4.

[13] Boudreau 2016, pp. 298–381; De Graaf and Dimitriu 2015, pp. 3–13; Betz 2015b, pp. 37–56; Briant 2015, pp. 206–211.

'trickery' has acquired a derogatory undertone, "just as 'Gentlemen' do not open each other's mail', so decent people should not engage in what is sometimes seen as an indecent activity".[14] In addition, as Betz and Rid and more recently Giles have observed, acting in the information domain is for Western governmental institutions an asymmetric struggle as they cannot be caught spinning and spreading false information. Western media moreover, and their governments, harp on the need for balanced reporting of events and avoid accusations of spinning and deliberate framing at all costs.[15] Russia and other authoritarian regimes have no such compunction, indeed, the Russian annexation of Crimea has served as a rude awakening that non-Western regimes exploit the information domain as an integral instrument in foreign policy and their military doctrine.[16]

Western analysts use the term 'hybrid warfare' to explain Russia's recent warfare. The Russians, however, prefer to use the phrase 'non-linear warfare',[17] first used by Vladislav Surkov. He was one of Putin's key political advisors, and he published a science fiction short story *Without Sky* under his penname Natan Dubovitsky. The story, told by a child, discusses a futuristic way of warfare in which not only the military but also other power elements are involved.[18] Sooner rather than later, the term 'non-linear warfare' was taken over by the Russian military top brass.

The Chief of the General Staff of the Armed Forces of the Russian Federation, General Valery Gerasimov explained the essence of non-linear warfare as 'a tendency toward blurring lines between the status of war and peace', and he accentuated the important role of information warfare.[19] Gerasimov's article became well-known in the West; it was seen as the source document for the famous Russian Gerasimov-doctrine.[20] Later, General Gerasimov denied that his article represented a new Russian way of warfare. He only described how the West, particularly the United States, had conducted operations for the last 25 years.[21] Nevertheless, Gerasimov's article is a clear indication that Russia not only wants to dominate the information domain by looking for opportunities to exploit this man-made domain, but also to deliberately create situations to deceive their opponents and public opinion.

---

[14] Gooch and Perlmutter 1982, p. 1.

[15] Giles 2016a, pp. 37–40; Giles 2016b, p. 6; Betz 2015a, pp. 117–130; Rid 2007, pp. 7–8 and 118–119; and Rid and Hecker 2009, pp. 207–217.

[16] Ven Bruusgaard 2016, pp. 7–26.

[17] Perry 2014.

[18] Dubovitsky 2013.

[19] Gerasimov 2013.

[20] Western analysts used the term Gerasimov-doctrine to explain what had happened in The Crimea and Ukraine in 2014. A few of the main characteristics of the Gerasimov-doctrine are: (1) a vague line between war and peace, (2) a blurred line between uniformed and non-uniformed persons involved in the conflict, (3) use of information warfare and propaganda, (4) The ratio between civilians and military service men is 4:1.

[21] Bartles 2016, pp. 30–38; McDermott 2016, pp. 97–105.

8.3 Deception Warfare

Deception plays a key role in information warfare. It should be seen as a survival mechanism in order to gain an advantage in an exhaustion game.[22] Deception can be a smart and cheap solution. The crux is to deploy existing resources in a more unpredictable manner. In many cases, deception can be seen as the exertion of trickery or astuteness. It aims to deliberately induce misperception in the human mind.[23] Deception is often neither a result of chance nor the spin-off of another undertaking.[24] Michael Handel defines deception as "the process of influencing the enemy to make decisions disadvantageous to himself by supplying or denying information".[25] Others call deception the 'art of the bluff', which is already as old as warfare itself.[26]

Perhaps the best definition of deception is given by Barton Whaley, an expert on stratagem. In 1982 he described deception as "information designed to manipulate the behaviour of others by inducing them to accept a false or distorted presentation of their environment—physical, social, or political".[27] Even today, deception is considered a form of forgery in which false elements are shown and people tend to believe it. They accept it is as reality, and consequently it becomes their perception of reality. In sociology this acceptance is called the Thomas theorem, which was already formulated in 1928 by Dorothy and William Thomas: "If men define situations as real, they are in their consequences".[28] It means that no human interpretation is objective. Human actions are always affected by subjective perceptions of situations. Whether there even is an objectively correct interpretation is not important for the purpose of helping guide individuals' behaviour.[29] A social construct always exists, in contrast to deception. Deception is a deliberate effort of an initiator to create a perception of reality that is favourable to him for a certain period of time. Later Peter Berger and Thomas Luckmann researched the way reality is perceived. The central concept in their 'Social Construct of Reality' is that people and groups interact in a social system and create, over time, a mental representation of each other's actions. Both knowledge and people's belief of what reality is become part of their institutional fabric, and also of their perception of society.[30]

Three generic goals can be discovered in every form of deception. The prompt purpose is to condition a target's beliefs; the intermediate purpose is to affect the

[22] Kopp 2005, p. 81.

[23] Gerwehr and Glenn 2000, p. 15.

[24] McClesky 1991.

[25] Handel et al. 1987, p. 1.

[26] Latimer 2016, pp. 1–43; Keneally 2016.

[27] Whaley 1982, pp. 188–190.

[28] Thomas and Thomas 1928, pp. 571–572.

[29] Merton 1995, pp. 379–424.

[30] Berger and Luckmann 1967, pp. 47–91 and 129–172.

target's behaviour and actions, and the overall purpose is that the deceiver profits from these actions. The comprehensive goal of deception is frequently not measured very well. In those cases deception is praised with success when only the first goal is achieved, while the overall goal is never realized.[31] Deception needs to have a deeper impact than only superficially touching a human belief system. Deception per se has no impact; it only becomes significant when used as a means of achieving surprise. So surprise will meet all three mentioned objectives. Surprise is something that moves the human mind deeply. It can cause subjectivity in judgment, psychological perceptions and misperceptions, wishful thinking or opposing thoughts. These characteristics can abuse the most intelligent actor who innocently, inadvertently and reluctantly plays into his opponent's hands and helps to set his own ruse.[32]

The surprise element, however, is sometimes difficult to identify. Roberta Wohlstetter notices that it is rarely obvious, even the unintentional version. There is a strong human tendency, she writes, to pay primary attention to signals that support current existing expectations about the behaviour of an opponent. Most of the time, other contradictory signals are ignored or not well understood.[33] This misunderstanding is also known as 'Cry Wolf' syndrome: If one cries wolf too often, people will stop believing him.[34] Kass and London make a distinction between the initiator and target of surprise. For the initiator, surprise is the outcome of a deliberate effort. It is a plan coming together in a concentrated burst of activity, and everything works just right to produce the expected result. For the target, surprise is an event: sudden, stunning, sometimes traumatic, and humiliating. Surprise catches the victim at his weakest, exposing and exploiting his failings. It is only in hindsight that the victim becomes aware of what caused the event to happen.[35]

On the other hand, surprise occurs to the degree that the opponent is uncertain whether, 'when', 'where' and 'how' an actor will strike. Surprise is then considered as a force multiplier. Surprise can keep the opponent stumbling when his plans dictate that he should be reacting, preventing him from using his capabilities efficiently and effectively while the deceiver is exploiting his own to the maximum.[36] Astonishment caused by surprise is thus a main outcome of deception, and the deceiver is hoping that it will lead to a state of shock, leading to paralysis and indecisiveness among his opponent. Therefore, surprise should be a main element in every military planning process. And in order to catch Major Vronsky, our man Captain Morland, should make an unpredictable plan with surprise elements.

---

[31] Daniel and Herbig 1982a, p. 5; Daniel and Herbig 1982b, p. 157.

[32] Handel 1976, pp. 7–8.

[33] Wohlstetter 1962, pp. 392–393.

[34] Cambridge Dictionary 2017.

[35] Kass and London 2013, p. 68.

[36] Betts 1982, pp. 4–5.

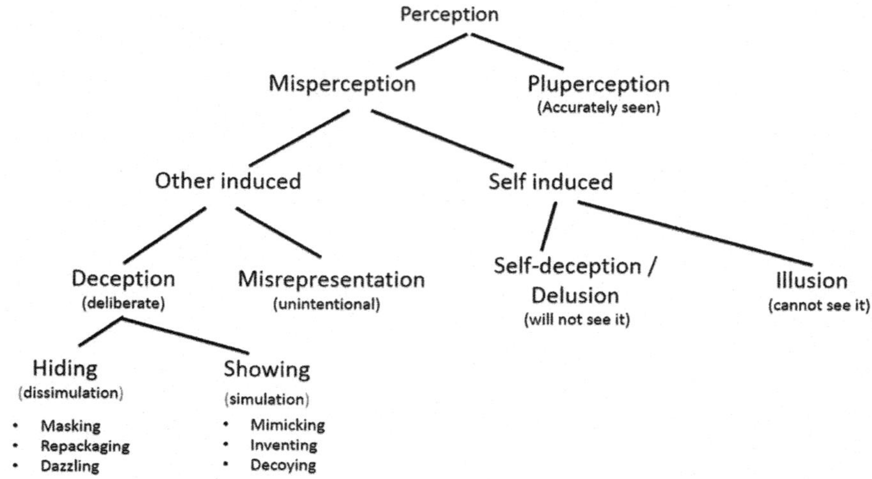

Figure. 8.1 A typology of perception (Whaley 1982, p. 180.)

8.4 Structure of Deception Warfare

Deception operations are structured along two ways: dissimulation and simulation. Both elements are quite simple. Dissimulation is about 'hiding the real'; it is covert. Half of the deception is concealed from the target. The foremost purpose is to hide or at least obscure the truth. Simulation, contrastingly, is 'showing the false'. It is overt. Half of the deception is presented to the target. The drive is to pretend, portray, and profess an intended lie. Today it can be said that both simulation and dissimulation are always present together in a single deception activity. "Nothing is ever 'just' hidden, something is always shown instead, even if only implicitly".[37] Barton Whaley went more into detail, when he designed a typology for deception.[38] The typology starts with perception, since deception is the alteration of perceived reality. More specifically, deception falls in the category of perception that is called misperception. It is a psychological phenomenon, which takes place in the 'eye of the beholder'. Whaley is convinced that men are not deceived by others, but merely by themselves. The deceiver is only attempting to provoke deception, by revealing a false picture of reality. In order to be deceived, one must both perceive this attempted portrayal and accept it in the terms intended and projected. Misperception can be self-induced in two ways. First, there is a form of self-deception in cases in

[37] Rothstein and Whaley 2013, pp. 19–20.
[38] Whaley 1982, p. 179.

which one can see the deception, but refuses to see it. This is also called a delusion. Second, an illusion occurs when one cannot see, nor discover the deception due to his own shortfalls[39] (Fig. 8.1).

Situations in which others cause the misperception are of much greater importance for this chapter. Misperception, induced by others, can be divided into deception, when it is intended, and misrepresentation, when it is unintentional. Whaley uses the earlier discussed two ways of deception to make a further distinction: (1) dissimulation or 'hiding the real', and (2) simulation or 'showing the false'. There are three ways of 'hiding'. The first one is by 'masking', making reality invisible. It either interposes a screen to cover it, or integrates it with its environment so it is overlooked. The second is by 'repackaging'; i.e. hiding reality by disguising it and changing its appearance. The third way is causing confusion through 'dazzling', which bewilders, confounds, baffles, and perplexes people, reducing certainty about the real nature of anything.[40]

There are also three ways of 'showing'. The first is 'mimicking', which shows falseness by having one thing imitate another. It copies an adequate number of aspects of the other to represent an acceptable imitation. Perfect examples are doppelgangers or look-alikes. The second one is 'inventing', which aims at showing another reality. Unlike mimicking, which duplicates an existing thing, inventing creates something completely new, nevertheless false. The third way is 'decoying', which distracts attention. It offers a misleading option, and is therefore a matter of feints and misdirection. Deceivers using decoy create alternative characteristics that give an extra pattern. Ideally, this alternative pattern bares more confidence and carries more faith than the real one. Just as simulation and dissimulation are opposites, their separate sub-categories also stand opposing one another. Thus, masking has its counterpart in mimicking, repackaging in inventing and dazzling in decoying. Despite the contradictions, both simulation and dissimulation are always present together in any single deception operation.[41]

8.5 How Deception Works

The next step in understanding deception is to determine how it works. What are the mechanisms behind these two ways of deception? Many researchers have come to the conclusion that the aforementioned ways of 'hiding' and 'showing' are based on cognitive and psychological principals. Humans are constantly deceived, although they may not realise it. Psychologists have identified several biases, which make the human mind peculiarly receptive to being betrayed. The brain is an associative network that develops over time during a process known as 'learning'. Thinking and trying to be rational is no more than testing incoming information against an

[39] Whaley 1982, p. 180.
[40] Whaley 1982, pp. 183–184.
[41] Whaley 1982, pp. 184–186.

existing network of associations. If the perceived information conforms to the associative network, humans usually accept it. Therefore, the mind relies to a great extent upon predetermined ideas.[42]

Most people find it difficult to cope with ambiguous circumstances. Humans dislike uncertainty, and they make every effort to reduce and stop uncertainty. Individuals can only process bits of information one at a time. They are forced to select information and to allocate its priorities. The temptation is to jump to conclusions, and the tendency is to focus on new and interesting information while routine receives a much lower indication. Another inclination for the human mind is to be lulled by monotony and predictability. It tends to pay less attention to occasions that occur again and is not good at recognizing minimal or gradual changes. The mind can thus be lured into a desired direction.[43]

In contrast to computers, the human brain is poorly equipped to perform logical or analytical tasks. Human rational, or 'higher-order' cognitive abilities are from an evolutionary viewpoint recently acquired abilities. Therefore humans typically resort to simple heuristic rules, the so-called 'mental shortcuts', during rational decision-making situations. Although heuristics can result in quite acceptable outcomes, they may yield suboptimal decisions, also known as 'cognitive biases'. [44] Biases can be understood by Kahneman's dual thinking mode, representing two different modes of information processing in decision making; the first operates automatically and fast, based on intuition, while the second is more rational, it is slow and deliberate and allocates attention to effortful mental activities.[45] Biases mostly occur when people only use system 1 without utilising system 2 for making judgments and decisions.[46] Although the enumeration is not complete, prominent biases and heuristics are: (1) anchoring bias: a tendency to make judgments or decisions coloured by previously and often first presented information (the 'anchor'), (2) representativeness bias: the probability that an entity belongs to a certain category is assessed by its similarity to the typical features of that category instead of its simple base rates, (3) blind spot bias: a tendency to recognize biases in others but not in oneself, (4) hindsight bias: the tendency to erroneously perceive events as inevitable or more likely once they have occurred, (5) confirmation bias: a tendency to search for, interpret, focus on and remember information in a way that confirms one's preconceptions, (6) framing bias: the way information is presented gives direction to human judgement, (7) negativity bias or 'loss aversion': a tendency to consider negative information more important than positive information, and (8) availability heuristic: a tendency to judge the frequency or likelihood of an event by the ease with which relevant information comes to the mind.[47]

[42] Dewar 1989, p. 9.

[43] Dewar 1989, pp. 9–10.

[44] Korteling et al 2016.

[45] Kahneman 2011, pp. 20–24.

[46] Toet et al. 2016, Appendix A/p. 1.

[47] Toet et al. 2016, Appendix A/pp. 1–3.

Daniel and Herbig distinguish two variants of deception, producing different effects and operating in different directions. The first is 'ambiguity-increasing' or A-type deception, also known as the less elegant variety, which confuses an opponent so that he is unsure as to what to believe. By ensuring an impact, A-type deception requires that the deceiver's lies are plausible and consistent enough to the target's comfort that he cannot ignore them. Enhancing uncertainty by pending extra information, a target may delay decision-making, thereby giving the deceiver wider freedom to arrange resources and take or remain the initiative. By guaranteeing a high level of ambiguity concerning the deceiver's intentions, the target is forced to spread his resources 'to cover all important contingencies' and thereby reducing the resistance the deceiver can expect at any time. The other variant, labelled 'misleading' or M-type deception, is more complicated. M-type deception is just designed to reduce ambiguity by offering attractiveness of one wrong alternative. It causes a target to concentrate its resources on a single eventuality, maximizing the deceiver's chances for prevailing all others.[48] The essence of M-type deception is captured by the phrase that "the ultimate goal of stratagem is to make the enemy quite certain, very decisive and wrong".[49]

More in practice, a deceiver must first of all scrutinise what his opponent will understand and accept. As soon as one has received an appreciation, one needs to condition the opponent's behaviour by reinforcing his expectations. These expectations are related to the opponent's values and norms, and will affect his judgment. Again a deceiver must create dissimulation (hiding reality) and present simulations (showing falsehood) by using a mix of physical and information techniques, which can reinforce one another. If there is a deception narrative, make sure that the story is credible, consistent, verifiable and executable. The ultimate aim is to influence the opponent's behaviour and not only his thinking.[50] Explaining opportunities to deceive an opponent, the 'Jones Dilemma'[51] indicates that deception becomes more difficult as the number of sources, which are available to the deception target to confirm the 'real situation', increases.[52] Instead, the greater the number of sources that are deceptively manipulated, the greater the chance the deception will be believed.[53]

Whaley describes different types of deception and ruse. He starts with 'diversion', which he defines as any movement of military units intended to imply a main attack. Its purpose is to distract the opponent's attention and strength away from the

[48] Daniel and Herbig 1982a, pp. 5–7; Daniel and Herbig 1982b, pp. 157–159.

[49] Whaley B, op. cit. in: Handel 1982, p. 131.

[50] Verrall 2016.

[51] The Jones Dilemma maxim is named after Professor R.V. Jones, one of the main researchers in British scientific intelligence during World War II. Jones' idea became well-known and was further developed when he discussed the detection of targets during a defence and attack: "that as many different physical means of detection as possible should be used in parallel. It may, therefore, be better in some circumstances to develop two or three independent means of detection, instead of putting the same total effort into the development of one alone." In: Jones 1957, p. 7.

[52] CIA 1980, pp. 5–41.

[53] Verrall 2016

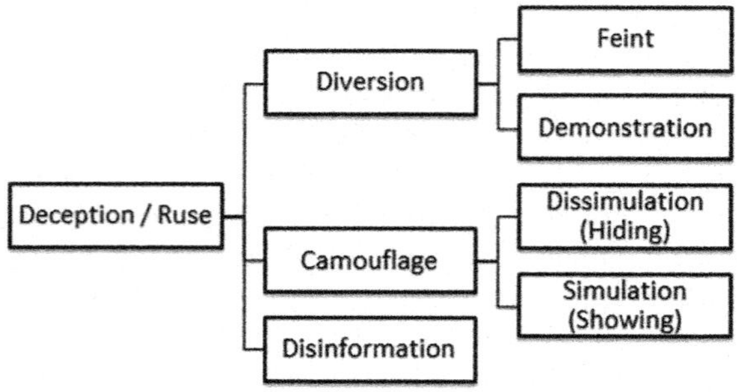

Figure 8.2 Different types of deception and ruse (Whaley 2007, pp. 8–12.)

real operation. Diversion takes two forms. First of all, a 'feint'; this is a mock attack or a simulation of a build-up to an attack. Secondly, a 'demonstration', which involves commitment to battle. Its purpose is to fix the opponent, locking his forces in combat and making the opponent's reserves irrelevant. Another type of ruse and deception is 'camouflage', which also has two sides: (1) dissimulative camouflage, in which military objects are concealed and (2) simulative camouflage, in which dummy military objectives are displayed to cheat the opponent. The third and last type of ruses and deception is 'disinformation'. This type includes all intended false information that is fed into another's information system in order to deceive him. While any type of communication channel can be used for spreading disinformation, the commonly used ones are: word-of-mouth rumours, written media, broadcast media, social media, diplomacy and espionage (Fig. 8.2).[54]

Unfortunately, Captain Tom Morland, a real exponent of Western thoughts, does not like the use of disinformation. He is straightforward and does not want to be tricked by his government. In his unit values such as honesty and mutual respect are much more important. Soldiers need to trust each other and their military leadership. On the other hand, Captain Tom Morland recognises the importance of having an understanding of the opponents' culture and preferences. He needs to fathom Major Vronsky's insights and experiences, which will contain the key to eliminate him.

8.6 Russian Deception Warfare

Shimon Naveh, a retired IDF-general,[55] is convinced that in the Russian way of warfare with its deep operations,[56] deception must attain the highest degree of operational shock, the so-called '*udar*'. This Soviet concept was developed in the

[54] Whaley 2007, pp. 8–1212.

[55] IDF stands for Israel Defence Forces.

[56] After World War I, Mikhail Tukhachevsky came with the idea of 'deep battle', breaking through the first defence lines of the opponent. During the Interbellum, his friend Vladimir Kiriakovitch Triandafillov made a next step in the development of the Soviet's military doctrine. Triandafillov

wake of World War I. While recognizing that military systems cannot only be destroyed by physical means, Russian theorists began to develop alternative approaches for defeating an opponent's system effectively. According to Naveh, the main role of deception is to create a 'state of mind'. It is rather a psychological mechanism that leads to paralysis of an opponent.[57] Today, the Russians still conduct deep operations and create '*udar*', especially in the information domain. Recent Russian information warfare has been divided into two parts: information-technical and information-psychological. The former concentrates on the technical creation of bits and bytes, while the latter concerns the 'battle of the wills'.[58] Current Russian doctrine for the application of force in the battle space indicates that information war is merely a confrontation for damaging information systems, processes, and resources to undermine the political, economic, and social systems, as well as brainwashing the populace for destabilizing the opponent's society and state.[59]

Deception, whether military or political, has been an integral characteristic in Slavic tradition. Mongol methods of warfare masterfully deflected enemy attention toward false threats; a lesson copied by their Muscovite vassals and, in turn, their Czarist and Soviet successors.[60] Russian deception became more known to the Western public after World War II. Paradoxically during the Cold War, the KGB[61] and the GRU,[62] the two most secret institutions of the Soviet state, were among the best-known Soviet organizations in the Western World. But in the Soviet Union, deception was not only allocated to secret services. Almost the entire government apparatus was committed to deception operations, and Soviet intelligence was only a part of it; however, an important one.[63] The central components in strategy, not only during the Czarist and Soviet period but also in the Russian Federation tradition, are cunning: indirectness, operational ingenuity, addressing weakness and avoiding strength.[64] General Makhmut Gareev, president of the Russian Academy

(Footnote 56 continued)

introduced the 'operational level' and the notion of 'shock armies', which were able to break through the opponent's resistance lines. And Georgii Isserson improved Triandafillov's thoughts by creating time/space factors in depth and organizing units in echelons. This is how the concept of 'deep operations' came into being. See Harrison 2001, pp. 169–217; Harrison 2010, pp. 60–151.

57 Naveh 1997, pp. 164–192.

58 Jonsson and Seely 2015, p. 12.

59 Ministry of Defence of the Russian Federation 2011, pp. 4–12.

60 Dziak 1987, p. 3. See also Baczkowski 1947; Hingley 1977, Chap. 3.

61 KGB stands for *Komitet Gosudarstvennoy Bezopastnosti*, which was the main security agency for the Soviet Union from 1954 until 1991.

62 GRU stands for *Glavnoye Razvedyvatel'noye Upravleniye*, which has been the 'Main Intelligence Agency of the Staff of the Armed Forces of the Russian Federation' since 1918.

63 Heuer 1987, pp. 22–23.

64 Adamsky 2015, p. 25.

of Military Science and a decorated World War II veteran,[65] has stated that "deceit of the adversary and cunning stratagem, dissemination of disinformation, and other deception methods have been, historically seen, integral parts of the military profession".[66]

Russian deception was primarily created in the information realm. Soviet propaganda and disinformation has worked like a well-oiled machine with significant resources and a network of front organizations, media outlets and even 'agents of influence' and agitators at its disposal. Some of these elements have been activated again in Russia, throughout the recent years.[67] More in general, modern Russian information warfare stems from the so-called '*spetspropaganda*' (special propaganda), which was first taught as a subject at the Russian Military Institute of Foreign Languages in 1942.[68] *Spetspropaganda* was used for blocking influence and the application of pressure. The Soviets conducted this type of propaganda in line with the socio-technical principals of successful propaganda, which were: (1) the principle of massive and long-lasting impact, (2) the principle of believing desired and manipulated information, (3) the principle of supposed obviousness,[69] and (4) the principal of emotional agitation.[70] The so-called 'agitprop'—the combination of agitation[71] and propaganda[72]—originates from the period immediately following the Russian Revolution. While agitprop together with *dezinformatsiya* did not stop after the collapse of the Soviet Union, these forms of influencing are still used by the Russian government. Together with the creation of main media outlets, such as RT and Sputnik to spread the message,[73] and the foundation of a new news agency *Rossiya Segodnya*[74] to merge the Russian story lines,[75] it is one of the vital parts of the current system of Russian information warfare.[76]

---

[65] The Russians prefer to speak of the 'Great Patriotic War' instead of 'World War II'.

[66] Gareev 2013, pp. 3–9. See also Adamsky 2015, p. 25.

[67] Pynnöniemi and Rácz 2016, p. 311.

[68] Lukas and Pomeranzev 2016, p. 6; Grigas 2016, pp. 45–46.

[69] Darczewska 2014, p. 25. The principle of 'supposed obviousness' causes the propaganda thesis to be associated with created political myth: e.g. the Soviet population equals determined and hard-working labourers, farmers and soldiers.

[70] Emotional agitation brings the recipient of the information to a condition in which they will act without much thought, and in an irrational way. See Kenez 1985, pp. 1–17 and 251–260.

[71] Agitation indicates 'emotion and speech', it refers to the mental state of the receiver.

[72] Propaganda refers to words and messages: What is the content? And how is disseminated?

[73] Both RT and Sputnik are relatively new media outlets, using the Internet. RT was formerly known as Russia Today, but because the word Russia has a negative connotation in the West Russia Today was reframed to RT.

[74] 'Rossiya Segodnya' means actually 'Russia Today', to make things complicated and create confusion.

[75] Van Herpen 2016, pp. 76–77.

[76] Lukas and Pomeranzev 2016, p. 6.

8.7 'Maskirovka'

Most Russian deception warfare is based on *maskirovka* and reflexive control mechanisms. During the Ukraine conflict in 2014, journalist Oestron Moeler wrote that President Putin was using Russia's traditional way of deception: '*maskirovka*'. In his point of view *maskirovka* stands for 'deliberately misleading the opponent with regard to own intentions causing the opponent to make wrong decisions thereby playing into your own hand'. Moeller adds that current *maskirovka* is mainly done through 'cunning use of networks to shape perceptions blurring the picture and opening up for world opinion to see this view as the correct one intending to legitimize the policy steps'.[77]

The origin of this phenomenon is disputed. Thomas explains that *maskirovka* was merely a military idea dating back to the Czar's Imperial Army.[78] Some Russian scholars trace it back to the year 1380 to the Battle of Kulikovo Field, some 120 miles south of Moscow, where the Russian leader Prince Dmitry Ivanovich Donskoy of Moscow divided his mounted fighters into two groups and fooled the Mongol Golden Hordes of Mamai.[79] Glantz, however, denies these views. He believes that the *maskirovka* concept flourished during the Soviet epoch. After studying the field regulations of the Red Army in the Interbellum,[80] Glantz concludes that *maskirovka* is part of a Marxism-Leninism tradition, directed to a class struggle. All measures that accelerate the change process during this struggle are permitted, if not crucial. War, in its different forms, is a normal element of that process, and therefore *maskirovka* is tolerated both in peacetime and during war. Glantz marks the difference between Russian and Western thinking: *Maskirovka* assumes a moral idiosyncrasy somewhat alien to the West that view deception as immoral, akin to dishonesty. Marxist-Leninist theory embraces all forms of deception and it defines morality differently than the west.[81]

The concept of *maskirovka* permeates the entire nation. It is not just a military term, it is a broad concept throughout Russian society that encompasses many English terms like 'camouflage', 'concealment', 'deception', 'imitation', 'disinformation', 'secrecy', 'security', 'feints', 'diversions', and 'simulation'. While these terms overlap to a great extent, a complication is that the Russian term is greater than the sum of these English terms. Thus, understanding the concept means grasping the entire concept rather than its components.[82] *Maskirovka* as well as the entire deception concept in Russia, has undergone a change that dominates the discourse of deception at the moment. The terms '*obman*' (deception) and '*vvedenie v zabluzhdenie*' (mislead) appear to have replaced the term '*maskirovka*' with the

---

[77] Oestron Moeler 2014.

[78] Thomas 2011, p. 107.

[79] Thompson 2014.

[80] Red Army 1937, p. 52.

[81] Glantz 1989, pp. 3–4.

[82] Smith 1988.

latter now understood to be a part of those two terms. Today, '*voennaya khitrost*' (military cunning or stratagem) is most used as a deception tool.[83] The *maskirovka* concept, although included in Soviet rationale, did not stop after the Cold War. The Soviet successor states, especially Russia, still use passive and active measures of *maskirovka* in varying degrees in order to manage their interests. Shea called *maskirovka* a unilaterally imposed 'fog of war' that distorts the truth for both external and domestic consumption clouds. It used to counter Western peacetime military engagement.[84] Also today, *maskirovka* is seen as a major component in Russia's security strategy. A special form of *maskirovka* is the use of reflexive control mechanisms.

8.8 Reflexive Control

The 'reflexive control' theory is a refinement of *maskirovka.* Reflexive control is considered as a 'manipulator of thought processes'.[85] It is similar to the Western concept of perception management, except that it attempts to control more than to manage a subject. Reflexive control involves creating a pattern or providing partial information that causes an opponent to react in a predetermined fashion without realizing that he is being manipulated.[86] Reflexive control appears to be 'an outgrowth of historical and cultural soil of Russia, germinated by the emphasis on control, nourished by psychological aspects of the Russian mindset such as dependence on a leader, awareness of external input the importance placed on cognition and reflection, and a society characterized by suspicion and deception'.[87] Vladimir Lefebvre developed the concept in the 1960s,[88] and he elucidates that reflexive control is simply the process of conveying the basis for decision-making from one person to another. The reflexive aspect refers to the subject's self-images and the action as perceived through these self-images, which is part of the theory. Any deceitful movement, provocation, intrigue, even calumny, or more generically, a falsehood of any type, creates reflexive control. Thus, a difference between a lie and reality can become almost fluid; for example, sending truthful information to an opponent, with the intention that the opponent considers it false, may deceive the opponent and lead to a decision advantageous to the sender. The trick is that the truth is considered a fib; all is mixed up, and at the end the opponent cannot see the wood for the trees.[89]

---

[83] Thomas 2011, pp. 107–108.

[84] Shea 2002, p. 63.

[85] Thomas 2011, p. 118.

[86] Thomas 2005, p. 241.

[87] Chotikul 1986, pp. 77–78.

[88] Chotikul 1986, pp. 77–78.

[89] Lefebvre 2015, pp. 42–43.

Reflexive control can be seen as a means of conveying specially prepared information to a partner or an opponent to incline him voluntarily making predetermined decisions desired by the initiator of the action.[90] Reflexive control is also seen as a whole of systems, methods, and tasks to influence the perception and behaviour of the enemy, population, and international community on all levels. It is a system approach based on perception management to target enemy leadership and to alter their orientation in such a way that they make decisions favourable to Russia and take actions that lead to a sense of despair within their leadership.[91] Likewise, reflexive control targets the psychological character of human beings, and it involves deliberate manipulation of their decision-making.[92]

Russian Major General (ret) Ionov, a specialist on information warfare, identified four basic methods for reflexive control, which serves as a checklist for commanders at all levels:

- **Power pressure**, which includes: the use of superior force, force demonstrations, psychological attacks, ultimatums, threats of sanctions, threats of risk, combat reconnaissance, provocative manoeuvers, denying the opponent access to certain areas, destabilizing the situation in the opponent's rear, demonstrating ruthless actions and showing mercy toward an opponent ally that has stopped fighting.
- **Measures to present false information about the situation** ('*dezinformatsiya*'), which includes: disguise, creation of mock installations, weapons bluffing, changing a mode of operation, concealing true relationship between units or creating false ones, or deliberately losing critical documents.
- **Affecting the opponent's decision-making algorithms**, which includes: the systematic conduct of games of what is perceived as routine plans, publishing a deliberately unrecognized doctrine, dissemination of false background data, and taking action to neutralize the opponent's operational thinking.
- **Modifying the decision-making time**, which can be done by surprise and suddenly starting combat actions, transferring information so that the opponent, when working out what seems feasible and predictable, makes a hasty decision that changes the mode and character of its operation.[93]

Colonel Komov named the ideas of reflexive control 'intellectual methods of information warfare'.[94] He created a general list of reflexive control mechanisms that can cause a range of psychological effects (see Table 8.1). If one of these mechanisms fails, the overall reflexive control method needs to employ another mechanism, or its initial effect might degrade soon.[95]

---

[90] Thomas 2011, p. 118.

[91] Selhorst 2016, p. 151.

[92] Shemayev 2007, p. 35.

[93] Ionov 1995, pp. 46–48; Thomas 2005, pp. 251–252.

[94] Thomas 2005, p. 255.

[95] Thomas 2011, p. 131; Selhorst 2016, pp. 151–152.

Table 8.1 Mechanism of reflexive control and their effects (Komov 1997, pp. 18–22. See also Thomas, 2011, pp. 129–130; Selhorst 2016, p. 152.)

Mechanism	Effect
Deception	Forcing the opponent to reallocate forces to a threatened region during the preparatory phases of combat operations
Deterrence	Creating the perception of insurmountable superiority
Distraction	Creating a real or imaginary threat to one of the opponent's most vital locations during the preparatory phases of combat operations, thereby forcing him to reflect on the wisdom of his decisions to operate along a certain axis
Division	Convincing the opponent that he must operate in opposition to coalition interests
Exhaustion	Compelling the opponent to carry useless operations, thereby joining combat with reduced resources
Overload	Frequently sending the opponent a large amount of conflicting data leading to indecisiveness
Pacification	Leading the opponent to believe that preplanned operational training is ongoing, rather than offensive preparations, and therefore reducing his attentiveness
Paralysis	Creating the perception of a specific threat to a vital interest or a weak spot
Pressure	Offering information that disgraces the government of the opponent in the eyes of its population
Provocation	Force him into a mode of taking action, which is advantageous for the instigator's side
Suggestion	Offering information that affects the opponent legally, morally, ideological, or in other modest areas

8.9 'Dezinformatsiya'

A relevant part of reflexive control is the use of *dezinformatsiya*. The 'science' of *dezinformatsiya,* as Russian specialists on information warfare framed it, is deeply rooted in the Russian soil and in the Russian history.[96] In the eighteenth century, the French Marquis de Custine remarked that "everything is deception in Russia. [...] Russian despotism not only counts ideas and sentiments for nothing but remakes facts. It wages war on evidence and triumphs in the battle".[97] Later Jozef Stalin decided that disinformation should look as if it were historically French. He organized an information campaign that the word *dezinformatsiya* derived from the French language. It was a composition of the French words '*des*' and '*information*'. It was a meaningless expression, but another form of Russian ruse. Stalin pretended that *dezinformatsiya* was a French capitalist tool targeted against the peaceful people of the Soviet Union.[98]

[96] Pacepa 2013, p. 36.

[97] De Custine 1987, pp. 14 and 161.

[98] Pacepa 2013, p. 39.

A vital element in *dezinformatsiya* is the way how the message is framed. Frames are the mental structures that organize all our information and shape the way we see the world. The world is thus a reflection of how humans frame it by constructions of semantics and images, including the omission of information to emphasize a certain meaning.[99] A frame is a prominent element of a narrative or a message.[100] German cognitive lingual scientist Elisabeth Wehling clarifies that frames evoke associations. These associations not only put words and observations in a framework, but make how humans perceive the world. Framing gives direction to human judgment of social, societal, economic, and political developments.[101] These judgments can affect the public. It can create psychological effects, while actions and decisions are based on it. It can also be seen as an element of the 'social construct of reality', as explained in Sect. 8.3.

Soviet circles used the term '*dezinformatsiya*' to refer to what the Soviet leaders called 'active measures'.[102] Ladislav Bittman, a former Czechoslovakian intelligence officer, has explained that disinformation is a carefully constructed false message leaked into an opponent's communication system to deceive the decision-making elite or the public. To succeed, every disinformation message must at least partly correspond to reality or generally accepted views. Without a considerable degree of plausible, verifiable information, it is difficult to gain the victim's confidence.[103] Today the concept of disinformation is still in use in the Russian Federation and is reframed by Western specialists as 'Kremlin's Weaponization of Information'[104] or earlier as the 'fabrication of information'.[105]

A special form of '*dezinformatsiya*' is '*kompromat*'. '*Kompromat*', its literal translation is 'comprising material', refers to discrediting information that can "be collected, stored, traded, or used strategically across all domains: political, electoral, legal, professional, judicial, media, or business." Russian '*kompromat*' conflicts are machinations exercised through the dissemination of often unsubstantiated or unproven information" (documents, messages, files, etc.), which are destructive for all those involved.[106]

[99] Lakeoff 2014, pp. xi–xv.
[100] Maley 2015, p. 84.
[101] Wehling 2016, pp. 21–41.
[102] Shultz and Godson 1984, p. 39.
[103] Bittman 1984, p. 49.
[104] Pomerantsev and Weiss 2014.
[105] George 1987, p. 11.
[106] Ledeneva 2006, pp. 58–59.

8.10 Findings in Ukraine and Russia

NATO's STRATCOM Centre of Excellence (COE) has discovered that the Russian *dezinformatsiya* frames during the Crimea and Ukraine conflict were largely built on association mechanisms linked to historical memories, such as 'The Great Russian Empire', 'The Great Patriotic War', 'Nazi atrocities' and 'The fascists in the Ukraine'.[107] The specificity of Russian collective memory of the 'Great Patriotic War' was a key factor in shaping the Russian public opinion toward the 'fascism' media frame. It can be considered as an ultimate form of 'agitprop'. The fascism frame helped building popular support for combatting these existential threats. The term 'fascism' is emotionally charged in Russia. It is not just a term describing distant events or stories from long ago in history books; it is a highly reminiscent term calling up a series of intense images that are saturated with typical meanings and associations and with deep relevance for the Russian population.[108]

Similarly, the concept of '*Russkiy Mir*'[109] was often used to underpin a narrative, such as the concept of the 'Russian World and Compatriots Abroad', and 'Ukrainians and Russians – one nation united under the Russian World'. Russians also used other narratives like 'The West is weak and divided', 'Russia's actions are legitimate', 'Western Ukrainians are Neo-Nazis' and 'The West is evil'.[110] NATO's STRATCOM COE learned that Russia's disinformation campaign has eroded over time, as more and more evidence and information was revealed to refute the lies and falsifications.[111] Nevertheless, the Russian involvement in Crimea and Ukraine was a huge 'laboratory' situation for Russian reflexive control concepts. At the strategic and operational level the Russian government and secret services created confusion about their political-military intentions and the deployed troop numbers, while at the tactical level it provided "a critical cover of deception for Russian forces for executing deployments and manoeuvers to take control of key facilities and positions, as well as penetrating deeply to paralyze a possible Ukraine response".[112]

The Western liberal media was no match for the unity of messages originating from Russia. Western media repeated Russian narratives, even the fraudulent ones, to their audiences, and therefore validated and reinforced the messages. Trying to portray balanced news, Western news editors were still striving for the two sides to every story, even though in reality there was only one. The Kremlin had a wide range and variety of different dissemination tools to spread the message, from the

---

[107] NATO Strategic Communication Centre of Excellence 2014, and NATO Strategic Communication Centre of Excellence 2016, pp. 71–82.

[108] Gaufman 2015, pp. 144–149.

[109] '*Russkiy Mir*' is a Russian view of the world and especially the Russian population and the Compatriots Abroad; the Russian identity, Russian history and culture, and the Russian Orthodox Church form a main part of it. See Petro 2015, pp. 3–7; Laruelle 2015, pp. 3–15.

[110] NATO Strategic Communication Centre of Excellence 2014.

[111] NATO Strategic Communication Centre of Excellence 2014.

[112] Kasapoglu 2015, p. 6.

president, diplomats, print and broadcast media to the Kremlin troll army,[113] a fancy name for Putin's bloggers on social media. RT played the most visible and vital role in spreading the message.[114] Paul and Matthew, RAND specialists on strategic communication and social science, called the Russian way of spreading information 'the firehose of falsehood', because of its two distinctive features. First, there were many channels and messages, and second, the Russian government had "a shameless willingness to disseminate partial truths or outright fictions".[115]

Michal Shavit, lecturer at the Interdisciplinary Center Herzliya in Israel, discovered another new generic element in the media representation of conflicts. Nowadays, a high degree of performativity and a blurring line between entertainment and news information have become part of the daily life in the media realm. This development generates new meanings and perceptions for the public.[116] It is exactly what the Russian state-owned channels were doing during the Crimea and Ukraine crises. Russian TV Channel 1, instigated by the Kremlin, attracted viewers by making Russian versions of popular Western shows and reality TV. During the shows hidden messages on the conflict were broadcasted. Likewise, such 'infotainment' brought in viewers, who then stay tuned for serious television programmes on current affairs.[117]

Remarkably the majority of the Russians believed their leaders, although the Ukrainian population tends to be sceptical of all elements of the Russian narrative.[118] In 2015, Russia's Levada Centre, one of the last neutral polling organizations in Russia, presented a research that underscored the very different attitude to the truth of Russian operations. Almost 37 percent of the investigated Russians said that their leaders were telling the truth about no Russian troops in Ukraine, while 38 percent said 'no'.[119] Meanwhile, Pew Research Center examined the populations of many NATO member states,[120] and they came to the conclusion that Russia and its president, Vladimir Putin, were held in low regard. Especially the median favourability whether Putin was doing the right thing and telling the truth regarding world affairs dropped down from 53% in 2003 to 28% in 2007 and to 16% in 2015.[121]

Maria Snegovaya, a columnist for Russian-language business daily *Vedomosti*, explained that the Russian government and Russian secret services used the

---

[113] Known as the 'Kremlin troll army', the men and women work 12-hour shifts around the clock, flooding the Internet with propaganda aimed at stamping President Vladimir Putin's specific vision on Russia, and the entire world. See Associated Press/New York Post 2015.

[114] Giles 2016a, b, pp. 32–39.

[115] Paul and Matthew 2016, p. 1.

[116] Shavit 2017, p. 178.

[117] Lukas and Pomeranzev 2016, p. 9.

[118] Gerber and Zavisca 2016, p. 94.

[119] Gerber and Zavisca 2016, p. 94.

[120] NATO member states like the United States, United Kingdom, Canada, Germany, France, Spain, Italy and Poland.

[121] Pew Research Center 2015, pp. 15–16.

reflexive control in Ukraine by obscuring its main goals and repeatedly denying its military presence in Ukraine's Donbass region, despite overwhelming evidence to the contrary. But there was more, the absence of a declaration of war against Ukraine in 2014 allowed Russia to limit its enemy's decision-making time, another aspect of reflexive control.[122] Thomas endorses Snegovaya's statement. He describes that even though the reflexive control theory was developed long ago, it is still undergoing refinement in Russia. During the Crimea and Ukraine crisis, the theory has been used in conjunction with other concepts, such as friendly embraces with Ukraine politicians, simulacrums and manipulated pictures, analogies with past conflicts, stimulated reactions among local populations, as well as creative and destructive applications.[123] The overall goal was to put Ukraine decision-makers under pressure in order to make quick and inaccurate judgments and decisions.

One of the first targets of Russian '*kompromat*' was Victoria Nuland, an Assistant Secretary of State for European Affairs at the U.S. State Department. In February 2014, at the peak of the crisis in Ukraine and just before the Russian invasion in Crimea, Nuland was secretly taped while speaking on the phone with Geoffrey Pyatt, the U.S. ambassador in Kiev. Frustrated with Europe's posture towards the Ukraine crisis, Nuland said literally: "F*ck the EU." Shortly after, a special assistant to the Russian deputy prime minister Dmitry Rogozin tweeted a link to the recording of the intercepted phone call, which was published on YouTube. The Nuland leak provoked a minor diplomatic hitch between the European Union and the United States.[124]

8.11 Closing Remarks

By and large, the Russian Federation has used different methods to generate a form of misperception among the decision-makers and the public of the west and their neighbouring countries during their actions in Crimea and Ukraine. It deliberately created deception, using both ways: 'hiding the real' with repackaging information and dazzling techniques, as well as 'showing the false' displaying a new reality. The Russians obviously violated the 'golden' practical rule that a deceiver must have an understanding of the opponent's values and preferences. Superficially, this would indicate an anomaly, but the Russians' campaign was sophisticatedly set. They had frequently used the ambiguity-increasing form of deception, the A-type: sending many different messages, and causing a lot of confusion. Meanwhile, they accepted collateral damage, impaired respect, a loss of credibility and 'information fratricide' caused by falsehoods and conflicting messages, as long as they could influence a certain percentage of the population of Ukraine and Russia and their

[122] Snegovaya 2015, p. 14.
[123] Thomas 2015, p. 118.
[124] Rid 2016; BBC News 2014.

respective public opinion. The Russia government and the Russian secret services were swiftly with the dissemination of their strategic narratives containing several dezinformatsiya frames. They tried to create anchoring and cognitive biases among the Western public, but they were not very successful. While Putin is still acclaimed in the Russian Federation, he is not very glorified in the Western world, partly due of what happened in Crimea and Ukraine.

This chapter has tried to paint a compelling picture on how the Russians can manipulate reality. Let us return to Captain Tom Morland, the hero at the start of this chapter, and see how this affects him. While observing riots and mortally wounded victims in the streets of Riga, Morland knew that he had to come with a cunning plan to capture Major Anatoly Vronsky. Only thinking in physical options was obsolete, as we know now. Vronsky was part of a Russian unit, the 45th Guards Spetsnaz Brigade, an operational part of the *GRU*, and they have grasped all the common tactics, techniques and procedures of the Western armies. No, Morland's plan should impact in the information domain, the relatively new man-made domain in which current and future conflicts are fought. It was thus imperative for Captain Morland to get a message out to his Headquarters back in the United Kingdom about what had really happened in the streets of Riga. Hopefully the headquarters can feed the press rapidly with my information, the only truthful information, were Morland's thoughts. Speed is paramount for news feeds. The first story-line published in any form of media will create the perceptions of the people world-wide and the public opinion. After creating such a perception, it is difficult to change such an understanding. It is known that the first message sticks in the mind of the people. Captain Morland quickly drafted an account of his impression, while his sergeant was setting up his SATCOM radio.[125] In the blink of an eye, they were ready. They looked at each other, nodding approvingly. The only thing Morland could say: "Lo and behold, let the truth be told!"

References

Adamsky D (2015) Cross-Domain Coercion: The Current Russian Strategy, Proliferation Papers, No. 54, November 2015. Institute Français des Relation Internationales (Ifri) (French Institute for International Relations), Paris, France

Associated Press/New York Post Online (2015) Russia's Online 'Kremlin Troll' Army (29 May 2015) http://www.nypost.com/2015/05/29/russias-online-kremlin-troll-army/ Accessed 28 October 2016

Baczkowski W (1947) Toward an Understanding of Russia: A Study in Policy and Strategy. Lipshutz Press, Jerusalem, Israel

Bartles C (2016) Getting Gerasimov Right. Military Review (Jan–Feb), pp. 30–38. http://www.usacac.army.mil/CAC2/MilitaryReview/Archives/English/MilitaryReview_20160228_art009.pdf Accessed 4 March 2017

[125] SATCOM stands for 'SATellite COMmunications' system, which is used as a world-wide military secure satellite communications system.

BBC News (2014) Victoria Nuland: Leaked Phone Call 'Impressive Tradecaft'. BBC Online (7 February 2014) http://www.bbc.com/news/world-europe-26079591 Accessed 7 March 2017

Betts R (1982) Surprise Attack: Lessons for Defense Planning. The Brookings Institution, Washington DC, USA

Betz D (2015a) Carnage & Connectivity: Landmarks in the Decline of Conventional Military Power. Oxford University Press, Oxford, UK

Betz D (2015b) Searching for El Dorado: The Legendary Golden Narrative of the Afghanistan War. In: De Graaf B, Dimitriu G (eds) Strategic Narratives, Public Opinion, and War. Routledge, Taylor & Francis Group, London, UK

Bittman L (1984) The KGB and Soviet Disinformation: An Insider's View. Pergamont-Brassey's International Defense Publishers, McLean VA, USA

Boudreau B (2016) We Have Met the Enemy and He Is Us: An Analysis of NATO Strategic Communications, The International Security Assistance Force (ISAF) in Afghanistan, 2003–2014. NATO Strategic Communications Centre of Excellence, Riga, Latvia

Briant EL (2015) Propaganda and Counter-Terrorism: Strategies for Global Change. Manchester University Press, Manchester, UK

Brown AC (2007) Bodyguard of Lies: The Extraordinary True Story Behind D-Day. Lyonn Press/Rowman & Littlefield, Lanham MD, USA

Cambridge Advanced Learner's Dictionary & Thesaurus (2017) Cry Wolf Definition. Cambridge University Press, Cambridge (UK). http://www.dictionary.cambridge.org/dictionary/english/cry-wolf Accessed 8 January 2017

Central Intelligence Agency (CIA) (1980) Deception Maxims: Facts and Folklore, A Research Paper (release date: 2 December 2015), Washington DC, USA http://www.governmentattic.org/18docs/CIAdeceptionMaximsFactsFolklore_1980.pdf Accessed 27 October 2016

Chotikul D (1986) The Soviet Theory of Reflexive Control in Historical and Psychocultural Perspective: A Preliminary Study. Naval Postgraduate School, Monterey CA, USA http://calhoun.nps.edu/bitstream/handle/10945/30190/soviettheoryofre00chot.pdf?sequence=1 Accessed 30 August 2015

Daniel D, Herbig K (1982a) Propositions on Military Deception. In: Daniel D et al (eds) Strategic Military Deception. Pergamon Press Inc., New York NY, USA

Daniel D, Herbig K (1982b) Propositions on Military Deception, In: Gooch J, Perlmutter A (eds) Military Deception and Strategic Surprise. Frank Cass and Co. Ltd, London, UK

Darczewska J (2014) The Anatomy of Russian Information Warfare: The Crimean Operation, A Case Study. Point of View No 42. May 2014, Centre for Eastern Studies, Warsaw, Poland

De Custine A (1987) Journey for Our Time: The Russian Journals of the Marquis de Custine, Gateway Edition. Regnery, Phyllis Penn Kohler Edition & Translation, Washington DC, USA

De Graaf B, Dimitriu G (2015) Introduction: Shaping Societies for War, Strategic Narratives and Public Opinion. In: De Graaf B, Dimitriu G (eds) Strategic Narratives, Public Opinion, and War. Routledge, Taylor & Francis Group, London, UK

Dewar M (1989) The Art of Deception in Warfare. David & Charles Publishers, Newton Abbot, Devon, UK

Dziak J (1987) Soviet Deception: The Organizational and Operational Tradition. Daily B, Parker P (eds) Soviet Strategic Deception. Lexington Books, Washington DC/Health and Company, Lexington MA, USA

Dubovitsky N (2013) *Bez Neba* (Without Sky). In: *Russkiy Pioneer* (Russian Pioneer, a Russian Magazine), No. 46, 12 March 2014. English translation by Bowler B, Bewildering Stories. http://www.bewilderingstories.com/issue582/wihout_sky.html Accessed 06 August 2015

Dylan B (1999) Things Have Changed. Song text of the official soundtrack of the movie 'Wonder Boys' (2000) http://www.bobdylan.com/songs/things-have-changed Accessed 17 October 2016

Gaufman E (2015) Memory, Media, and Securitization: Russian Media Framing of the Ukrainian Crisis - Russian Media and the War in Ukraine. Journal of Soviet and Post-Soviet Politics and Society 1(1):141–ff.

Gareev M (2013) *Voennaia nauka na somvremenonnom etap* (Military Science at the Present Stage). *Voyenno-Promyschlennyy Kuryer* (Military Industrial Courier), No. 13(481) April 2013

George R (1987) A Historical Investigation of Soviet Strategic Deception, Student Report. Air Command and Staff College, Maxwell Air Force Base, AL, USA http://www.dtic.mil/dtic/tr/fulltext/u2/a179921.pdf Accessed 27 October 2016

Gerasimov V (2013) *Tsennost nauki v predvidenii* (The Value of Science in Prediction). A Lecture given at a closed Voroshilov General Staff Academy Conference, later reprinted in: *Voyenno-Promyschlennyy Kuryer* (Military Industrial Courier), No. 12(481), 27 February 2013

Gerber T, Zavisca J (2016) Does Russian Propaganda Work? The Washington Quarterly, 39 (2):79–98

Gerwehr S, Glenn R (2000) The Art of Darkness: Deception and Urban Operations, RAND Monograph Report M1132-1. RAND Corporation, Arroyo Center, RAND Army Research Division, Santa Monica CA, USA

Giles K (2016a) Russia's 'New' Tools for Confronting the West: Continuity and Innovation in Moscow's Exercise of Power. Research Paper, Russia and Eurasia Programme. March 2016, Chatham House, The Royal Institute of International Affairs, London, UK http://www.chathamhouse.org/sites/files/chathamhouse/publications/research/2016-03-21-russias-new-tools-giles.pdf Accessed 27 October 2016

Giles K (2016b) The Next Phase of Russian Information Warfare. NATO Strategic Communication Centre of Excellence, Riga (LTV) http://www.stratcomcoe.org/next-phase-russian-information-warfare-keir-giles Accessed 1 March 2017

Glantz D (1989) Soviet Military Deception in the Second World War. Frank Cass and Company, Totowa, NJ, USA

Gooch J, Perlmutter A (eds)(1982) Military Deception and Strategic Surprise. Frank Cass and Co. Ltd, London, UK

Grigas A (2016) Beyond Crimea: The New Russian Empire. Yale University Press, New Haven, CT, USA

Handel M (1976) Perception, Deception and Surprise: The Case of the Yom Kippur War, Jerusalem Papers on Peace Problems. The Hebrew University, Leonard Davis Institute for International Relations, Jerusalem, Israel

Handel M (1982) Intelligence and Deception, In: Gooch J, Perlmutter A (eds) Military Deception and Strategic Surprise. Frank Cass and Co. Ltd, London, UK

Handel M et al (1987) Strategic and Operational Deception in Second World War. Frank Cass and Co. Ltd, London, UK

Harrison R (2001) The Russian Way of War: Operational Art, 1904–1940. University Press of Kansas, Lawrence KS, USA

Harrison R (2010) Architect of Soviet Victory in World War II: The Life and Theories of G.S. Isserson. Mc Farland & Company, Inc., Publishers, Jefferson NC, USA

Heuer R (1987) Soviet Organization and Doctrine for Strategic Deception, In: Daily B, Parker P (eds) Soviet Strategic Deception. Lexington Books, D.C. Health and Company, Lexington MA, USA

Hingley R (1977) The Russian Mind. Charles Scribner's Sons, New York NY, USA

Ionov M (1995) On Reflexive Control on the Enemy in Combat. Military Thought - A Russian Journal of Military Theory and Strategy (English version), No. 1, January 1995, pp. 46–48

Jones R (1957) The Theory of Practical Joking – Its Relevance to Physics. Bulletin of the Institute of Physics, 8(6):193–201

Jonsson O, Seely R (2015) Russian Full-Spectrum Conflict: An Appraisal After Ukraine. The Journal of Slavic Military Studies 28(1):1–22

Kahneman D (2011) Thinking, Slow and Fast. Penguins Books, London, UK

Kasapoglu C (2015) Russia's Renewed Military Thinking: Non-Linear Warfare and Reflexive Control, Research Paper No. 121. November 2015, Research Division – NATO Defence College, Rome, Italy http://www.ndc.nato.int/news/news.php?icode=877 Accessed 30 October 2016

Kass L, London J (2013) Surprise, Deception, Denial and Warning: Strategic Imperatives. Orbis 57(1):59–82

Keneally A (2016) The Art of the Bluff: How Presidents Can Leverage Deception. Lawfare Blog (1 June 2016) http://www.lawfareblog.com/art-bluff-how-presidents-can-leverage-deception Accessed 24 October 2016

Kenez P (1985) The Birth of the Propaganda State: Soviet Methods of Mass Mobilization, 1917–1929. Cambridge University Press, Cambridge, UK

Kitzen M (2012) Western Military Culture and Counter-Insurgency: An Ambiguous Reality. Scientia Militaria, South African Journal of Military Studies 40(1):1–24

Komov S (1997) About Methods and Forms of Conducting Information Warfare. Military Thought - A Russian Journal of Military Theory and Strategy (English version), No. 4 July–August 1997, pp 18–22

Kopp C (2005) Classical Deception Techniques and Perception Management vs. the Four Strategies of Information Warfare. In: Pye G, Warren M (eds) Conference Proceedings of the 6th Australian Information Warfare & Security Conference (IWAR 2005), School of Information Systems, Deakin University, Geelong VIC, Australia

Korteling J (2016) Neurowetenschappelijke Mechanismen van Cognitieve Bias (Neuroscientific mechanisms of cognitive bias). TNO Rapport, no R11451, TNO Eart, Life & Social Sciences, Soesterberg (NLD)

Lakeoff G (2014) Don't Think of an Elephant: Know Your Values and Frame the Debate. Chelsea Green Publishing, White River Junction VT, USA

Laruelle M (2015) The "Russian World": Russia's Soft Power and Geopolitical Imagination. Center on Global Interests (May 2015) http://www.globalinterests.org/wp-content/uploads/2015/05/FINAL-CGI_Russian-World_Marlene-Laruelle.pdf Accessed 24 March 2016

Latimer J (2016) Deception in War, first published in 2001 by John Murray. Thistle Publishing, London, UK

Ledeneva A (2006) How Russia Really Works: The Informal Practices that Shaped Post-Soviet Politics and Business. Cornell University Press, Ithaca NY, USA

Lefebvre V (2015) Conflicting Structures, a Translation of '*Konfliktuyushchiye Struktury*' (1967) by '*Vysshaya Shkola*'. Leaf & Oaks Publishers, Los Angeles CA, USA

Leipold JD (2015) Milley: Russia No.1 threat to US, Official Home Page of the United States Army (9 November 2015) http://www.army.mil/article/158386/milley_russia_not_threat_to_us Accessed 17 October 2016

Lukas E, Pomeranzev P (2016) Winning the Information War: Techniques and Counter-strategies to Russian Propaganda in Central and Eastern Europe. A Report by CEPA's Information Warfare Project in Partnership with the Legatum Institute. Centre for European Policy Analysis & Legatum Institute (August 2016) http://www.cepa.ecms.pl/files/?id_plik=2773 Accessed 29 September 2016

Maley W (2015) The War in Afghanistan: Australia's Strategic Narratives. In: De Graaf B, Dimitriu G (eds) Strategic Narratives, Public Opinion, and War. Routledge/Taylor & Francis Group, London, UK

McClearly P (2015) More Pentagon Generals Line Up to Proclaim Russia's 'Extential' Threat to U.S. Foreign Policy, The Cable (14 July 2015) http://www.foreignpolicy/2015/07/14/more-pentagon-generals-line-up-toproclaim-russia-existential-threat-to-u-s/ Accessed 23 August 2015

McCleskey E (1991) Applying Deception to Special Operations Direct Action Missions. Defense Intelligence College, Washington DC, USA

McDermott R (2016) Does Russia Have a Gerasimov Doctrine? Learning From Today's Wars. Parameters 46(1):97–105

Merton R (1995) The Thomas Theorem and the Matthew Effect. Social Forces 74(2):379–424

Ministry of Defence of the Russian Federation (2011) *Konseptual'nyye Vzlyady Na Deyatel'nost Vooruzhennykh Sil Rossiyskoy Federatsii v Informationnom Prostrastve'* (Conceptual Vision on the Activities of the Armed Forces of the Russian Federation in the Information Space). http://www.pircenter.org/media/content/files/9/13480921870.pdf Accessed 23 October 2016

NATO Strategic Communication Centre of Excellence (2014) Analysis of Russia's Information Campaign Against Ukraine. NATO Strategic Communications Centre of Excellence, Riga, Latvia. http://www.stratcomcoe.org/~/media/SCCE/NATO_PETIJUMS_PUBLISKS_29_10.ashx Accessed 30 August 2015

NATO Strategic Communication Centre of Excellence (2016) Framing of the Ukraine-Russia Conflict in Online and Social Media. NATO Strategic Communications Centre of Excellence, Riga, Latvia. http://www.stratcomcoe.org/framing-ukraine-russia-conflict-online-and-social-media Accessed 23 June 2016

Naveh S (1997) In Pursuit of Military Excellence: The Evolution of Operational Theory, reprinted in 2004. Frank Cass Publishers, Abington, Oxon, UK

North Atlantic Council (2014) Wales Summit Declaration, issued by the Heads of State and Government participating in the meeting of the North Atlantic Council in Wales (5 September 2014, last update: 26 September 2016). http://www.nato.int/cps/en/natohq/official_texts_112964.htm Accessed 27 February 2017

North Atlantic Council (2016) Warsaw Summit Declaration, issued by the Heads of State and Government participating in the meeting of the North Atlantic Council in Warsaw 8–9 July 2016 (9 July 2016, last update: 03 August 2016). http://www.nato.int/cps/en/natohq/official_texts_133169.htm Accessed 27 February 2017

Oestron Moeler J (2014) Maskirovka: Russia's Masterful Use of Deception in Ukraine. Website of The World Post, a Partnership of Huffington Post and Berguruen Institute (23 April 2014). http://www.huffingtonpost.com/joergen-oerstroem-moeller/maskirovka-russias-master_b_5199545.html Accessed 30 August 2015

Pacepa I (2013) Disinformation: Former Spy Chief Reveals Secret Strategies for Undermining Freedom, Attacking Religion, and Promoting Terrorism. WND Books, Inc., Washington DC, USA

Paul C, Matthew M (2016) The Russian "Firehose of Falsehood" Propaganda Model: Why It Might Work and Options to Counter It. RAND Corporation Report (11 July 2016) RAND Corporation, Santa Monica CA, USA. http://www.rand.org/pubs/perspectives/PE198.html Accessed 24 October 2016

Perry B (2014) Non-Linear Warfare in Ukraine: The Critical Role of Information Operations and Special Operations. Small Wars Journal (14 August 2015). http://www.smallwarsjournal.com/jrnl/art/non-linear-warfare-in-ukraine-the-critical-role-of-information-operations-and-special-operations Accessed 24 August 2015

Petro N (2015) Russia's Orthodox Soft Power. U.S. Global Engagement, Carnegie Ethics Online, Carnegie Council for Ethics in International Affairs (23 March 2015). http://www.carnegiecouncil.org/publications/articles_papers_reports/727 Accessed 29 August 2015

Pew Research Center (2015) NATO Publics Blame Russia for Ukrainian Crisis, but Reluctant to Provide Military Aid (released 10 June 2015). Pew Research Center, Washington DC, USA

Pomerantsev P, Weiss M (2014) The Menace of Unreality: How the Kremlin Weaponizes Information, Culture and Money. A Special Report Presented by The Interpreter (November 2014). http://www.interpretermag.com/wpcontent/uploads/2014/11/The_Menace_of_Unreality_Final.pdf Accessed 30 August 2015

Pynnöniemi K, Rácz A (eds) (2016) Fog of Falsehood: Russian Strategy of Deception and Conflict in Ukraine. The Finnish Institute of International Affairs, Helsinki, Finland. http://www.fiia.fi/en/publication/588/fog_of_falsehood/ Accessed 13 October 2016

Red Army of the Soviet Union (1937) *Vremennyi Polevoi Ustav RKKA, 1936* (Temporary Field Regulations of the Red Army, 1936), translated by U.S. Army War College, Translation Section, Washington DC, USA/Voenizdat, Moscow, Soviet Union

Rid T (2007) War and Media Operations: The US Military and the Press from Vietnam to Iraq. Routledge, Taylor & Francis Group, London, UK

Rid T (2016) How Russia Pulled Off the Biggest Election Hack in U.S. History: Putin, Wikileaks, the NSA and the DNC Email Fiasco that Gave Trump and Clinton Another Reason to Be at Odds. Esquire Online (20 October 2016) http://www.esquire.com/news-politics/a49791/russian-dnc-emails-hacked/ Accessed 7 March 2017

Rid T, Hecker M (2009) War 2.0: Irregular Warfare in the Information Age. Praeger Security International, Westport CT, USA

Rothstein H, Whaley B (2013) Section II Principles and Other Fundamentals, Introduction, In: Rothstein H et al (eds) The Art and Science of Military Deception. Artech House, Boston MA, USA

Selhorst T (2016) Russia's Perception Warfare. Militaire Spectator (Military Spectator) 184(4): 148–164. http://www.militairespectator.nl/thema/strategie-operaties/artikel/russias-perception-warfare Accessed 27 October 2016

Shavit M (2017) Media Strategy and Military Operations in the 21st Century: Mediatizing the Israel Defence Forces. Routledge, Taylor & Francis Group, London, UK

Shea T (2002) Post-Soviet Maskirovka, Cold War Nostalgia, and Peacetime Engagement. Military Review 82(3):63–67

Shemayev V (2007) Cognitive Approach to Modelling Reflexive Control in Socio-Economic Systems. Information & Security, An International Journal, 22:28–37. http://www.procon.bg/system/files/22.04_Shemayev.pdf Accessed 27 October 2016

Shirreff R (2016) 2017 War With Russia: An Urgent Warning from Senior Military Command. Coronet, Hodder & Stoughton Ltd, London, UK

Shultz R, Godson R (1984) Dezinformatsia: Active Measures in Soviet Strategy. Pergamont-Brassey's International Defense Publishers, McLean VA, USA

Smith C (1988) Soviet Maskirovka. Airpower Journal 2(1):28–37 http://www.airpower.maxwell.af.mil/airchronicles/apj/apj88/spr88/smith.html Accessed 13 October 2014

Snegovaya M (2015) Putin's Information Warfare in Ukraine: Soviet Origins of Russia's Hybrid Warfare. Institute for the Study of War, Washington DC, USA

Stavridis J (2016) Foreword. In: Shirreff R (2016) 2017 War With Russia: An Urgent Warning from Senior Military Command. Coronet, Hodder & Stoughton Ltd, London, UK

Thomas T (2005) Cyber Silhouettes: Shadows over Information Operations. Foreign Military Studies Office (FMSO), Fort Leavenworth KS, USA

Thomas T (2011) Recasting the Redstar: Russia Forges Tradition and Technology Through Toughness. Foreign Military Studies Office (FMSO), Fort Leavenworth KS, USA

Thomas T (2015) Russia Military Strategy: Impacting 21st Century Reform and Geopolitics. Foreign Military Studies Office (FMSO), Fort Leavenworth KS, USA

Thomas W, Thomas D (1928) The Child in America: Behavior Problems and Programs. Knopf Publishers, New York NY, USA

Thompson M (2014) The 600 Years of History Behind Those Ukrainian Masks: Maskirovka Has Been A Part of the Russian Military Since Before There Was a Russia. TIME Online (17 April 2014) http://www.time.com/67419/the-600-years-of-history-behind-those-ukrainian-masks/ Accessed 18 October 2016

Toet A et al (2016) Effects of Personal Characteristics, Mental State and External Factors on Susceptibility to Bias: a Literature Study, TNO Report, no R11452, TNO Earth, Life & Social Sciences, Soesterberg, The Netherlands

Tolstoy L (2016) Anna Karenina, 2nd edn. A New Translation by Rosamund Bartlett. Originally published as a book in 1878. Oxford World Classics, Oxford University Press, Oxford, UK

U.S. Department of Defense (2015) NATO Commander Breedlove Discusses Implications of Hybrid War. DoD News (23 March 2015) http://www.defense.gov/News/Article/604334 Accessed 23 August 2015

Van Herpen MH (2016) Putin's Propaganda Machine: Soft Power and Russian Foreign Policy. Rowman & Littlefield, Lanham MD, USA

Ven Bruusgaard K (2016) Russian Strategic Deterrence. Survival 58(4):7–26

Verrall N (2016) Future of Military Deception Study Day: Context and Setting. A Presentation for RUSI's Future of Military Deception Study Day, 6 September 2016, Royal United Services Institute, Whitehall, Westminster, London, UK

Wehling E (2016) Politische Framing: Wie Eine Nation Sich Ihr Denken Einredet – Und Daraus Politik Macht (Political Framing: How A Nation's Thinking Is Foist – And How Politics Are Based On It). Herbert von Halem Verlag, Cologne, Germany

Whaley B (1982) Towards a General Theory of Deception. In: Gooch J, Perlmutter A (eds) Military Deception and Strategic Surprise. Frank Cass and Co. Ltd, London, UK
Whaley B (2007) Stratagem: Deception and Surprise in War, originally published in 1969 by the Center for International Studies at MIT. Artech House Boston MA, USA
Winning N (2015) U.K. Foreign Secretary Philip Hammond Warns Russia Is Potential Big Threat to Security. The Wall Street Journal (10 March 2015) http://www.wsj.com/articles/u-k-foreign-secretary-philip-hammond-warns-russia-is-potential-threat-to-security-1425997164 Accessed 17 October 2016
Wohlstetter R (1962) Pearl Harbor: Warning and Decision. Stanford University Press, Stanford CA, USA

Author Biography

Colonel Han Bouwmeester, M.Sc., MMAS is Associate Professor of Military Strategy and Land Warfare at the Department of War Studies, Netherlands Defence Academy. He holds a Master of Science degree in Public Administration & Political Sciences, is a graduate of the U.S. Army School of Advanced Military Studies and is preparing his Ph.D. thesis on Russia's information campaigns and deception.

Chapter 9
Manoeuvring and Generating Effects in the Information Environment

Paul A.L. Ducheine, Jelle van Haaster and Richard van Harskamp

Is this the real life? Is this just fantasy?.
Queen—Bohemian Rhapsody.

Abstract This chapter aims to offer a framework for States in general and armed forces in particular for generating effects in or through the information environment by answering the question: "How to generate effects in or through the information environment and therefore, how to manoeuvre in this information environment?" This environment is part of the larger operational environment and consists of three dimensions: the cognitive, virtual and physical. These dimensions in turn host certain layers and these hold targetable entities. States can create effects in this environment by wielding the instruments at their disposal (diplomacy, informational, military and economic) for various purposes (anticipation, prevention, deterrence, protection, intervention, stabilization and normalization). In order to be able to do so, States must organise and equip themselves for manoeuvring in the information environment. To show that indeed States can use this environment, this chapter highlights some cases where an actor uses the information environment to great effect, being: the U.S. Election Information Campaign (2016) and the BlackEnergy Operation (2015) in Ukraine. These cases are indicative of the potential of manoeuvring in the information environment by States and their armed forces.

Keywords Cyber Operations · Cyberspace · Cyber Warfare · Information Operations · Information Warfare · Psychological Operations · Non-Kinetic Effects · National Interests · Strategic Interests

P.A.L. Ducheine (✉) · J. van Haaster · R. van Harskamp
Faculty of Military Sciences, Netherlands Defence Academy, Breda, The Netherlands
e-mail: p.a.l.ducheine@uva.nl

P.A.L. Ducheine and F.P.B. Osinga (eds.), *Netherlands Annual Review of Military Studies 2017*, NL ARMS, DOI 10.1007/978-94-6265-189-0_9

Contents

9.1 Introduction

Early 2016, the US Secretary of Defense Ashton Carter affirmed that he had given US' Cyber Command "its first wartime assignment" in the war with ISIS.[1] Likewise, the UK's Secretary of State for Defence Michael Fallon MP confirmed that his country's offensive cyber capabilities are being deployed in this campaign against ISIS.[2] These statements combined with NATO's pronouncement to recognise cyberspace as a domain of operations,[3] have spearheaded the use of cyber operations during conflict. It is evident that the information environment, including cyberspace, is increasingly being used to project power, in war and peacetime. Current cyber capabilities, however, almost exclusively focus on the logic (software) and infrastructure (network) to be exploited for military purposes. For a brief period (2000s to early 2010s), these 'hard' cyberspace logic-focused operations have overshadowed the use of cyberspace for primarily influencing human cognition. Tantamount to this is that various cyberspace doctrines only briefly mention the use of cyber operations to affect the humans.[4] With the rise of so-called 'fake'

[1] Military.com 2016.

[2] Fallon 2016.

[3] NATO, Warsaw Summit Communiqué, 9 July 2016, paras 70–71, http://bit.ly/29wBtNW.

[4] See for instance The Joint Chiefs of Staff 2013.

news, election hacks, the hybrid warfare hype and regained attention for 'maskirovka',[5] 'soft' influence activities are once more in the limelight.[6]

Whether for strategic ends as part of a so-called 'hybrid threat' posed by States or terrorist campaigns,[7] or serving operational goals in a so-called comprehensive approach in recent counterinsurgency or stabilization operations,[8] it seems fair to conclude that information as a source of power, after its successes in the east,[9] enjoys a reappraisal in the Western Hemisphere after 2000.[10] Increasingly, States anticipate, or at least research the use of other than kinetic means, ways and strategies to solve its conflicts.

This chapter therefore aims to offer a framework for States in general and armed forces in particular for generating effects in or through the information environment by answering the question: "How to generate effects in or through the information environment and therefore, how to manoeuvre in this information environment?" It thereby contributes to this publication's overarching purpose to address the strategic and operational issues related to the various uses of non-kinetic capabilities by armed forces in pursuance of a State's goals, or as used in this volume: *to fight without killing*. This is not limited to times of war, but also involves the public acknowledgement of the potential of cyber capabilities to be used to affect other states in times of peace. The use of capabilities in the information environment blurs the dichotomy of peace and war to the point of the distinction becoming obsolete.

To this end, first, the information environment itself will be analysed. Next, the purpose (the end) of manoeuvring in the information environment will be discussed, and, closely connected to the why of manoeuvring in the information environment, when this is envisioned. Thirdly, the means required to do so will be addressed by subsequently analysing the physical, conceptual and moral component of fighting power. After that, the ways to manoeuvre will then be discussed by providing a generic model, as well as by reviewing recent examples of manoeuvres in the information environment.

Before starting, three points of reference should be clarified. First, although initially departing from a military point of view, this chapter will not restrict itself in this way. Therefore, the analysis presented will use the State as its point of departure, and will include civilian as well as military capabilities alike. Thus, apart from the fact that information as a part of instruments of power will be addressed,

[5] Russian military deception, sometimes known as *Maskirovka*, is a military doctrine developed from the start of the twentieth century. The doctrine covers a broad range of measures for military deception, ranging from camouflage to denial and deception. See Chap. 8 by Han Bouwmeester in this volume.

[6] Heickerö 2010, p. 20; Wirtz 2014, p. 21; Giles 2016, p. 26.

[7] Inter alia: Selhorst 2016, pp. 148–164; see also Chap. 8 by Han Bouwmeester in this volume.

[8] Inter alia: Ducheine 2015a b, c, pp. 201–220; see also Chap. 5 by Thijs Brocades Zaalberg in this volume.

[9] Kragh and Åsberg 2017; Lee 2014.

[10] See Herrick 2016. pp. 99–112; Farrell et al. 2013, pp. 308–309; Rid and Hecker 2009, pp. 132–133.

"information related capabilities", whether military or civilian, will be taken into consideration as well. Secondly, this chapters aims to contribute to strategic as well as military operational notions of the subject of manoeuvring in the information environment. In doing so, references to state practice, inter alia that of the Netherlands armed forces, will be used. This practice will serve as an example only, as the aim is to cover the themes in a generic international manner. Thirdly, although the requirement of the availability of intelligence is evident, this prerequisite will not be addressed in this chapter.

9.2 Where? Manoeuvring in the Information Environment

An important aspect contributing to this chapter's aim—to discern the opportunities of manoeuvring in the information environment—is delineating the information environment. Therefore, this section will first briefly outline the background of the information environment. Next, the dimensions and contents of the information environment will be derived from technological and military doctrinal analysis. Lastly, a conceptual model of the information environment involving the different dimensions (or layers), entities (or components) and (target) actors will be presented. This model will subsequently be used in this chapter's analysis.

9.2.1 Doctrinal History of the Information Environment

The information environment is a construct often employed in publications regarding information operations. The capacities nowadays included under the ambit of information operations are long established in warfare. However, the integral coordination and application of these capacities is relatively new, that is, the early-90s.[11] Early doctrine described information operations as a new concept with "five core capabilities", being Electronic Warfare, Psychological Operations, Operational Security, military deception and Computer Network Operations.[12] The concept followed after the Second Gulf War (Desert Storm, 1991) was dubbed information war, later information warfare and finally replaced with the construct information operations.[13] These operations are conducted in the information

[11] Office of the Chief of Naval Operations 1995; U.S. Army Training and Doctrine Command 1995; The Joint Chiefs of Staff 1998.

[12] U.S. Department of Defense 2003, p. 9.

[13] Bemis and Morgan 2008, pp. 19–21.

environment, which is described as "the aggregate of individuals, organizations, and systems that collect, process, disseminate, or act on information.[14]

Terminological issues regarding the concepts of scope and content of information operations, information activities, information domain and the information environment have plagued many discussions in this matter. These are not 'right or wrong' discussions, they revolve around being correct or doctrinally correct. Hence, these concepts are easily reconciled. The information domain, as often used in discussions, is referred to in doctrines as the information environment.[15] Although used interchangeably, from a doctrinal perspective they have different implications. An environment is a larger/higher concept than a domain, the US doctrine uses two environments: the traditional and the informational. Adjoined these two make up the operational environment.[16] In military doctrine, a domain is 'reserved' for acknowledged domains such as air, land, sea, space and cyberspace (see Fig. 9.1).[17] In other words, the term information environment as will be used here, is doctrinally more precise; noting however, that amongst the military, the terms domain and environment are often used as synonyms.

9.2.2 *A Model of the Informational Environment*

Despite terminological differences, contemporary military doctrine agrees upon the dimensions of the information environment (see Fig. 9.1), namely: physical, cognitive (psychological) and informational (virtual).[18] These "three interrelated dimensions [...] continuously interact with individuals, organizations, and systems".[19] The physical dimension is "composed of command and control (C2) systems, key decision makers, and supporting infrastructure that enable individuals and organizations to create effects".[20] The cognitive (US and UK) or psychological (NATO) dimension "encompasses the minds of those who transmit, receive, and respond to or act on information".[21] Lastly the virtual (UK and NATO) or

[14] The Joint Chiefs of Staff 2014, pp. I-2, I-3.

[15] North Atlantic Treaty Organization 2009, p. 1-1; the Joint Chiefs of Staff 2014, p. ix; UK Development, Concepts and Doctrine Centre 2010, p. 3-7; British Army 2010, p. 4-14.

[16] The Joint Chiefs of Staff 2014, p. x.

[17] An exception to this line of reasoning is the new doctrine for information operations that has replaced dimensions with domains, see: An exception to this line of reasoning is the new doctrine for information operations that has replaced dimensions with domains. See North Atlantic Treaty Organization 2015, p. 1-2.

[18] UK Development, Concepts and Doctrine Centre 2010, p. 2-5; The Joint Chiefs of Staff 2014, p. I-1; North Atlantic Treaty Organization, 2015, p. 1-2.

[19] The Joint Chiefs of Staff 2014, p. I-1.

[20] The Joint Chiefs of Staff 2014, p. I-2.

[21] The Joint Chiefs of Staff 2014, p. I-3.

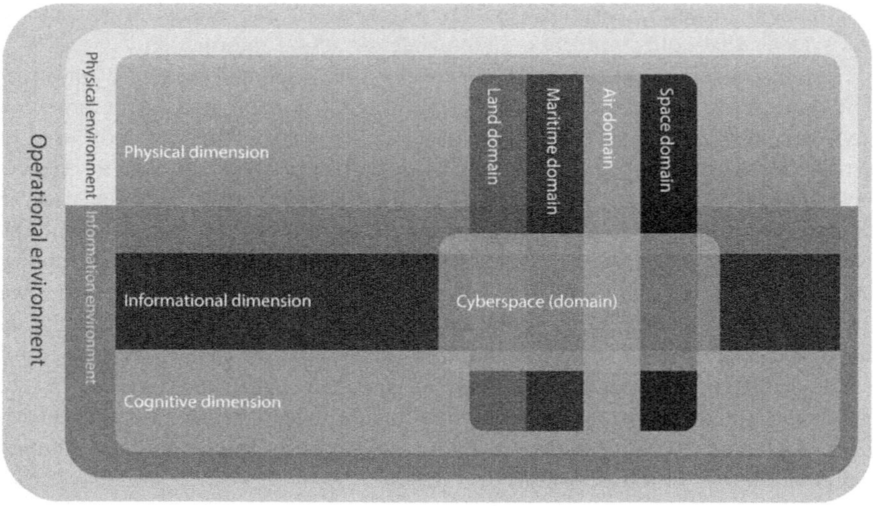

Figure 9.1 Relation between environments, dimensions and domains

informational (US) dimension "encompasses where and how information is collected, processed, stored, disseminated, and protected".[22]

The three dimensions making up the operational environment comprise seven layers (see Fig. 9.2) that in turn comprise certain entities (see Fig. 9.3). The physical dimension includes the geographical, physical network and physical layer; the virtual dimension envelops the logical and virtual persona layer and the cognitive dimension encompasses the cognitive and social layer.

The layers subsequently hold various entities that are needed to operationalize the layer concept (see Fig. 9.3); these entities can be affected via activities and/or operations.

The geographical layer includes geographical locations, the physical layer physical objects and persons, the physical network layer physical network infrastructure (i.e. routers, cables, computers), the logical virtual objects (e.g. software), the virtual persona layer virtual personas (e.g. social media profiles and mail accounts), the cognitive layer encompasses human psyche (e.g. will, perception and behaviour) and the social layer networked/interacting groups or audiences.[23] These entities are interconnected and interrelated; affecting one will affect others too.

A current surge towards human-machine interfaces has further blurred the distinct differences between virtual and cognitive/physical aspects of humans.[24]

[22] The Joint Chiefs of Staff 2014, p. I-3.

[23] For a more detailed description of these entities, see Ducheine and Van Haaster 2014, pp. 303–327; [Dutch] Ducheine and Van Haaster 2013, pp. 368–387; UK Development, Concepts and Doctrine Centre 2010, p. 2-9; U.S. Army 2010, pp. 8–9.

[24] See, for example, Anthony 2014; Hewitt 2014; Regalado 2014.

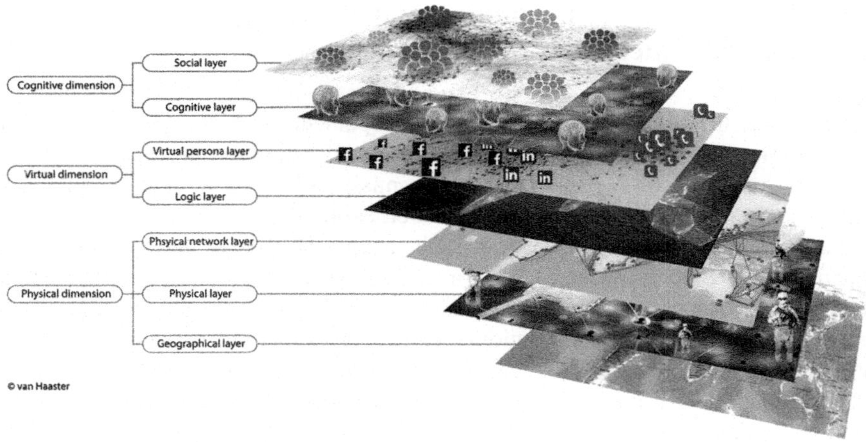

Figure 9.2 Dimensions and layers (Based on: UK Development, Concepts and Doctrine Centre 2010, p. 2-9; The Joint Chiefs of Staff 2013, p. I-3; Ducheine and Van Haaster 2014; Dekkers et al. 2016.)

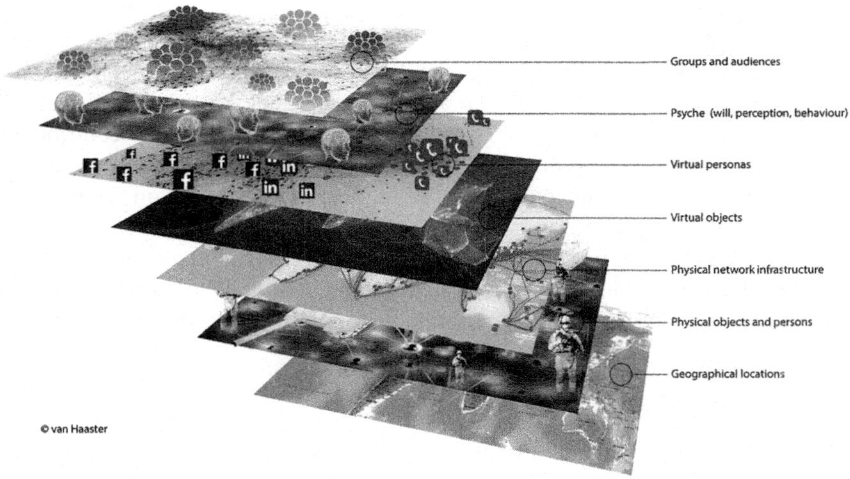

Figure 9.3 Entities

Similarly, the rise of artificial intelligence (AI), although still nascent, is starting to blur the difference between physical or virtual objects and a virtual 'AI' persona.[25] In the present authors' opinion, society will increasingly be confronted with hybrid entities, humans linked with hardware and software, virtual objects with an individual virtual persona.

[25] See, for instance, Abadi and Andersen 2016; Rayner 2016; Suzuki et al. 2005.

9.3 Why? Pursuing, Promoting and Securing National Interests

The purpose, the 'why' of manoeuvring, is to pursue, promote, secure or defend national interest in times of peace and in war. For the military, this may too easily sound as "winning wars" or to "accomplish missions", the real strategic goal is a higher one. This actual purpose may be stated explicitly in a Grand Strategy,[26] in so-called White Papers,[27] or in policy documents issued by various functional departments.[28] Sometimes, purpose is implied in or can be deduced from ad hoc documents or statements. In the Dutch situation, two complementary policy documents define the Netherlands' national security ambition, encompassing five vital and three strategic interests.[29] Jointly, they cover territorial integrity, physical security, economic security, ecological security, social and political stability and the international legal order.

Whether to promote or advocate national vital and strategic interests in peacetime, or to secure or defend them in times of war or conflict, instruments of power provide the means enabling States to do so. These instruments of power are quite often referred to by the acronym DIME: diplomacy, information, military and economy.[30] The synchronized and harmonised application of these instruments of power is caught in terminology such as the 'whole of government' or 'comprehensive approach'.[31] Ideally, a proper and well-coordinated use of DIME offers synergic advantages compared to the stand-alone or side-by-side employment of separate instruments. Jointly (or separately), the DIME-instruments contribute to strategic functions in international relations.[32] In the Dutch situation these strategic functions encompass anticipation, prevention, deterrence, protection, intervention, stabilization and normalization.[33]

Though much is expected from cyber capabilities and manoeuvring in the information environment, given the scarce number of publicly known demonstrations available for research, the contribution, if any, to these strategic functions is

[26] Inter alia The White House 2015.

[27] See the French White Paper [Livre Blanc] on Defence and National Security 2013.

[28] Such as the various National Cyber Security Strategies, see https://ccdcoe.org/cyber-security-strategy-documents.html.

[29] Netherlands National Security Strategy 2006; Netherlands Ministry of Foreign Affairs 2013.

[30] US Joint Forces Staff College/National Defense University.

[31] UK Ministry of Defence 2014, p. 4-1.

[32] South African Defence Review 2015; in a similar way: US Department of Defense 2013, pp. I-10–I-11.

[33] [In Dutch] Netherlands Ministry of Defence 2010, p. 193; Jordan et al. 2008; Netherlands Ministry of Defence 2013; Based on the strategic functions defined in the French White Papers [Livre Blanc] on Defence and National Security as of 2007.

subject of much speculation, and—luckily also—research.[34] In the meantime, the effectiveness of military cyber capabilities and manoeuvres in the information environment may be derived by learning from experiences and lessons drawn in other fields where these capabilities are more widely used. This might prove problematic as this involves quite different paradigms,[35] however, extracting (or copying) means, methods, modus operandi and concepts from other fields of application may end up illustrative.[36]

The general idea is that manoeuvring in the information environment will offer advantages and reveal capabilities for States to be used alongside or in combination with other instruments, thus supplementing kinetic military capabilities.

9.4 Doing What in the Information Environment

This notion is reflected in Fig. 9.4, where physical as well as non-physical activities and operations (Fig. 9.4: white arrows) are aimed at objects, persons, network infrastructure, virtual objects, virtual personas, psyche and audiences in order to influence via the cognitive dimension to create an effect in a target actor (Fig. 9.4: actor B). These activities or operations are directed vis-à-vis or via these intermediate entities seeking premeditated and designated effects (Fig. 9.4: red arrow), which may influence other actors or interests—in turn other actors may respond. Despite Western military culture and its prevalence to physical action,[37] NATO's information operations doctrine acknowledges that "the cognitive/psychological domain is the most important as it consists of cognition and emotions, which affect an individual's decision-making".[38]

The activities specifically tailored to affecting the cognitive dimension are known as "information operations" and/or "information activities". From a NATO perspective information operations (InfoOps or IO) are "a staff function to analyse, plan, assess and integrate information activities"; information activities being "actions designed to affect information or information systems".[39] The US perspective to information operations is that they entail the "integrated employment, during military operations, of information-related capabilities in concert with other lines of

[34] For a constructive attempt: Siedler 2016, pp. 23–36; Also: Healey 2013; Raitsasalo 2015; Geers 2010, pp. 298–303; Libicki 2009; Lindsay 2013, pp. 365–404; Kuehl 2009.

[35] Ducheine 2015a, pp. 211–232.

[36] See, for instance, Ducheine and Van Haaster 2014, pp. 303–327; Learning from intelligence services, law enforcement agencies, criminal organisations, marketing, psychology and social behaviour studies, and of course from international relations and strategic studies, offers a temporary indication of the utility of operations and activities in the information environment.

[37] See Ducheine 2015b, pp. 201–220.

[38] North Atlantic Treaty Organization 2015, p. 1-2; See also Manoeuver warfare doctrine and the Air/Land battle concept.

[39] North Atlantic Treaty Organization 2015, p. 1-5.

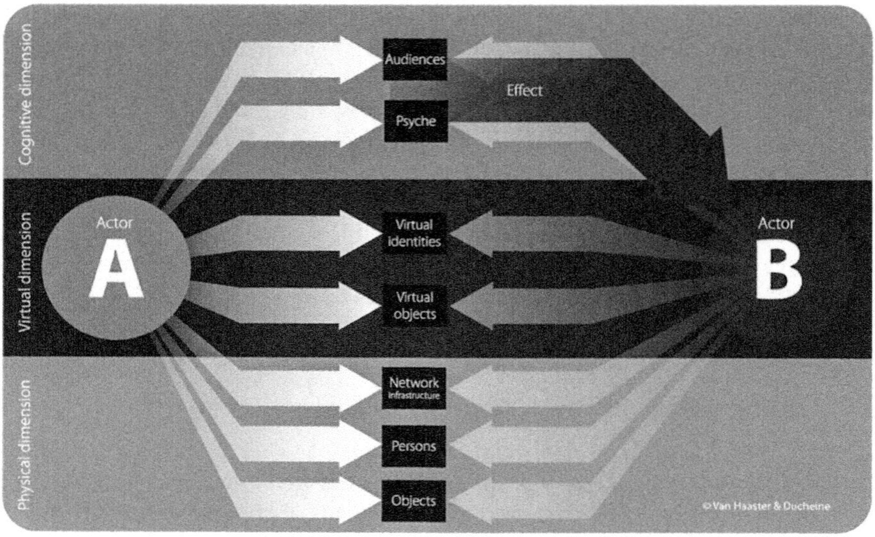

Figure 9.4 Creating effects (for a more detailed description of these entities, see Ducheine and Van Haaster 2014, pp. 303–327; UK Development, Concepts and Doctrine Centre 2010, p. 2-9; U. S. Army 2010, pp. 8–9.)

operation".[40] The UK has a similar terminological approach to information operations, namely as a "coordinated military activity undertaken to affect decision-makers".[41] As NATO's information activities are labelled information operations in UK and US doctrine, it depends per State what the scope of information operations and/or activities is. The following section will discuss the requirements needed to actually do so, for manoeuvring in the information environment.

9.5 Requirements for Manoeuvring in the Information Environment

Before being able to manoeuvre in the information environment, States as well as forces require the capability to project military power into this environment. One of the ways of thinking about capability is through the concept of power as it is used in military doctrine, namely: fighting power. Fighting power consists of three components, the conceptual, physical and moral.[42]

---

[40] The Joint Chiefs of Staff 2014, p. GL-3.

[41] UK Joint Doctrine & Concepts Centre 2002, p. 1-2; UK Development, Concepts and Doctrine Centre 2007, p. 1-2.

[42] British Army 2010, p. 2-2; Netherlands Ministry of Defence 2013, p. 66.

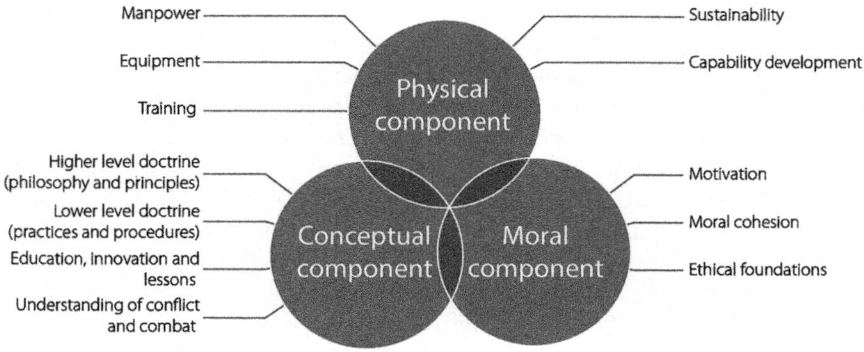

Figure 9.5 (Fighting) power components (based on Fig. 2.1 in British Army 2010, p. 2-2; Netherlands Ministry of Defence 2013, p. 66.)

Although a Western view of fighting power, it also reflects how other, non-Western, actors reflect on military capabilities.[43] The conceptual component encompasses "the ideas behind how to operate and fight". The moral component comprises "people's will and ability to get people to operate and fight". And the physical component entails the "means to operate and fight".[44] The aspects of these components required for projecting fighting power are depicted in Fig. 9.5. In the present authors' opinion, these three components apply to other instruments of power as well. Power of whatever kind requires means, a proper concept as well as the will and dedication to use those means accordingly. The next subsections will discuss the ramifications on the components of fighting power when organising for manoeuvring in the information environment.

9.5.1 Conceptual Component

Apart from the proper mind-set as an essential part of the moral component and the capacities as part of the 'physical' component just described, capacities wilfully employed also require ideas, concepts, to generate credible capabilities. The conceptual component comprises, amongst other, higher- and lower-level doctrine; education, innovation and lessons; and understanding of conflict and combat.[45] It also encompasses strategies on the use of power and the pursuance of national interest.

---

[43] See, for instance, Ministry of Defence of the Russian Federation 2003, pp. 69–82; Security Council of the Russian Federation 2010, pp. 40–44; Li 1999, pp. 146–174.

[44] British Army 2010, p. 2-2.

[45] British Army 2010, pp. 2-3 to 2-10.

States and non-state actors have promulgated grand strategies or issued ad hoc strategic guidance, expressing strategic or operational concepts of thinking regarding the use of instruments of power. These ideas often apply to regular geopolitical settings to the use of power, whether in or outside crises and war. Concepts can be found on the political-strategic, military-strategic, or on the operational and tactical military level of deployments. The idea is that the purpose of the use of instruments of power, including the military, will be guided, explicitly or implied, by a stated or deducted concept on the use of power.

It is evident that the synchronised and harmonised wielding of the instruments of power to create synergy is more complicated than using them separately. Despite criticism related to the effectiveness this comprehensive approach,[46] States already employ instruments in this manner; or, when confronted with a comprehensive (and or hybrid) application of adversaries' power, are forced to or starting to adhere a similar approach. This goes for the constituent elements of the military as well. Full spectrum and joint operations, using all dimensions of the Operational Environment are once more becoming more accepted and used.[47] Increasingly, military targeting procedures developed for classical physical engagement, are being used for operations in the information environment.[48]

Despite this progress, doctrine regarding the information environment is sometimes immature or even non-existent.[49] It is telling that NATO is trying to close the conceptual gap in this respect by pushing hard for its first allied joint publication on cyber operations. The Tallinn Manual on the international law of cyber warfare, supported by NATO's Cooperative Cyber Defence Centre of Excellence (CCDCOE),[50] was the first to shed light on the legal ramifications of the deployment of cyber capabilities in war. The follow-up process generated a second manual applicable to cyber operations in peacetime too.[51] In addition, numerous reports on strategic communication, information operations and hybrid threats have been published by others.[52] The challenge will be to orchestrate and deconflict the overlaps in doctrine and concepts.

Yet, despite familiarities and old wine in new bottles, those planning and executing operations, military or otherwise, will be faced with biases and preferences. These can partly be addressed in the conceptual component, for instance through training and education. However, organisational culture will have a say too, as will rivals or even opponents that are the subject (targets) of operations, of the use of

[46] See for instance Clingendael Conflict Research Unit 2012, pp. 2–3.

[47] For a Dutch example of this process, see National Coordinator for Security and Counterterrorism, 2016.

[48] Ducheine and Van Haaster 2014, pp. 303–328; Ducheine 2015b, pp. 201–220; Pratzner 2015, pp. 78–95.

[49] See for example North Atlantic Treaty Organization 2015; The Joint Chiefs of Staff 2014.

[50] Schmitt 2013; also Ducheine 2015c, pp. 456–475.

[51] Schmitt 2017; also Ziolkowski 2013; for an early Dutch effort, Ducheine 2012, p. 116.

[52] See for instance Echevarria II 2016.

power. In addition, organisations and people have their conceptual comfort zones. Of course this applies to both sides of the model just presented (Fig. 9.4: Actor A and B).

Preferences, when properly exploited, may turn into weaknesses. This applies to issues of overlapping competences, seems, or lacunas too. Seems and overlapping competences will cause friction, therefore coordination is required. In addition to these doctrinal and organisational issues, legal competences, including its seems, overlaps and voids, may also offer opportunities to be exploited. In this respect, one of the characteristics of a holistic approach is demonstrated by the Russian Federation. Whilst manoeuvring below the threshold of armed conflict as stated in the UN Charter, Russia embraces 'bespredel' (denial), 'maskirova', or proxies and non-military instruments, exploiting these seems, overlaps and gaps, to avoid responses from actors targeted and affected.

9.5.2 Physical Component

The physical component comprises manpower, equipment, training, sustainability and capability development. Although the concept of fighting power is sufficiently general to be equally applicable to all operations, both traditional and information, the scope of these elements will change.

Manpower comprises the people engaged in information operations. As with most operations, various roles or tasks exist, depending on the purpose of operations. Given the numerous types of information operations,[53] the manpower required is increasingly diverse. Depending on the goal of an information operation, manpower could include regular soldiers (e.g. for physical destruction or deception), cyber operators (e.g. developers, hackers, forensics), press affairs officers, legal advisors, linguists, cultural advisors, financial advisors, political advisors, psychologists, analysts, etcetera. In addition, many other specialties such social-media experts, web designers, content developers, web-care, marketers will be needed to manoeuvre in the information environment.

This non-kinetic expertise could be present in peacetime military organisations, although it will take some effort to orchestrate and reinforce the specialities before being able to create effects in the information environment. For instance, the social media experts, content developers, public affairs officers, psychologists and marketers reside in the non-operational, often civilian, parts of the defence organisation,

[53] NATO and US doctrine earmark the following operations to be capabilities to be employed under the ambit of information operations: Psychological operations; presence, posture and profile; operations security; information security/information assurance; deception; electronic warfare/joint electromagnetic spectrum operations; physical destruction; key-leader engagement; computer network operations/cyberspace operations; civil-military cooperation/civil-military operations; strategic communication; joint interagency coordination; public affair; space operations; military information support operations and intelligence.

whilst the psychological operations teams and cyber operators are located in the operational, military parts of the military. Currently, in the Dutch context, a single authority that can wield these capacities although being envisioned, is still lacking. Thus, the issue is less about the manpower; instead, it is about organising the manpower into an organisation that can be orchestrated.

On the equipment part, all essential parts can be found in the physical network layer of the information environment, ICT-infrastructure and hardware, as well as in the virtual layer. The latter comprising virtual persona (digital accounts), ICT-software (operating systems, firmware, applications), protocol and scripts, as well as content (information and data). Information and data may be available in the physical dimension or in the cognitive and informational/virtual dimension. Hence, the classical physical component comprises truly physical as well as virtual elements such as software, data, virtual identities. Seemingly hard to reconcile, this concept is demonstrated by the value of datasets and information, often coined the 'new gold' in modern information societies.[54]

9.5.3 Moral Component

The moral component, which epitomises the moral qualities needed to conduct (military) operations, comprises three important elements: "ethical foundations, moral cohesion and motivation".[55] The elements are equally, if not more important when manoeuvring in the information environment important. The pinnacle, however, is adopting the mind-set and willingness of affecting actors in situations of war and peace. This is even more important as soon as it becomes evident that States are being targeted or affected by all instruments of power by other actors and States 24/7, not only during conflict and rivalry, but in this respect predominantly in peacetime as well.

Manoeuvring in the information environment blurs the lines between armed conflict and peace in a geographical and temporal sense. As states and non-state actors can, enabled by information technologies, increasingly and more easily affect others from within their borders, the geographical delimitation of in and outside theatre will vanish, that is from a mind-set perspective and not from the legal perspective. Also militaries often still talk of and use 'lines of departures', marking the temporal division between pre-operation and the start of an operation (on 'D-day' at 'H-hour').[56] Manoeuvring in the information environment requires a different attitude, which may involve "stop [thinking] in terms of D-days and lines of departure at all".[57] Operations aimed at shaping the environment should not be preserved for the small timespan before an operation; instead shaping should be

[54] Smit-Kroes 2011; Singh 2013.

[55] British Army 2010, p. 2-11.

[56] North Atlantic Treaty Organization 2014, p. 2-D-5.

[57] Van Haaster and Roorda 2016, p. 185.

done continuously, taking into account all instruments of power in a synchronised and harmonised manner: before, during and after conflict.[58] These activities should involve aspects such as "developing situational understanding; developing options to influence audiences; persuading and empowering other actors to make choices that are advantageous; and conducting limited offensive actions in order to keep adversaries off-balance."[59] Governments wishing to employ their military instrument of power in the information environment should prepare itself and its military for doing so. In addition, on the strategic level, society and government should be prepared to actively promote their strategic and vital interests comprehensively throughout the whole spectrum of peace and war.

At the same time, governments, society and militaries should be resilient too and beware being targeted 24/7.[60] Traditionally military resilience is established with security training aimed at creating understanding and awareness.[61] These measures are, however, primarily focused on military personnel and facilities and less on their surroundings (family, friends and other connections). As social-media permeates every aspect of real and virtual life and society, adversaries can easily uncover the network surrounding individual soldiers and hurt them where they can be hurt most.[62] The so-called 'attack surface' of soldiers extends as they can be influenced more easily on different fronts and dimensions: physically, cognitively and virtually. Traditionally, military personnel primarily faced physical peril on a battlefield, whilst within the paradigm of information manoeuvre the will face many *Dantean* risks on moral, conceptual and physical levels, not only personally, but in their social environments such as their family as well.[63] Although this already happened on a small scale in modern warfare,[64] recent developments suggest an increase in these types of activities.

9.6 Examples

This section will discuss two recent examples of manoeuvres in the information environment, in doing so this section will reflect on the model forwarded in Fig. 9.2. It will discuss the context of the case, the purpose and ways of trying to achieve an effect and whether or not it was successful in doing so. The first example, the U.S. election information campaign of 2016, was selected for its distinct strategic purpose

---

[58] See Chap. 8 by Han Bouwmeester in this volume.

[59] British Army 2010, p. 5-21.

[60] See for example National Cyber Security Centre 2016.

[61] Often referred to as operations (OPSEC), personal (PERSEC), communications (COMSEC) and information security (INFOSEC).

[62] Van Haaster and Roorda 2016, p. 183.

[63] Alighieri 1982.

[64] Hughes 2015; Herridge 2014.

and envelops not only cyber means and methods but also other information activities. The second is an operation with a disruptive physical effect induced by social engineering and cyber activities, namely the Black Energy case.

9.6.1 Information Campaign 'U.S. Election Hacks' (2016)

Before, during and in the wake of the 2016 U.S. elections there were many allegations regarding Russian interference in the democratic process, amongst others from the U.S. Intelligence Community (IC). Following his election, the president-elect's sceptical approach as to the IC's allegations resulted in the Office of the Director of National Intelligence (ODNI) releasing a declassified report on Russian activities during the elections, which offers a glance into the allegedly Russian activities during the election.[65] Using open-source information, other sources arrived at the same conclusion: It is very likely that Russian agencies have interfered in the U.S. elections.[66]

The "influence campaign" encompassed the following goals: "[…] to undermine public faith in the US democratic process, denigrate Secretary Clinton, and harm her electability and potential presidency."[67] The campaign was carefully staged; starting with cyber operations aimed at collecting intelligence in March 2016, directed at, amongst other, "both major US political parties."[68] Having obtained the requisite intelligence, selected content was publically disclosed in data dumps and "exclusives to media outlets" via the Romanian hacker 'Guccifer 2.0', DCLeaks.com and WikiLeaks.[69] The compromising material, also known as "*Kompromat*",[70] was used to target various individuals involved in the election, for example the Democratic National Committee's (DNC) chair Debbie Wasserman Schultz,[71] the chairman of the 2016 Hillary Clinton presidential campaign John Podesta,[72] and most prominently Hilary Clinton.[73] Although the definitive answer as to the impact of the campaign on the result remains to be seen, some have called this campaign—somewhat alarmist—the "political equivalent of 9/11".[74]

---

[65] U.S. Office of the Director of National Intelligence 2017.

[66] See for instance Rid 2016; Gilsinan and Calamur 2017; Taub 2017.

[67] U.S. Office of the Director of National Intelligence 2017, p. 1.

[68] U.S. Office of the Director of National Intelligence 2017, p. 2.

[69] U.S. Office of the Director of National Intelligence 2017, pp. 2–3; Rid 2016.

[70] Rid 2016.

[71] Martin and Rappeport 2016; Shear and Rosenborg 2016.

[72] WikiLeaks 2016a.

[73] WikiLeaks 2016b.

[74] Morell and Kelly 2016.

The information campaign aimed at the 2016 U.S. election is an example of how states can contribute to the strategic function 'prevention' (see para 9.3), which encompasses "active steps intended to prevent a threat occurring to [State] interests".[75] It shows how States can manoeuvre in the information environment to achieve a strategic goal, namely: a preferential outcome at the political-strategic level. The means to influence the target audience, that is, the U.S. electorate, are diverse. As with all activities, the prime target is to influence psyche, in this case the perception as to which candidate is qualified for presidential office. Influencing psyche in this case involved using and exploiting virtual personas (e.g. using Guccifer 2.0, DCLeaks, and trolls whilst exploiting the Clinton, Podesta and Wasserman Schultz mail accounts) and virtual objects (e.g. exploiting databases and mail servers whilst using websites and blogs to disclose the finds). As to the effectiveness of the campaign there is much yet unresolved, apart from a general state of confusion as to what has actually occurred before, during and after the elections. Considering that the ODNI has deemed undermining public faith in the US democratic process and denigrating Secretary Clinton to be the goals of the campaign, for now it seems that the influence campaign was successful. As such, it is a well-documented example of manoeuvring in the information environment to achieve strategic ends.

9.6.2 'BlackEnergy' Operation (2015)

Whereas the U.S. elections case was primarily aimed at discrediting persons via virtual personas and in the preparatory stages included 'hard' cyber operations, the Black Energy example could be characterised as a 'hard' cyber operation aimed at affecting software and hardware. This sub-section will briefly discuss the context of the case, the purpose, the means involved and the effectiveness.

BlackEnergy is the name of malware previously used by criminals (from 2007 on) and later by advanced persistent threat (APT) groups (2014).[76] Currently BlackEnergy has become synonymous with the power outage in Ukraine December 23, 2015. Some use BlackEnergy to refer to the APT group using BlackEnergy whilst generally it is used to refer to the specific strand of malware and/or the Ukraine outage.[77] The BlackEnergy operation of 2015 was significant as it heralded "ICS [industrial control system] attacks going mainstream".[78]

---

[75] Netherlands Ministry of Defence 2010, p. 15.

[76] F-Secure 2014; Kaspersky 2016a.

[77] Kaspersky 2016a.

[78] Van Haaster et al. 2016, p. 72.

The operation was conducted amidst of geo-political tensions between Russia and Ukraine regarding the Crimean annexation (2014). Within that geopolitical context there are series of events that could be considered relevant for this operation,[79] however, there is one event that can be considered a catalyst for conducting this operation: the Crimean blackout November 21 (2015). The Crimean blackout was caused by the destruction/sabotage of "four electricity transmission towers" on Ukrainian territory resulting in a blackout in Crimea.[80] There is speculation "that the subsequent blackouts [by the BlackEnergy operations] in Ukraine were retaliation for the attack on the [towers]"—albeit not the original motivation as the operation started six months before the Crimean blackout.[81]

The operation consisted of a preparation stage of obtaining access to networks via spear-phishing from 2014 to mid-2015; the mails contained malicious Office files (first Excel, later Power Point and Word).[82] Using macros in Office, an "old-school method from the 1990s",[83] the 2015 BlackEnergy operation targeted railway, mining, media and power sectors in Ukraine[84] and ICS/SCADA and energy companies worldwide.[85] After rising tensions in the region, from mid-2015 the operation has geared towards obtaining control over the regional *Prykarpattya Oblenergo* and *Kyivoblenergo* energy providers in Ukraine.[86]

After gaining access to the corporate networks, the attackers "conducted extensive reconnaissance, exploring and mapping the networks" and, as a textbook example, targeted the Windows Domain Controllers.[87] By doing so the attackers "acquired legitimate credentials" that facilitated remote access via a virtual private network (VPN) to industrial control systems—including, amongst other, the electrical breakers.[88] After laying the groundwork, the attackers took additional steps in order to extend the blackout period for maximum effect. In order to thwart the recovery operation the attackers had overwritten firmware on "serial-to-Ethernet converters"—the link between logical (SCADA) and physical control systems—in order to "prevent [company] operators from sending remote commands to re-close breakers once a blackout occurred."[89] Besides that, they "reconfigured the

---

[79] See Zetter 2017. She lists other causes such as the Ukrainian parliament considering privatisation of privately owned power companies in Ukraine of which some are owned by powerful Russian oligarchs.

[80] BBC 2015; Radio Free Europe/Radio Liberty 2015; Russia Today 2015.

[81] Zetter 2017.

[82] Cys-Centrum 2016; Kaspersky 2016b.

[83] Zetter 2017.

[84] Trend Micro 2016.

[85] Kaspersky 2016b.

[86] Zetter 2017.

[87] Van Haaster et al. 2016, p. 62.

[88] U.S. Department of Homeland Security ICS-CERT 2016.

[89] Zetter 2017.

uninterruptible power supply [UPS]" for the companies control centres, resulting in the company operators manning the control centre would also lack power in the event of an outage.[90]

At around 15:30 on December 23 the attackers "launched a telephone denial-of-service attack [from Moscow] against customer call centres to prevent customers from calling in to report the outage", "entered the SCADA networks through the hijacked VPNs and sent commands to disable the UPS [and] began to open breakers".[91] After completion the attackers used KillDisk, "a disk wiping malware",[92] to erase files and corrupt the master boot record, "rendering systems inoperable" and deleting tracks.[93] The attack resulted in a population of "around 1.4 million [...] left without electricity for a few hours".[94]

The BlackEnergy operation of 2015 could be seen as an example of how States can manoeuvre in the information environment. Assuming Russia's involvement, focussing on the virtual dimension, the author-State could have sought a strategic goal, namely: retaliate against a target-State in order to dissuade it from future activities harming the interests of the author-State. This is an example of the strategic function 'intervention' as the author-State aims to "enforce a change in the behaviour of one or more parties that threaten the interests of the [State]".[95] In this case, although speculative, Russia responding in kind to Ukraine for allowing the sabotage of Crimea's power supplies. By exploiting virtual personas (mail accounts targeted in the spear-phishing campaign) the attackers gained access to the network and were able to escalate access to the domain controller, control systems and SCADA systems (virtual objects) and gain control over physical network infrastructure (serial-to-Ethernet converters) and 'regular' physical energy infrastructure (objects). By doing so the author-State creates a blackout, aimed to send the message: "you think you can take away the power [in Crimea]? Well I can take away the power from you."[96]

9.7 Conclusion

This chapter has sought to answer the question: "How to generate effects in or through the information environment and therefore, how to manoeuvre in this information environment?" First this chapter has discussed the information environment construct. This environment, a part of the larger operational environment

[90] Zetter 2017.

[91] Zetter 2017.

[92] Trend Micro 2016.

[93] U.S. Department of Homeland Security ICS-CERT 2016.

[94] ESET 2016.

[95] Netherlands Ministry of Defence 2010, p. 15.

[96] Zetter 2017.

consist of three dimensions, the cognitive, virtual and physical. These dimensions in turn host certain layers and these hold targetable entities. Departing from that conceptual model of the information environment, this chapter has turned to how to manoeuvre in that environment. First the 'why' was discussed, considering the potential there are many incentives for a State to use this environment. A State can create effects in the information environment by wielding the instruments at its disposal (diplomacy, informational, military and economic) for various purposes, for instance strategic functions (anticipation, prevention, deterrence, protection, intervention, stabilization and normalization). In order to be able to do so, it must organise and equip itself for manoeuvring in the information environment, this chapter has used the components of fighting power in order to highlight critical elements required for creating effects. Lastly, this chapter has highlighted the U.S. Election Information Campaign (2016) and the BlackEnergy operation (2015) as examples of manoeuvring in the information environment in order to achieve strategic goals. These cases demonstrate that the use of and manoeuvring in the information environment might indeed contribute to achieving effects and strategic goals.

References

Abadi M, Andersen DG (2016) Learning to Protect Communications with Adversarial Neural Cryptography. https://arxiv.org/pdf/1610.06918.pdf Accessed 4 March 2017

Alighieri D (Dante) (1982) The Inferno, translated by Ciardi J. Signet Classics, London

Anthony S (2014) The First Human Brain-to-Brain Interface has been Created in the Future, Will we all be Linked Telepathically? (3 September 2014) http://www.extremetech.com/extreme/188883-the-first-human-brain-to-brain-interface-has-been-created-in-the-future-will-we-all-be-linked-telepathically Accessed 14 January 2015

BBC (2016) Crimea Power Blackout: Russia Accuses Ukraine of Sabotage (30 Nov 2015) http://www.bbc.com/news/world-europe-34967093 Accessed 10 February 2017

Bemis J, Morgan G (2008) Exposing the Information Domain Myth. Air & Space Power Journal 22(3):19–21

British Army (2010) Army Doctrine Publication: Operations. Development, Concepts and Doctrine Centre, Shrivenham

Clingendael Conflict Research Unit (2012) CRU Policy Brief: How to make the Comprehensive Approach Work. Clingendael, The Hague

Cys-Centrum (2016) Киберугроза Blackenergy2/3. История Атак На Критическую Ит Инфраструктуру Украины (Cys-Centrum February 10) http://www.cys-centrum.com/ru/news/black_energy_2_3 Accessed 4 March 2017

Dekkers PAP, Benten C, Dijkstra H (2016) Advisory Report: Information as Weapon, Vector and Target. Ministry of Defence, The Hague

Ducheine P, Osinga F, Soeters J (2012) Cyber warfare: Critical perspectives. TMC Asser Press, The Hague

Ducheine P (2015a) Non-Kinetic Capabilities: Complementing the Kinetic Prevalence to Targeting. In: Ducheine P, Schmitt M, Osinga F (eds) Targeting: Challenges of Modern Warfare. TMC Asser Press, The Hague, pp 201–220

Ducheine P (2015b) The Notion of Cyber Operations. In: Tsagourias N, Buchan R (eds) Research Handbook on International Law and Cyber Space. Edward Elgar Publishing, Cheltenham, pp 211–232

Ducheine P (2015c) Military Cyber Operations. In: Gill TD, Fleck D (eds) Handbook of the International Law of Military Operations, 2nd edn. Oxford University Press, Oxford, pp 456–475

Ducheine P, Van Haaster J (2013) Cyber-Operaties en Militair Vermogen. Militaire Spectator 182 (9):368–387

Ducheine P, Van Haaster J (2014) Fighting Power, Targeting and Cyber Operations. In: Brangetti P, Maybaum M, Stinissen J (eds) Proceedings of the 6th International Conference on Cyber Conflict, NATO CCD COE, Tallinn, pp 303–328

Fallon M (2016) Secretary of State for Defence Michael Fallon MP at The Second International Cyber Symposium (21 October 2016) http://www.youtube.com/watch?v=6TqQTVkYHYc Accessed 16 February 2017

Farrell T, Osinga F, Russel JA (2013) Military Adaptation in Afghanistan. Stanford University Press, Stanford CA

French White Paper [Livre Blanc] on Defence and National Security (2013) http://www.defense.gouv.fr/content/download/215253/2394121/file/White%20paper%20on%20defense%20%202013.pdf Accessed 4 March 2017

F-Secure (2014) BlackEnergy & Quedagh: The Convergence of Crimeware and APT Attacks. F-Secure, Helsinki https://www.f-secure.com/documents/996508/1030745/blackenergy_whitepaper.pdf Accessed 4 March 2017

Geers K (2010) The Challenge of Cyber Attack Deterrence. Computer Law & Security Review 26 (3):298–303

Giles K (2016) Russia's New Tools for Confronting the West: Continuity and Innovation in Moscow's Exercise of Power. Chatham House, London

Gilsinan K, Calamur K (2017) Did Putin Direct Russian Hacking? And Other Big Questions. The Atlantic (6 January 2017) http://www.theatlantic.com/international/archive/2017/01/russian-hacking-trump/510689/ Accessed 6 January 6 2017

Healey J (ed) (2013) A Fierce Domain: Conflict in Cyberspace, 1986 to 2012. Cyber Conflict Studies Association

Heickerö R (2010) Emerging Cyber Threats and Russian Views on Information Warfare and Information Operations. FOI, Stockholm

Herrick D (2016) The Social Side of 'Cyber Power'? Social Media and Cyber Operations. In: Pissanidis N, Rõigas H, Veenendaal M (eds) Proceedings of the 8th International Conference on Cyber Conflict Cyber Power. NATO CCD COE Publications, Tallinn, pp 99–112

Herridge C (2014) Army Warns US Military Personnel on ISIS Threat to Family Members. Fox News (2 October 2014) http://www.foxnews.com/politics/2014/10/02/army-warns-us-military-personnel-on-isis-threat-to-family-members.html Accessed 15 November 2016

Hewitt J (2014) New Brain Implant Tech from Blackrock is Making 'Mind Over Matter' a Reality. extremetech.com (18 December 2014) http://www.extremetech.com/extreme/194935-new-brain-implant-tech-from-blackrock-is-making-mind-over-matter-a-reality Accessed 14 January 2015

Hughes C (2015) Jihadis Threaten to Slaughter British Soldiers' Wives and Families as Police Issue Social Media Warning. Mirror (31 July 2015) http://www.mirror.co.uk/news/uk-news/jihadis-threaten-slaughter-british-soldiers-6173859 Accessed 15 November 2016

Jordan D et al (2008) Understanding Modern Warfare. Cambridge University Press, Cambridge

Kaspersky (2016a) BlackEnergy APT Attacks in Ukraine Employ Spearphishing with Word Documents. Kaspersky Lab (28 January 2016) https://securelist.com/blog/research/73440/blackenergy-apt-attacks-in-ukraine-employ-spearphishing-with-word-documents/ Accessed 10 February 2017

Kaspersky (2016b) Newly Discovered BlackEnergy Spear-Phishing Campaign Targets Ukrainian Entities. Kaspersky Lab (28 January 2016) http://www.usa.kaspersky.com/about-us/press-center/press-releases/2016/newly-discovered-blackenergy-spear-phishing-campaign-targets-uk Accessed 10 February 2017

Kragh M, Åsberg S (2017) Russia's strategy for influence through public diplomacy and active measures: the Swedish case. Journal of Strategic Studies (5 January 2017) http://dx.doi.org/10.1080/01402390.2016.1273830. Accessed 4 March 2017

Kuehl D (2009) From cyberspace to cyberpower: Defining the problem. In: Kramer F, Starr S, Wentz L (eds) Cyberpower and National Security. Potomac Books, Lincoln

Lee S (2014) China's 'Three Warfares': Origins, Applications, and Organizations. Journal of Strategic Studies 37(2):198–221. http://dx.doi.org/10.1080/01402390.2013.870071

Li N (1999) The PLA's Evolving Campaign Doctrine and Strategies. In: Mulvenon JC, Yang RH (eds) The People's Liberation Army in the Information Age. RAND, Santa Monica, pp 146–174

Libicki M (2009) Cyberdeterrence and Cyberwar. RAND, Santa Monica CA

Lindsay JR (2013) Stuxnet and the Limits of Cyber Warfare. Security Studies, 22(3):365–404

Lipovsky R, Cherepanov A (2016) BlackEnergy Trojan Strikes again: Attack Ukrainian Electric Power Industry. ESET (4 January 2016) http://www.welivesecurity.com/2016/01/04/blackenergy-trojan-strikes-again-attacks-ukrainian-electric-power-industry/ Accessed 10 February 2017

Martin J, Rappeport A (2016) Debbie Wasserman Schultz to Resign D.N.C. Post. The New York Times (25 July 2016) http://www.nytimes.com/2016/07/25/us/politics/debbie-wasserman-schultz-dnc-wikileaks-emails.html Accessed 8 February 2017

Military.com (2016) Cyber Command Gets First Wartime Assignment in Fight ISIS (5 April 2016) http://www.military.com/daily-news/2016/04/05/cyber-command-gets-first-wartime-assignment-in-fight-isis.html Accessed 16 February 2017

Ministry of Defence of the Russian Federation (2003) Aktual'nyye Zadachi Razvitiya Vooruzhënnykh Sil Rossiyskoy Federatsii [The priority tasks of the development of the Armed Forces of the Russian Federation] Ministry of Defence of the Russian Federation, Moscow

Morell M, Kelly S (2016) Former CIA Acting Director Michael Morell: "This is the Political Equivalent of 9/11". The Cipher Brief (11 December 2016) http://www.thecipherbrief.com/article/exclusive/fmr-cia-acting-dir-michael-morell-political-equivalent-911-1091 Accessed 8 February 2017

National Coordinator for Security and Counterterrorism (2016) Magazine Nationale Veiligheid en Crisisbeheersing, 14(5) https://www.nctv.nl/binaries/MagazineNationaleVeiligheidenCrisisbeheersing20165_7interactief_tcm31-234692.pdf Accessed 4 March 2017

National Cyber Security Centre (2016) Cyber Security Assessment Netherlands. https://www.ncsc.nl/english/current-topics/Cyber+Security+Assessment+Netherlands/cyber-security-assessment-netherlands-2016.html Accessed 4 March 2017

Netherlands Ministry of Defence (2010). Future Policy Survey: Summary and Conclusions. Ministry of Defence, The Hague

Netherlands Ministry of Defence (2013) Netherlands Defence Doctrine. Ministerie van Defensie, The Hague

Netherlands Ministry of Foreign Affairs (2013) International Security Strategy - A secure Netherlands in a secure world. http://www.government.nl/latest/news/2013/06/21/a-secure-netherlands-in-a-secure-world [Kamerstukken II 2012-13, 33 694, no. 1, Internationale Veiligheidsstrategie]

Netherlands National Security Strategy (2006) [Strategie nationale veiligheid]. Parliamentary Papers II [Kamerstukken II] 2006-07, 30 821, no. 2

North Atlantic Treaty Organization (2009) Allied Joint Publication 3.10: Allied Joint Doctrine for Information Operations. North Atlantic Treaty Organisation, Brussels

North Atlantic Treaty Organization (2014) Allied Administrative Publication 06: NATO Glossary of Terms and Definitions (English and French). NATO Standardization Agency, Brussels

North Atlantic Treaty Organization (2015) Allied Joint Doctrine for Information Operations (Edition A Version 1 ed). NATO, Brussels

Office of the Chief of Naval Operations (1995) OPNAV Instruction 3430.26: Implementing Instruction for Information Warfare/Command and Control Warfare (IW/C2 W)

Pratzner P (2015) The Current Targeting Process. In: Ducheine P, Schmitt M, Osinga F (eds) Targeting: Challenges of Modern Warfare. TMC Asser Press, The Hague, pp 77–97

Radio Free Europe/Radio Liberty (2016) Pylon 'Blown Up' in Ukraine, Causing Crimea's Blackout. Radio Free Europe/Radio Liberty (22 November 2015) http://www.rferl.org/a/ukraine-crimea-black-out-state-of-emergency/27379758.html Accessed 10 February 2017

Ragalado A (2014) Military Funds Brain-Computer Interfaces to Control Feelings. MIT Technology Review (29 May 2014) http://www.technologyreview.com/news/527561/military-funds-brain-computer-interfaces-to-control-feelings/ Accessed 14 January 2015

Raitasalo J (2015) Hybrid Warfare: Where's the Beef? War on the Rocks (23 April 2015) http://warontherocks.com/2015/04/hybrid-warfare-wheres-the-beef Accessed 4 March 2017

Rayner A (2016) Can Google's Deep Dream Become an Art Machine? The Guardian (28 March 2016) http://www.theguardian.com/artanddesign/2016/mar/28/google-deep-dream-art Accessed 9 November 2016

Rid T (2016) How Russia Pulled Off the Biggest Election Hack in U.S. History. Esquire (20 October 2016) http://www.esquire.com/news-politics/a49791/russian-dnc-emails-hacked/ Accessed 8 February 2017

Rid T, Hecker M (2009) War 2.0: Irregular Warfare in the Information Age. Praeger Security International, London

Russia Today (2016) State of Emergency, Blackout in Russia's Crimea After Transmission Towers in Ukraine Blown Up. http://www.rt.com/news/323012-crimea-blackout-lines-blown-up/ Accessed 10 February 2017

Schmitt MN (2013) Tallinn Manual on the International Law Applicable to Cyber Warfare. Cambridge University Press, Cambridge

Schmitt MN (2017) Tallinn Manual 2.0 on the International Law Applicable to Cyber Operations. Cambridge University Press, Cambridge

Security Council of the Russian Federation (2010) Military Doctrine Russian Federation [ОЕННАЯ ДОКТРИНА РОССИЙСКОЙ ФЕДЕРАЦИИ]. Security Council of the Russian Federation, Moscow

Sellhorst AJC (2016) Russia's Perception Warfare - The development of Gerasimov's doctrine in Estonia and Georgia and its application in Ukraine. Militaire Spectator 185(4)148–164, http://www.militairespectator.nl/thema/strategie-operaties/artikel/russias-perception-warfare Accessed 4 March 2017

Shear MD, Rosenberg M (2016) Released Emails Suggest the D.N.C. Derided the Sanders Campaign. The New York Times (3 July 2016) http://www.nytimes.com/2016/07/23/us/politics/dnc-emails-sanders-clinton.html Accessed 8 February 2017

Siedler RE (2016) Hard Power in Cyberspace: CNA as a Political Means. In: Pissanidis N, Rõigas H, Veenendaal M (eds) Proceedings of the 8th International Conference on Cyber Conflict Cyber Power. NATO CCD COE Publications, Tallinn, pp 23–36

Singh A (2013) Is Big Data the New Black Gold? Wired.com (undated) https://www.wired.com/insights/2013/02/is-big-data-the-new-black-gold/ Accessed 1 March 2017

Smit-Kroes N (2011) Data is the New Gold - Opening Remarks, Press Conference on Open Data Strategy, European Commission, Brussels (12 December 2011) https://ec.europa.eu/digital-single-market/en/news/data-new-gold Accessed 1 March 2017

South African Defence Review (2015) http://www.dod.mil.za/documents/defencereview/Defence%20Review%202015.pdf Accessed 4 March 2017

Suzuki T, Inaba K, Takeno J (2005) Conscious Robot that Distinguishes between Self and Others and Implements Imitation Behavior. Springer, Berlin

Taub A (2017) 'Kompromat' and the Danger of Doubt and Confusion in a Democracy. The New York Times (15 January 2017) http://www.nytimes.com/2017/01/15/world/europe/kompromat-donald-trump-russia-democracy.html Accessed 8 February 2017

UK Development, Concepts and Doctrine Centre (2007) Joint Doctrine Publication 3-45.1: Media Operations. The Development, Concepts and Doctrine Centre, Shrivenham

UK Development, Concepts and Doctrine Centre (2010) Joint Doctrine Publication 04: Understanding. Ministry of Defence, Shrivenham

UK Joint Doctrine & Concepts Centre (2002) Joint Warfare Publication 3-80: Information Operations. Ministry of Defence, Shrivenham

U.S. Army (2010) Cyberspace Operations Concept Capability Plan 2016-2028. The U.S. Training and Doctrine Command, Fort Eustis

U.S. Army Training and Doctrine Command (1995) TRADOC Pamphlet 525-69: Military Operations Concept for Information Operations. Department of the Army, Fort Monroe

U.S. Department of Defense (2003) Information Operations Roadmap. Department of Defense, Washington DC

U.S. Department of Homeland Security (2016) Alert (IR-ALERT0H-16-056-01): Cyber-Attack Against Ukrainian Critical Infrastructure. ICS-CERT. http://www.ics-cert.us-cert.gov/alerts/IR-ALERT-H-16-056-01 Accessed 10 February 2017

U.S. Joint Chiefs of Staff (1998) Joint Publication 3-13: Joint Doctrine for Information Operations. The Joint Chiefs of Staff, Washington DC

U.S. Joint Chiefs of Staff (2013) Joint Publication 3-12 (R): Cyberspace Operations. The Joint Chiefs of Staff, Washington DC

U.S. Joint Chiefs of Staff (2014) Joint Publication 3-13: Information Operations (20 November 2014). The Joint Chiefs of Staff, Washington DC

U.S. Joint Forces Staff College/ National Defense University (no date) Elements of National Power. http://jfsc.ndu.edu/Portals/72/Documents/library/Bibliographies/Elements_of_National_Power.pdf Accessed 1 March 2017

U.S. National Security Strategy (2015) The White House (February 2015) http://nssarchive.us/wp-content/uploads/2015/02/2015.pdf Accessed 1 March 2017

U.S. Office of the Director of National Intelligence (2017) Background to "Assessing Russian Activities and Intentions in Recent US Elections": The Analytic Process and Cyber Incident Attribution. U.S. Office of the Director of National Intelligence, Washington DC

Van Haaster J, Gevers R, Sprengers M (2016) Cyber Guerilla. Syngress, Boston

Van Haaster J, Roordan MP (2016) D-Day's Demise: The Impact of Hybrid Warfare on Traditional Operational Rationale. Militaire Spectator 185(4):175–185

WikiLeaks (2016a) Hillary Clinton Email Archive. http://www.wikileaks.org/clinton-emails/ Accessed 8 February 2017

WikiLeaks (2016b) The Podesta Emails. http://www.wikileaks.org/podesta-emails/ Accessed 8 February 2017

Wilhoit K (2016) KillDisk and BlackEnergy are Not just Energy Sectory Threats. Trend Micro (11 February 2016) http://www.blog.trendmicro.com/trendlabs-security-intelligence/killdisk-and-blackenergy-are-not-just-energy-sector-threats/ Accessed 10 February 2017

Wirtz JJ (2014) Cyber War and Strategic Culture: The Russian Integration of Cyber Power into Grand Strategy. In: Geers K (ed) Cyber War in Perspective: Russian Aggression against Ukraine. CCD COE, Tallinn

Zetter K (2016) Inside the Cunning, Unprecedented Hack of Ukraine's Power Grid. Wired (3 March 2016) http://www.wired.com/2016/03/inside-cunning-unprecedented-hack-ukraines-power-grid/ Accessed 8 February 2017

Ziolkowski K (2013) Peacetime regime for state activities in cyberspace: International law, international relations and diplomacy. NATO CCD COE Publication,Tallinn

Author Biographies

Brigadier-General Professor Dr. P.A.L. Ducheine is Professor for Cyber Operations and Cyber Security at the Netherlands Defence Academy and a Professor of the Law of Military Cyber Operations at the University of Amsterdam. Trained at the Netherlands Royal Military Academy, he joined the Engineer Corps in 1988 and switched to the Army Legal Service in 1998. He holds a Master of Science degree in Public Administration & Political Sciences as well as a Master degree and a Ph.D. in Law.

Lieutenant J. van Haaster, LL.M. is a Ph.D. researcher for Cyber Warfare at the Department of War Studies, Netherlands Defence Academy and the Amsterdam Centre of International Law (ACIL).

Colonel R.H. van Harskamp is a Lecturer at the Department of War Studies, Netherlands Defence Academy. He commanded 42 Mechanized Infantry Battalion during deployment in Iraq (2004), and the Task Force Uruzgan in Afghanistan (2008).

Chapter 10
Strategic Counter-Marketing to Fight ISIL

Jan van Lieshout and Robert Beeres

Abstract This chapter discusses ISIL's marketing techniques. We think that, due to its strong and successful branding and communication strategies, from a marketing perspective ISIL can be considered a leader on the market of terror produced by violent non-state actors (VNSAs). We propose ISIL can be fought by using counter-marketing strategies, as a significant part of non-kinetic warfare. Counter-marketing can be done in one of two ways. Either by preventing your opponents from adhering to their marketing strategies or by developing the best marketing strategy yourself. Any victories gained by counter-marketing will result in ISIL losing in various respects, such as supporters, revenues and combat power.

Keywords (Counter) Marketing Strategies · Marketing Mix · Violent Non-State Actors (VNSAs) · ISIL · NATO

Contents

J. van Lieshout (✉) · R. Beeres
Faculty of Military Sciences, Netherlands Defence Academy, Breda, The Netherlands
e-mail: jmm.v.lieshout@mindef.nl

R. Beeres
e-mail: RJM.Beeres@mindef.nl

P.A.L. Ducheine and F.P.B. Osinga (eds.), *Netherlands Annual Review of Military Studies 2017*, NL ARMS, DOI 10.1007/978-94-6265-189-0_10

10.1 Introduction

Considering violent non-state actors (VNSAs) as brands is nothing new under the sun.[1] Within the 'market of VNSAs', ISIL, in particular, can be said to position itself as a highly recognizable brand. From a strategic marketing perspective, the VNSA appears successful in the market of a continuous threat to different identities, cultures and religions.[2] By means of (the threat of) coercion, taking place in various arenas, forms, alliances and at multiple locations, efforts are made to counter violent non-state actors. However, as of yet, attempts to counter their specific branding strategies remain scarce. As these organizations appear experts in luring and recruiting novices to choose their brand,[3] the question could be put forward whether they could be countered by means of strategic marketing, much as in everyday business life, in which commercial brands use marketing to compete on consumer markets.

In this chapter we aim to study counter-marketing as a strategic, non-kinetic, tool to fight violent non-state actors, using ISIL as a case study. To do so, we apply commercial business marketing, both in theory and practice, to gain an insight into the key strategies and techniques ISIL uses, as well as to point out potential ways in which to fight these by using strategic counter-marketing techniques.[4]

Essentially, throughout commercial business, marketeers strive to seduce consumers to purchase their brands. To this end, branding strategies are used to create a coveted identity, thereby distinguishing their products from those of competitors. The so-called marketing mix, consisting of four instruments: product, price, place and promotion (4Ps) is applied to convince target groups in the market that the product seamlessly fits their needs, even such needs they were not yet aware of having.[5] As to *product*, this looks at the physical product or service to be bought, and includes all additional factors that determine what product will be chosen by the consumer. The second instrument pertains to *pricing*, encompassing both the money to pay for the product as well as the product-related costs for suppliers and consumers. Thirdly, *place* refers to the ways in which the product is distributed and, moreover, to the location at which the consumer may obtain the product. Lastly, *promotion* is about the organization handling communications with the market, thereby promoting sales. This fourth instrument involves all marketing communication strategies to focus consumers on the specific product and its advantages.

As to the quality of the product that is being branded, this does not have to actually surpass products offered by competitors. Branding strategies are about making promises regarding the fit between one specific brand and the target group's

---

[1] Archetti 2015, p. 50; Ligon et al. 2015, p. 27.

[2] Farwell 2014, p. 49.

[3] Farwell 2014, p. 50; Ingram 2015, p. 730.

[4] See also Helmus et al. 2007.

[5] McCarthy 1964; Floor and Van Raaij 2010, p. 11; Verhage 2013, p. 9.

situation, attitudes, convictions, desires and needs. The deeper such a connection is felt, the more desirable the product will become.

Based on the above, it is to be expected that consumers wavering between products will select the one that, on account of how it is being marketed, they believe they covet most. If the product or brand performs as expected or even better, purchases will be repeated and chances are consumers will develop an actual preference for the selected brand. This holds all the more true for product groups in which brands and images have more impact than the physical product or its tangible characteristics. In such cases, loyal customers may turn into regular brand groupies, much to the crowning glory of the marketeer and a prosperous growth of the product's market share.

About the same approach to the marketing mix is noticeable in non-commercial marketing strategies of VNSAs such as Boko Haram, al-Qaeda and ISIL. Their names have become strong brands, imprinted in collective memories, distinguishable from one another by the use of logos, consistent communications and promises about the perfect fit between the VNSA's raison d'être and the desires and needs of their target groups.[6] In addition to security and economic realms, VNSAs should, therefore, also be studied from a marketing perspective, because as they behave like brands they will be facing the same choices, decisions, issues and developments as commercial product brands. In the end, any brand will have to be marketed to be sold and this requires in-depth knowledge of marketing strategies.

In this chapter, ISIL serves as a case study to gain an insight into VNSA marketing strategies. ISIL is regarded a key player and leader in the market for VNSAs, having created a strong brand based on the appliance of differing marketing instruments.[7] We aim to put forward the ways in which applying a strategic marketing perspective may supplement today's deployment of military and financial means to fight terrorism. To do so, first we will analyse the four Ps as ISIL puts these into practice in its strategic marketing mix. Next, based on the workings of these four Ps we will offer a tentative vision as to how to use counter-marketing to fight ISIL.

10.2 P for Product

Often, in comparing all four instruments the marketing mix consists of, the product will be the most visible. A product can be regarded as an array of elements, such as quality, brand, image, service and guarantees offering value to the buyer. Additionally, a product's success will depend on available creativity and expertise for developing a strong brand.[8] It appears some VNSAs, and ISIL especially, can

[6] Winkler and Dauber 2014, p. 5.

[7] Winter 2015, p. 6.

[8] Verhage 2013, p. 306.

call on excellent craftsmanship to develop their strong brand. In all its communications the organization is clearly recognizable by a consistent use of brand, logo and message.[9] A product, or rather, a brand, will offer its consumer added value and, based on comparisons, consumers will choose the best value.[10] Added value may be either of a tangible nature, as in physical products, and/or non-tangible. As far as the product ISIL goes, it is mainly its non-tangible nature that provides added value and appeal to buyers.

ISIL, as a product and a brand, is purchased by the buyer at the moment he joins ISIL. Therefore the decision to buy equals the decision to join and this can be regarded the most physical part of the product ISIL. To a far greater extent, ISIL, as a product, consists of mental elements, such as image, reputation, brand values, meanings and benefits. The visible part of the product mainly consists of conceptualisations of the brand (e.g., the ISIL flag or the use of its logo in videos).

ISIL's branding strategy is grounded in six *narratives* representing the pinnacles on which the product is built. Actually, these consist of six promises on *brutality*, *mercy*, *victimhood*, *war*, *belonging* and *utopianism*[11] to be received upon purchase (joining as a member). These narratives form the brand's core and therefore the product ISIL. Synthesizing its narratives, ISIL, as a product, proclaims to stand for *a better world*. ISIL's novice members, its buyers, buy the product and its benefits, as verbalized in the narratives mentioned above. According to Alberto Fernandez, coordinator of the U.S. state department's Centre for Strategic Counterterrorism Communications (CSCC), ISIL's core message and promise should be understood as follows:

> Muslims are being killed and ISIL is the solution. Besides the call to violently retaliate, there is also a call reaching out for the best side of mankind, their aspirations, hopes and dreams, their deepest longing for identity, faith and self-actualization.[12]

Especially to those living their daily life in circumstances filled with doubt, poverty, despair, anger, frustration and hope the product ISIL may appear very desirable to buy into. ISIL, as a product, is not only appealing because of the tempting and promising vision of a better future, but simultaneously the brand is being developed continuously.

Choosing between brands is a process highly influenced by image and reputation, aspects of both product and promotion. Image consists of all meanings and associations consumers foster regarding a specific brand.[13] Reputation follows image. One's image of a brand is shaped by all associations attributed to it, ranging from personal experiences and convictions, opinions of others, perceptions induced by communication and branding strategies. A brand's image may suffer or prosper

---

[9] Winkler and Dauber 2014.

[10] Verhage 2013, p. 306.

[11] Winter 2015.

[12] Cottee 2015.

[13] Verhallen 1988; Pruyn 1994, p. 13.

on all components it is made up of and the same goes for the brand's commercial success.

In the case of ISIL, by consistently using various means to communicate identical messages and promises, the VNSA's image and reputation appear to be continually bolstered. The message in each issue of Dabiq is the same as in their tweets, videos, Facebook messages and press releases.[14] As a result, ISIL's image remains strong and stable, whether embraced as something positive or not. The rift separating ISIL lovers and haters poses no problems to the brand, for lovers create new followers and jihadists and ISIL branding strategies are not aiming to win over the haters anyway.[15] Moreover, even when supporters decide not to go all the way and become ISIL members, they will not turn into opponents of the brand. This, also, can be considered a positive marketing result. It appears linking a positive message and promise of a better world to a consistent promotion of this message leads to a competitor that will be hard to beat.

10.3 P for Price

Price often concerns the money asked for a product or service. As to the VNSAs, they do not ask their followers to pay to join the organization, and in their case, price and costs are beyond monetary characteristics.[16] VNSA membership prices may instead be expressed in acts of terror.[17]

As all producers, VNSAs incur costs to put their product on the market. Next to organizational costs, there will be procurement costs and personnel to pay. In a commercial business, to compensate for such costs, generate production and make a profit, consumer prices are being calculated. Although VNSAs operate differently there will have to be a mechanism in place to compensate for costs, if only to ensure the VNSA's viability. This mechanism consists of members and supporters providing services and other deliverables to the VNSA, adding costs incurred by members committing an attack. It appears, the consumer price paid to consume the VNSA as a product or brand consists of (part of) the members' lives. Such extreme consumer prices are unheard of in any other organization. As to the price to pay, membership of NATO, EU or any other alliance comes much cheaper than any VNSA's. It could therefore be concluded appliance of the instrument Price is not a strong point in the VNSAs' marketing mix.

Having said so and despite the staggering price to pay new members keep joining ISIL. In marketing theory, this is explained by the concepts of price

[14] Colas 2017; Seo 2016.

[15] Ingram 2015, p. 730; Pons Soler 2016, p. 7.

[16] Khan and Estrada 2016, p. 1652.

[17] Mink 2016, p. 69.

perception and price acceptation.[18] Buyers will make up their minds to purchase any product based on an analysis of costs and benefits. The costs to join ISIL are weighed against the expected benefits, such as *a better world*, martyrdom and the glorification of performed acts of terror.[19] These benefits are perceived to be of all-encompassing value and any costs, therefore will be perceived acceptable, up front. If any price will be acceptable to its followers, consequentially, the counter-marketing strategies applied by alliances opposing ISIL or any other VSNA will not be successful if focused on this particular instrument of the marketing mix.

10.4 P for Place

Place refers to the paramount importance of the timely availability of products and services to all customers. Distribution is tasked to get the products at the right time to the right place (i.e. at the target group's location). Along these lines logistics are organized throughout the consumer market sector, whether this concerns supermarkets, auctions or web shops. However, VNSA products and brands are less tangible and less easily transported to a central collection point than most consumer goods.

Taking a closer look at how ISIL distributes its products to consumers, it becomes noticeable that physical locations open for purchase are scarce, with the exception of the caliphate itself. Buyers in search of ISIL's physical product therefore will have to travel. For the main part however, everything on offer by ISIL has to be purchased through other channels, rather remote from any central market. Joining the VNSA, by becoming its customer is possible by means of social media or by using already established customer communities. The internet offers a communication platform without geographical limitations which greatly enhances the opportunities for ISIL's distribution networks to operate at a large scale, supplied and supported via main arteries that spring from headquarters whilst, at the same time, remaining very tightly knit at a cellular level. Over time, increased popularity (i.e., market shares) will lead the cells to expand and create new cells at different locations to be fed and supported by the ones they originated from.

Because of this organizational set-up and ultimately, the digital and global availability of its product, even upon physically destroying its headquarters and the caliphate, ISIL cannot be expected to cease to exist overnight. For at a cellular level, ISIL's *raison d'être* remains alive and by means of its strong message and expertise to promote its brand ISIL will continue to be appealing to new and returning members anywhere. Branches of ISIL are viable to the extent to keep on fighting for their goal and the so-called *wireless caliphate* will keep to exist.[20]

[18] Verhage 2013.

[19] Zahedzadeh et al. 2015, p. 8.

[20] Nacos 2016.

As far as the P for place is concerned, Western organizations face severe competition regarding the delivery of their products and services. Competing against the ways in which ISIL executes their P for place still is to little avail. Instead, currently the counter-marketing focus is mainly on monitoring potential new supporters by scrutinizing planned trips to the caliphate, for people entering a shop can be expected to intend making a purchase.[21] As to the purchases made on social media, where ISIL remains highly active despite incurred losses competition remains tough. On closing Twitter accounts supporters switch easily to another social medium.[22]

10.5 P for Promotion

VNSAs, by the use of various media invest heavily in promotion. Counter-strategies focusing on the P for promotion are indicated for damage control, because this marketing tool in particular is what keeps these organizations afloat.[23] ISIL appears omnipresent promoting its objectives, specifically, regarding the recruitment of new members and bolstering its reputation. ISIL's activities on Twitter, YouTube and Facebook[24] not only result in thousands of daily followers but, also in placing over 40,000 tweets per day.[25] Besides, ISIL regularly issues Dabiq, an online VNSA glossy.[26] Apart from using social media, ISIL posts its video messages to traditional media and within its central organization boasts a broadcasting unit, al-Bayan Radio.[27] Another promotion platform, used by ISIL regards video games, as becomes apparent from the release of the adapted version of *Grand Theft Auto* (GTA) called "Salil al-Sawarem".[28] Throughout inaccessible regions the VNSA uses flyers and posters to communicate with remote villages.[29] To top it all, ISIL commands a network of so-called *fan boys* disseminating online propaganda.[30] In this article, we leave out news on ISIL reported by traditional media, as the VNSA does not direct these communications itself.

Every brand will communicate its core message using every media considered suitable. Generally, such messages entail promises as to the essential end benefits that will be experienced on purchase; in other words, the payoffs that render the

---

[21] Field 2016.

[22] Fisher 2015.

[23] Gartenstein-Ross 2015.

[24] Bartlett and Reynolds 2015, p. 13.

[25] Farwell 2014, p. 50.

[26] All editions of Dabiq can be downloaded on http://www.clarionproject.org.

[27] Schmid 2015, p. 3; Winter 2015.

[28] Al-Rawi 2016.

[29] O'Callaghan et al. 2014.

[30] Cottee 2015.

brand's promises explicit. ISIL, the brand acts in the same way. Its core message, wrapped in six narratives encompasses the "perfect state" and the fact that to every —new-member this perfect state will be achievable.[31] Also, in terms of brand's exclusiveness, it should be noted ISIL is the only VNSA to offer this perfect state, as theirs is based on the one and only Islam "the Prophetic methodology".[32] Muslims who do not adhere to the Prophetic methodology (e.g., members of Al Qaeda) are considered heretics or even opponents, much the same as non-Muslims are viewed.[33] ISIL's brand identity is strong and upon joining new recruits are offered immediate change as well as a life time chance for long run improvements.[34]

The brand's strength increases because its core message (including the narrative on the one and only Islam)[35] is consequently and consistently communicated in every act of speech regardless of the media used.[36]

Thus, in this respect ISIL can be said to behave like any regular brand and for this reason should be analysed and countered in the same way. Once again, it should be noted, ISIL is quite professional at branding- and communication strategies and excels in creating synergetic effects. All broadcasting of communiques is centralized and the VNSA can avail of its own departments for production and dissemination of all acts of speech.[37] In this way, both unity and effectiveness in brand communication are guaranteed.

10.6 Towards Strategic Counter-Marketing

When comparing ISIL's strategic marketing mix to counter-marketing strategies deployed by alliances and organizations fighting the VNSA, it can be concluded ISIL's marketing outranks its opponents.[38] According to Alberto Fernandez, this is because in the fight against ISIL its opponents lack the same communicative strong promises, messages and narratives.[39] Based on both quantity and quality with respect to the delivery, (unity of) content and the multimedia approach in communications, ISIL counts as a strong competitor that will not easily be defeated.

---

[31] Winter 2015.

[32] Wood 2015.

[33] Novenario 2016, p. 954.

[34] Winter 2015.

[35] Bartlett and Reynolds 2015.

[36] Wood 2015.

[37] Winter 2015.

[38] Also see Farwell 2014; Klausen 2015.

[39] Cottee 2015.

Moreover, as to its target group the overarching promise of the perfect state is rich in meaning and emotional affect,[40] the message itself is also strong.

Apart from recruiting fighters, ISIL is able to create devoted groupies and the already mentioned fan boys amongst its followers. Many of them are digitally active and volunteer to disseminate messages posted by ISIL on social media. Carefully orchestrated communications aim to strengthen brand sympathy amongst the target group, as can be witnessed in explanatory articles in Dabiq to clear misunderstandings and to report on how promises are kept, if only partially. Organizations fighting ISIL cannot avail of such fan groups and for this reason their messages and communications are less followed, disseminated and praised.[41]

In marketing terminology ISIL uses all sympathy and devotion to successfully develop brand loyalty. Brand loyalty is paramount for any brand to remain viable, because it induces customers to repeat their purchases.[42] In the case of ISIL, because of brand loyalty supporters can be expected to be there to stay and to continue their contributions to the brand's success.

At the other side, in countering ISIL the one message communicated on social media is composed of "Don't do this".[43] In marketing practice it is known positive messages will beat negative messages anytime and ISIL's positive messages turn out stronger than the opponents' negative ones.[44] Communicating one's own brand advantages rather than communicating the competitor's disadvantages will result in brand sympathy, loyalty and followers.[45]

By communicating their "Don't do this" message, ISIL's competitors are not attracting many new customers (followers), nor do they seem to be able to increase their market shares by brand loyalty.[46] Whereas ISIL avails of fan groups, the opposition does not.[47] Contrary to the worldwide dissemination of posts by ISIL fan boys and due to a lack of response, the US Department of State recently closed their Twitter account Think Again Turn Away. As its successor @The GEC counts no more than 26.8 K followers this attempt also appears meaningless.[48]

ISIL scores high on all four Ps. Even the P concerning their exaggerating price, in terms of paying with your life does not decrease the number of supporters. Effective counter-marketing strategies to fight ISIL, in which each partial result contributes to the ultimate effect, would have to aim for reputation damage, thereby decreasing new support to the VNSA and to a decrease in revenues and budgets for buying weaponry. This then would result in less terrorist attacks causing less victims.

---

[40] Cottee 2015.

[41] Cottee 2015.

[42] Mosmans 2013, p. 45.

[43] Bartlett and Reynolds 2015, p. 23; Nance 2016, pp. 422–423.

[44] Also see Cottee 2015.

[45] Mosmans 2013.

[46] Beutel et al. 2016; Mahadevan 2015, p. 2.

[47] Cottee 2015.

[48] See on Twitter @theGEC, Global Engagement; Nance 2016, p. 422.

However, to date, Western institutions such as EU, NATO and UN proclaiming to compete with VNSAs and despite their excellent means of distribution and affordable prices, are still offering a far less appealing product and brand. Besides, their products are promoted far less consistently, visible and effectively than ISIL's. In this respect, the counter-marketing battle appears almost lost. To strike an effective counter offensive, appropriate *marketing weapons* should be used. Next to the P for promotion, strategic marketing offers three additional weapons to fight the competition. Most prominently regarding the P for product, ISIL's opponents, including NATO,[49] are in dire need of a strong brand. NATO exists already as a brand in the market for international security. Based on a preliminary NATO brand analysis, however it appears that apart from a manual on logos, NATO does not pay a lot of attention to strategic branding and the same applies to its efforts in brand positions or propositions. NATO limits communications to providing information and communication projects undertaken by its strategic communications centre of excellence. Neither branding nor a comprehensive marketing strategy seem to be a subject of interest.[50] ISIL, on the contrary, heavily invests in transformational brand positioning and the ensuing end benefits, promises and narratives propel the VNSA into realms beyond mere logos.

As a brand, NATO appears rather empty and therefore the organization will have a hard time to become "alive" to the public at large, let alone to be experienced as sufficiently tempting and seducing to the extent people will prefer NATO over its opponents. Of course NATO membership is restricted to states instead of individual members, but even so the alliance could do better in terms of creating a clear image of its meaning, brand likeability and attractiveness.

Currently, as NATO does not want to present itself as a brand it will have a hard time standing up against a very strong transformational brand, such as ISIL. Neither will NATO be able to lure any wavering youngsters to their side.

To strengthen NATO as a brand its message and promises are in need of additional strength and in this respect white goods supplier Miele "Forever Better" could serve as an example. NATO, both as a product and in terms of supply is in need of a price that is perceived as the better one. In comparing products, consumers are sensitive to price advantages and when offers are indeed perceived to be comparable they will prefer the product that is priced most advantageously (in terms of costs). Although NATO's price seems reasonable, ISIL's price is better perceived and this is to be turned around. Similar to ISIL's, NATO's product is well distributed and apart from continuing to block ISIL accounts on social media, therefore competition in this respect seems to little avail.

Essentially, there appear two ways to counter-market ISIL. First, ISIL can be thwarted in executing its strategies, starting with the P for place, for instance by thwarting the system used for dissemination of messages. In doing so, the P for promotion will be threatened as well and as a result, ISIL's marketing mix will be

[49] See NATO 2016, para 7.

[50] See http://www.stratcomcoe.org.

weakened. This would then allow for a successful attack on ISIL's strategic marketing as a whole.

The second option in waging a counter-marketing war on ISIL necessitates NATO to execute the best strategy. To this end, NATO as a brand competing with ISIL has to develop a marketing mix and a strategy that aim for brand likeability, brand preference and increased market shares. In other words, to strengthen its brand NATO needs all four Ps to deliver added value in comparison to ISIL and together form the best proposition. NATO's marketing strategies should, amongst others, specifically target young people doubting whether or not to join ISIL.

10.7 Conclusion

From a marketing perspective, NATO and armed forces alike could gain from adding experienced brand constructors and—managers to their midst as their knowledge is needed to win the counter-marketing fight. Marketing has its place and use in an integrated effort applying both lethal and non-lethal means to beat the competition.[51] However, to date, literature on marketing strategies mainly focuses on commercial business practices. As to research into the use of (counter) marketing on behalf of the public sector, and particularly in international security, there remains a void that should be addressed to answer questions as to how, where and at what price specific products can be put on the market and how competitors are to be countered by using these same means.

We end this chapter by introducing a potentially strong opportunity for NATO branding. As NATO is an organization encompassing member states, we have to answer the question whether we want to position NATO as an umbrella brand and consequentially using different product brands for each individual product or as a corporate brand using its organizational name for all products it markets. In other words, is NATO to be marketed like Unilever containing many product brands or is NATO like Philips and should all products be branded under one corporate name?[52] It could be argued that the Unilever strategy should be preferred, as NATO is joined by 28 member states, so-called sub-brands. Endorsed by NATO each (member) state could then proceed to build its own brand. The result, however would be a kaleidoscope of small brands containing differing messages thereby preventing the creation of one recognizable competitor to prefer over ISIL. We therefore argue to market NATO as a corporate brand containing one message, one personality and one identity, no matter how many member states or sub-brands NATO includes. To be successful in counter-marketing ISIL with its strong narratives and attractive

---

[51] Nance 2016, p. 409.

[52] Riezebos 2011, p. 67.

future states on offer, NATO is in need of an equally strong message and end benefits. The NATO™ narrative springs naturally from Article 5 of the North Atlantic Treaty, that proclaims that an armed attack against one or more of its member states will be considered as an attack against all.[53] One for all and all for one. Translated into a brand promise NATO's message could be "solidarity" or "togetherness". This could well be the starting point both with respect to a NATO marketing strategy incorporating the development of all four instruments of the marketing mix as well as regarding the NAtions TOgether brand battle against ISIL.

References

Al-Rawi A (2016) Video games, terrorism, and ISIS's Jihad 3.0. Terrorism and Political Violence (5 Aug 2016 online):1–22. DOI:10.1080/09546553.2016.1207633

Archetti C (2015) Terrorism, communication and new media: explaining radicalization in the Digital Age. Perspectives on Terrorism 9:49–59

Bartlett J, Reynolds L (2015) The state of the art 2015: A literature review of social media intelligence capabilities for counter-terrorism. Demos, London

Beutel A et al (2016) Guiding Principles for Countering and Displacing Extremist Narratives. Journal of Terrorism Research 7(3):35–49. DOI:10.15664/jtr.1220

Colas B (2017) What Does Dabiq Do? ISIS Hermeneutics and Organizational Fractures within Dabiq Magazine. Studies in Conflict & Terrorism 40(3):173–190

Cottee S (2015) Why it's so hard to stop ISIS propaganda. The Atlantic (2 March 2015). http://www.theatlantic.com/international/archive/2015/03/why-its-so-hard-to-stop-isis-propaganda/386216/ Accessed 1 March 2017

Farwell JP (2014) The media strategy of ISIS. Survival 56(6):49–55

Field A (2016) The Dynamics of Terrorism and Counterterrorism: Understanding the Domestic Security Dilemma. Studies in Conflict & Terrorism (11 Aug 2016 online):1–14 DOI:10.1080/1057610X.2016.1221253

Fisher A (2015) Swarmcast: How Jihadist Networks Maintain a Persistent Online Presence. Perspectives on Terrorism 9(3):3–20

Floor JMG, Van Raaij WF (2010) Marketingcommunicatiestrategie. Noordhoff Uitgevers, Groningen/Houten

Gartenstein-Ross D (2015) Jihad 2.0: Social Media in the Next Evolution of Terrorist Recruitment. Hearing before the Senate Committee on Homeland Security & Governmental Affairs (7 May 2015). http://www.defenddemocracy.org/content/uploads/documents/Gartenstein_Ross_HSGAC_Testimony_Jihad2.0.pdf Accessed 1 March 2017

Helmus TC, Paul C, Glenn RW (2007) Enlisting Madison Avenue: The marketing approach to earning popular support in theaters of operation, Vol. 607. Rand Corporation, Santa Monica. http://www.rand.org/content/dam/rand/pubs/monographs/2007/RAND_MG607.pdf Accessed 1 March 2017

Ingram HJ (2015) The strategic logic of Islamic State information operations. Australian Journal of International Affairs 69(6):729–752

[53] North Atlantic Treaty, Article 5, http://www.nato.int/cps/en/natohq/official_texts_17120.htm.

Khan A, Estrada MAR (2016) The effects of terrorism on economic performance: the case of Islamic State in Iraq and Syria (ISIS). Quality & Quantity International Journal on Methodology 50(4):1645–1661

Klausen J (2015) Tweeting the Jihad: Social Media Networks of Western Foreign Fighters in Syria and Iraq. Studies in Conflict & Terrorism 38(1):1–22 DOI:10.1080/1057610X.2014.974948

Ligon GS, Harms M, Derrick DC (2015) Lethal brands: how VNSAs build reputations. Journal of Strategic Security 8(1):27–42

Mahadevan P (2015) Out-narrating the IS. Policy Perspectives 3(2):1–4 http://www.css.ethz.ch/content/dam/ethz/special-interest/gess/cis/center-for-securities-studies/pdfs/PP-2015-03.pdf Accessed 1 March 2017

McCarthy J (1964) Basic Marketing: a managerial approach. Irwin, Homewood IL

Mink C (2016) It's about the group, not God: social causes and cures for terrorism. Journal for Deradicalization 5:63–91

Mosmans A (2013) Branding. Adformatiegroep, Hoofddorp

Nacos BL (2016) Mass-mediated Terrorism: mainstream and digital media in terrorism and counterterrorism. Rowman & Littlefield Publishers, Plymouth

Nance M (2016) Defeating ISIS: Who They Are, How They Fight, What They Believe. Skyhorse Publishing, New York

NATO (2016) Warsaw Summit Communiqué, Issued by the Heads of State and Government participating in the meeting of the North Atlantic Council in Warsaw 8–9 July 2016 (9 July 2016) http://www.nato.int/cps/en/natohq/official_texts_133169.htm Accessed 1 March 2017

Novenario CMI (2016) Differentiating Al Qaeda and the Islamic State through Strategies Publicized in Jihadist Magazines. Studies in Conflict & Terrorism 39(11):953–967

O'Callaghan D, Prucha N, Greene D et al (2014) Online social media in the Syria conflict: Encompassing the extremes and the in-betweens. Advances in Social Networks Analysis and Mining (ASONAM). DOI: 10.1109/ASONAM.2014.6921619

Pons Soler M (2016) The analogy of Islamic radicalization and a successful marketing campaign, Documento Opinion 56. http://www.ieee.es/en/Galerias/fichero/docs_opinion/2016/DIEEEO56-2016_Radicalizacion_Marketing_ENGLISH_MPons.pdf Accessed 1 March 2017

Pruyn ATH (1994) Imago: een analytische benadering van het begrip en de implicaties daarvan voor onderzoek. In: Van Riel CBM (ed) Corporate communication. Bohn Stafleu Van Loghum, Houten

Riezebos R, Van der Grinten J (2011) Positioneren. Boom Lemma Uitgevers, The Hague

Schmid AP (2015) Challenging the Narrative of the Islamic State. International Centre for Counter-Terrorism, The Hague. http://www.icct.nl/wp-content/uploads/2015/06/ICCT-Schmid-Challenging-the-Narrative-of-the-Islamic-State-June2015.pdf Accessed 1 March 2017

Seo H, Ebrahim H (2016) Visual propaganda on Facebook: A comparative analysis of Syrian conflicts. Media, War & Conflict 9(3):227–251

Verhage BJ (2013) Grondslagen van marketing. Noordhoff Uitgevers, Groningen

Verhallen T (1988) Psychologisch marktonderzoek. Katholieke Universiteit Brabant, Tilburg

Winkler CK, Dauber CE (2014) Visual propaganda and extremism in the online environment. Strategic Studies Institute and US Army War College Press, Carlisle PA. http://www.dtic.mil/cgi-bin/GetTRDoc?AD=ADA607970 Accessed 1 March 2017

Winter C (2015) The Virtual 'Caliphate': understanding Islamic State's propaganda strategy. Quilliam, London. http://www.stratcomcoe.org/download/file/fid/2589 Accessed 1 March 2017

Wood G (2015) What IS really wants. The Atlantic (March) http://www.theatlantic.com/magazine/archive/2015/03/what-isis-really-wants/384980/ Accessed 1 March 2017

Zahedzadeh G, Barraza JA, Zak PJ (2015) Persuasive narratives and costly actions. Terrorism and political violence 29(1):160–172. http://dx.doi.org/10.1080/09546553.2015.1005077

Author Biographies

J.M.M. van Lieshout, M.Sc. is Assistant Professor of Defence Economics, Department of Military Management Studies at the Faculty of Military Sciences, Netherlands Defence Academy. He obtained a Master degree in Social Sciences at Tilburg University.

Professor Dr. R.J.M. Beeres is Professor of Defence Economics, Department of Military Management Studies at the Faculty of Military Sciences, Netherlands Defence Academy and an Associate Professor at Nyenrode Business University. He obtained a Ph.D. in administrative sciences at Radboud University Nijmegen.

Chapter 11
Delegitimising the Adversary: Understanding Actor and Audience Analysis as a Tool to Influence and Persuade

Christina van Kuijck

Abstract Influence and persuasion are gaining increased attention in warfare, but often social psychology approaches remain underused or neglected while this academic field has provided useful insights into this approach, especially as a non-kinetic capability. To effectively influence and persuade, we need to understand the adversary and the specific frame of reference of the relevant audience, which can be determined through Actor and Audience Analysis (AAA), a relatively new tool that has shown promising results. Based on the results of AAA, the communicator's characteristics and persuasion techniques must be tailored to the specific audience. This chapter will use delegitimisation of the adversary, specifically in the context of extremism and terrorism, as an example to demonstrate how influence and persuasion can be applied in a military context.

Keywords Non-kinetic influence · Influence · Persuasion · Social psychology · Actor and Audience Analysis · Communication · Extremism · Terrorism · Non-state adversaries

Contents

C.I.A. van Kuijck (✉)
Innovation Centre AIR, Royal Netherlands Airforce, Breda, The Netherlands
e-mail: cia.v.kuijck@mindef.nl; ciavankuijck@outlook.com

P.A.L. Ducheine and F.P.B. Osinga (eds.), *Netherlands Annual Review of Military Studies 2017*, NL ARMS, DOI 10.1007/978-94-6265-189-0_11

11.1 Introduction

"We haven't figured out how to reach that individual who's going down the path of radicalization" according to Seamus Hughes, deputy director of the Program on Extremism in Washington, who recently hired reformed ex-terrorist Jesse C. Morton to gain an understanding of the 'mind of the terrorist'.[1] Understanding non-state adversaries, especially extremist and terrorist groups, has become an important matter since 9/11 after the fall of the Twin Towers, both for the government and for various populations. In addition to the counterterrorism efforts by the international community, individual Muslims started to respond with the #NotInMyName campaign against the Islamic State (IS), which went viral in 2014.[2] In other words, Muslims are showing publicly their disagreement with the terrorist organisation, a form of delegitimisation.

Delegitimisation has been gaining increased attention as a coercive measure, but with the exception of Wilner, there has been no specific theoretical framework within coercion studies.[3] The reason is that coercion, the academic field in strategic studies that preoccupies itself with influencing adversaries, was initially focused on state actors, but non-state adversaries such as terrorist organisations tend to have another structure: the adversary is no longer a single entity, but a system or network composed out of different and complex levels of decision-making.[4] Later in the chapter we will see that influence requires a tailored approach, taking into account the specific frame of reference of the target we want to influence. The same applies to influence in war: if the goal is influencing adversary, neutral and friendly audiences to achieve a certain *end state* (ends), *how* we will do it depends on the specific frame of reference of the target (ways) and on the available *time, resources and capabilities* (means).[5] But what is influence exactly?

According to Schaefer, influence is the exercise of power through the process of persuasion.[6] Persuasion aims to change attitudes, beliefs and/or behaviour based on

---

1 Cohen 2016.

2 See for example isisnotinmyname.com. Accessed 7 December 2016.

3 See for example Knopf 2010, pp. 1–33; Knopf 2012, pp. 21–45; Wilner 2010a; Wilner 2010b; Wilner 2011; Wilner 2012; Wilner 2013; Long and Wilner 2014; Wilner 2015a, b.

4 Davis 2010, pp. 1–2, 19–25; Gross Stein 2012, pp. 51–52; Kroenig & Pavel 2012, pp. 23–24; Rid 2012, pp. 127–128; Miller 2013, pp. 134–139.

5 NATO AJP-01(D) 2010, nr. 0407; NATO AJP-5 2013, nr. 0211. See also Chap. 9 by Ducheine, Van Haaster and Van Harskamp in this volume.

6 Schaefer 2007, p. 357.

information received from others through intentional messaging.[7] Coercion attempts to influence an adversary's behaviour and decision-making processes through the use of credible and capable threats.[8] It is generally subdivided into deterrence (stopping an action or maintaining the status quo), compellence (causing or undoing an action) and coercive diplomacy (signalling a target through inducements and the limited use of force).[9] Authors agree that coercion and persuasion are two overlapping concepts within influence, where coercion is more focused on depriving the target from a measure of freedom or autonomy through threats in order to alter behaviour, and persuasion concentrates on the perception of free choice, which more broadly attempts to change beliefs, attitudes, and/or behaviour.[10]

Is delegitimisation therefore a method of coercion or a method of persuasion? And how does delegitimisation work? I argue that delegitimisation is one of the cross-points where the two concepts overlap, since a threat is being made to the adversary we want to influence by taking away their (potential) support, consent or recognition. This chapter seeks to understand how influence and persuasion can be developed in a military context through Actor and Audience Analysis (AAA) by using delegitimisation of the adversary as an example, with a special focus on extremism and terrorism. As such, Sect. 11.2 will set out what delegitimisation exactly is, both from a social-psychology and coercive perspective, specifically using insights from delegitimisation of terrorism and extremism. In order to effectively delegitimise the adversary, a clear objective is needed, in the case of coercion differentiating between deterring potential supporters or compelling active members of the adversary's group by stripping the justification for the support, consent or recognition of the adversary (Sect. 11.3). After having determined the relevant objective, an analysis of what makes the adversary attractive and of the frame of reference of the audiences we want to influence will be made in Sect. 11.4, explaining how the tool of Actor and Audience Analysis (AAA) works in this context, as well as some insights from (counter-) radicalisation and de-radicalisation to comprehend the relevant factors that make people join and leave extremist groups. Section 11.5 discusses the basic principles of influence and persuasion based on Cialdini's framework, describing on the one hand the characteristics a communicator must possess (e.g. credibility) and on the other hand techniques of influence (e.g. commitment and consistency), the use of these depending on the outcomes of AAA. Finally, a conclusion will be made in Sect. 11.6.[11]

---

[7] Rashotte 2007, p. 4427; Andersen and Taylor 2008, p.513; Perloff 2010, p.13; Northouse 2013, p. 9.

[8] Schelling 1966, pp. 2–6; Byman and Waxman 2002, p. 1; Bratton 2005, p. 99; Jakobsen 2010, pp. 279–281.

[9] Byman and Waxman 2002, pp. 3–9; Bratton 2005, p. 103; Jakobsen 2010, p. 279; Jakobsen 2011, p. 155; George and Smoke 1974, p. 11; Johnson et al. 2002, pp. 13–14; Thies 2003, p. 243; Bowen 2004, p. 58; Sperandai 2006, p. 253; Heuser 2010, p. 465.

[10] See for example Perloff 2010, pp. 17–19; Ledgerwood et al. 2014, p. 533.

[11] This chapter is based on the Yale Communication Model, which analyses the audience (whom), the role of the communicator (who) and the message (what and means). See Hovland et al. 1953. For modern applications, see for example Bordens and Horowitz 2002, pp. 217–231; Maio and Haddock 2007, pp. 569–570; Perloff 2010, pp. 126–127.

11.2 Delegitimisation

In social psychology delegitimisation can be defined as the process in which the support, consent or recognition for an entity is withdrawn, i.e. the loss of legitimacy by retracting the justification that lies within the socially constructed system of norms, values, beliefs and definitions of a target, undermining the entity's ability to exercise influence and control over others.[12] Kelman claims that delegitimisation is a social shift of norms, beliefs and values through influence, removing the acceptability of actions done by an entity or the entity itself, leading to, on the one hand, a change in the social beliefs and behaviour of the supporters, and, on the other hand, a denial of rights and powers of the delegitimised entity.[13] Academics have written on the delegitimisation of adversaries, more specifically terrorist organisations, but do not provide a specific framework on how to do this. However, some common factors are that delegitimisation is a form of preventing recruitment and radicalisation by changing perceptions, beliefs and behaviour; that former members belonging to the adversary, communities, neighbours and family can play a crucial role; that the contradiction between the organisation's ideology and messages must be stressed; and that alternatives must be provided.[14]

In the context of coercion, Wilner frames delegitimisation of terrorist organisations as a deterrence strategy that:

- Raises the costs of participating in terrorism by challenging the normative, religious, and socio-political justifications of the terrorist organisation;
- Manipulates the political, ideological, and religious rationale that informs terrorist behaviour;
- To reduce the achievement of terrorist goals;
- By attacking the legitimacy of beliefs that inform the terrorists' behaviour.[15]

There are, however, some problems with his approach. First, Wilner uses a mix between rational actor theory (all rational actors make the same cost-benefit calculus, and terrorists are not rational actors following this theory) and strategic culture (discover what a target values by studying a country's beliefs, geography, history, generational change, political structure, defence organisations, myths, and symbols).[16] The problem with these theories is that the former does not take into account the psychological and cognitive aspects in decision-making processes and

[12] See for example Suchman 1995, p. 574; Toros 2008, pp. 407–426; Beetham 2012, p. 123; Lamb 2014, p. 17.

[13] Kelman 2001.

[14] See for example McBride 2011; Benard 2011; Kalmonick 2012; Benard 2015.

[15] Wilner 2011, pp. 26–27; Long and Wilner 2014, pp. 127–128; Wilner 2015a, pp. 66–68; Wilner 2015b, pp. 445, 449.

[16] For rational actor theory, see for example Schelling 1960; Schelling 1966. For strategic culture, see Johnston 1995; Sheehan 2010, p. 171; Walton 2012, p. 321; Lantis and Howlett 2013, pp. 80–82.

presuppose that terrorists are not rational actors, while the latter ignores local and supranational cultures, and still holds on to cost-benefit calculations.[17]

Second, authors have pleaded for a broader approach than deterrence, as it does not function well with terrorist organisations and non-state adversaries due to the fact that states approach these adversaries based on actions and not motives, requiring a wider strategy of influence.[18] Even though Wilner does take into account motivation, he still remains within the traditional cost-benefit calculus. Moreover, Wilner recognises that delegitimisation can compel recruit to leave illegitimate violence behind, but still frames the measure within deterrence, which seems contradictory since deterrence is about maintaining the status quo and compellence about altering an existing course of action, inducing the target to do something.[19]

Third, although Wilner differentiates between individual, group and community levels, delegitimising a terrorist organisation requires a tailored approach that needs to differentiate between the audiences, and delegitimisation must be based on their frames of reference, which is a complex set of beliefs, schemas, values, culture, attitudes and other issues through which we form understanding and judgment and create perceptions and meaning.[20]

This doesn't mean that Wilner's framework is useless, but rather needs adjustment based on social psychology. The next sections will set out an adapted framework of delegitimisation in order to effectively influence terrorist beliefs and behaviour, using insights from counter-radicalisation, de-radicalisation and social influence.

11.3 Defining the Objective: Deterrence or Compellence?

The first step in influence is determining what one's objective is, followed by the identification of the target(s) one wants to influence and setting out their specific frame of reference to understand which influence and persuasion techniques are applicable.[21] For delegitimisation the goal is to withdraw support, consent and recognition of the adversary in order to weaken it and undermine its ability to inspire and persuade people, i.e. undermine its power by persuading the target to change its

[17] For rational actor theory see Lebow and Gross Stein 1989; Davis 2012, pp. 70–71; Gross Stein 2012, pp. 51–52; Miller 2013, pp. 139–140; Sauer 2015, p. 113. Rational actor theory has also been criticised as a psychological approach, see for example McAllister and Schmid 2011, p. 222; Rogers 2011, pp. 38–40. For strategic culture, see Booth 1981, p. 15; Coskun 2007; Bunn 2007, pp. 1–8; Long 2009; Toje 2009; Williams and Covarrubias 2009, p. 267; Echevarria 2011, pp. 431–445.

[18] Byman 1998, pp. 149–169; Davis and Jenkins 2004; Jenkins 2006; Fussey 2011, p. 174; McAllister and Schmid 2011, p. 260; Davis 2014.

[19] Wilner 2011, p. 4, 6; Long and Wilner 2014, pp. 152–153; Wilner 2015a, p. 17; Wilner 2015b, pp. 458–460.

[20] Sherif 1937; Eagly and Chaiken 1998, p. 304; Abrams and Levine 2012, pp. 62–63; Gavac et al. 2016, pp. 353–354.

[21] Van Gorp 2007; Seese 2014; Cwalina et al. 2015, pp. 90–91; Paul et al. 2015; Tatham 2015a.

Table 11.1 Goals of delegitimisation in academic literature

Target	Coercion	Prevention	Radicalisation
Friendly and neutral audiences: Risk of supporting, consenting and/or recognising the adversary	*Deterrence*: Maintaining The status quo—stop start of support	*Primary*: General group target	*Counter-radicalisation*: Prevention of radicalisation of youth, vulnerable groups and communities
		Secondary: audiences at risk	
Adversary audiences: Already actively supporting, consenting and/or recognising the adversary	*Compellence*: Stop or undoing course of action—stop current support	*Tertiary*: Active audiences	*De-radicalisation*: Active audiences to repent and disengage

attitudes, beliefs, and/or behaviour based on the received information through intentional messaging.[22] Learning from counterinsurgency, it can be translated into the reinforcement of the commitment of friendly audiences; gaining the support and commitment of neutral audiences; and weaken the will of (potentially) adversary audiences.[23] Conversely, in studies on the prevention of extremism authors differentiate between primary prevention (whole population groups), secondary prevention (vulnerable targets to radicalisation) and tertiary prevention (active members).[24] Translated to coercion the goal is to prevent friendly and neutral audiences to join and/or participate in the activities of the adversary (deterrence: maintaining the status quo; primary and secondary prevention), whereas persons already committed to the adversary's cause need to leave the organisation behind (compellence: stop or undoing a course of action; tertiary prevention).

Social psychology applied to extremism also differentiates between the targets that have to be influenced, as the mechanisms of influence are more or less applicable depending on the frame of reference of the target and the goal we want to achieve.[25] A difference is made between counter-radicalisation (the prevention of radicalisation of youth, vulnerable groups and communities) and de-radicalisation (the already radicalised targets that have to be influenced to repent and disengage from violence).[26] In other words: do we want deter potential support, consent and/or recognition, or do we want to compel those who are already involved to leave terrorism behind? Table 11.1 sets out the different goals of delegitimisation previously discussed.

---

[22] Perloff 2010, p. 13; Benard 2011; Northouse 2013, p. 9; Benard 2015.

[23] See for example AJP-3.4.4. nrs. 0217, 0446.

[24] See for example Bjorgo 2013, pp. 10, 26; Koehler 2015, p. 135.

[25] See for example Bjorgo 2013, pp. 6–9; Braddock 2014, p. 62; El-Said 2015, pp. 10–11; Koehler 2015, p. 135.

[26] Schmid 2013; El-Said 2015, pp. 10–11.

Extremist organisations, especially terrorist organisations, cannot survive without support, and it is therefore important to understand *why* people support, consent or recognise adversaries (frame of reference) and *how* we can change that in order for a non-state adversary to crumble (influence and persuasion), which can be done through Actor and Audience Analysis (AAA).[27]

11.4 Understanding the Frame of Reference

Having set out the relevant objective or end state (deter (potential) supporters of the adversary or compel active members to leave extremism behind), the next step is to determine what the relevant frame of reference of the audience is that we want to influence. This section will explain what Actor and Audience Analysis (AAA) is, a relative new measure developed by Psychological Operations (PsyOps) community that is gaining increased attention by defence organisations around the world.[28] Subsequently, important factors relevant for a specific the end-state we want to achieve will be considered and analysed. If we want to deter (potential) support, recognition and consent of the adversary, understanding why people join the group will be more useful to prevent support in the first place. However, if we want to compel active members to leave the adversary's organisation behind, other factors may play a more important role in influence and persuasion. Different processes and mechanisms lie behind the objective, but the difference does not make the factors mutually exclusive, rather that depending on the audience one measure will be more applicable than other: there is not a one-fits-all approach.

11.4.1 Actor and Audience Analysis

There are various methods to analyse a target, but one relatively new defence measure that has shown promising results is Actor and Audience Analysis (AAA—previously termed Target Audience Analysis[29]), a tool designed to identify the behaviour and beliefs of the target, and the reasons behind it. Based on this, triggers and circumstances are determined to change the behaviour of the

[27] Jacobson 2010; Fink and El-Said 2011; Sim 2013; Clubb 2014, pp. 842–861; Cronin 2009, pp. 61–62; Della Porta 2009, pp. 66–87; Haniff Bin Hassan 2012, pp. 358–384.

[28] See for example AJP-3.10.1(a) 2007 (NATO); FM 3-53 2013 (United States of America); AJP-3.10.1 2014 (United Kingdom). In the United States of America the term PsyOps has been replaced by Military Information Support Operations (MISO).

[29] Interview with Meike Wolf, Senior Actor and Audience Analyst at the Dutch Civil and Military Interaction Command (1 CMI-Commando).

target.[30] This model is useful because it takes into account the fact that communication must be tailored depending on the context and other environmental factors, as well as classifying different groups dealing with the problem of 'multiple audiences' (how to send different meanings in the same message to different audiences): the target audience (1); the hostile audience (2); the friendly audience (3); and the neutral audience (4).[31] The difference with classic social psychology concerning persuasion is the focus on multiple factors instead of only values, attitudes, beliefs and behaviour.[32] Furthermore, it allows for a tailored approach in the sense that it takes into account local dynamics and environment with respect to the behaviours that need to be changed, something that counter-extremism psychologists have pleaded for.[33] Another advantage is the possibility to influence the target tailored to a specific level: do we want changes in the overall population by using state policies (macro-level), look into the organisational dynamics and influence factions, communities and groups (meso-level), or persuade a specific individual, leader or special/small segments of the population (micro-level)?[34]

Tatham explains that behavioural grids are made on the basis of various parameters: descriptive, prognostic and transformative, which in turn contain 45 components (see Table 11.2).[35] Each of these parameters have different components which provide a detailed understanding, from attitudes, beliefs and values to motivations, common enemies and historic context.[36] These parameters should be seen as cyclical phases. The descriptive phase describes the characteristics of the population relevant to the objective, in our case, the target we want to influence. The prognostic phase concerns itself with the 'why' question: why do people support the adversary. In the transformative phase, possible manners of how we can influence the target are considered in a holistic manner.[37]

The relevant components are deduced from measurement, research and subsequently reliable knowledge of the audience, in sort a bottom-up approach. It must be noted that to understand who the relevant target is and how we can influence them, one must create a comprehensive understanding of the adversary in order to discover its appeal to potential recruits, and subsequently learn how that attractiveness can be turned beneficial to our goals.

---

[30] Tatham 2013, 2015a; Tatham and Giles 2015. For other models, see for example Boslaugh et al. 2005; Paul et al. 2015, pp. 171–172 (on audience segmentation), and Jackson 2009, pp. 216–217; Mac Ginty 2010; Paul et al. 2015, pp. 172–173 (on social network analysis).

[31] Bordens and Horowitz 2002a, b, pp. 231–232; Tatham 2015a, pp. 21–25.

[32] See for example Bordens and Horowitz 2002a, b; Maio and Haddock 2007; Hitlin and Pinkston 2013; Petty et al. 2013.

[33] See for example Archetti 2013, pp. 118–122; Bjorgo 2013, pp. 9–13; Medina and Hepner 2013; Bosi and Della Porta 2015, pp. 83–84.

[34] McAllister and Schmid 2011, pp. 214–260; Schmid 2013, p. 4; Bosi and Della Porta 2015, pp. 83–86; Ferguson 2016, p. 3.

[35] Tatham 2015a, pp. 27–29.

[36] Ibid. See also Tatham 2015b.

[37] Interview with Meike Wolf.

Table 11.2 Parameters actor and audience analysis (Tatham 2015a)

Descriptive	Prognostic	Transformative
Audience statistics	Informational	Message appeals
Language	Propensity for change	Common enemies
Literacy	Emic logic	Binary opposition
Current behaviour	Normative affiliation	Source credibility
Size	Locus of control	Reward structures
Leadership	Motivations	Decision path
Financing	Polarising motivating propensity	Filters
Relevant issues	Instrumental Behaviour	Noise
Historic context	Influenceability	Attitudes
Channel credibility	Beliefs	Initiating sets
Group composition	Intent	Rituals
Mission	Skill	Ability
Values	Power structures	Expectancy
Group membership	Ideology	Fears
Channel exposure	Fragility/Resistance	
Needs		

11.4.2 Deterring Potential Supporters: Learning from (Counter-)Radicalisation

Radicalisation is the process in which (1) an extremist belief system is adopted or promoted (2) through the complex interaction of various factors (3) to advance political, religious or social change (4) that can lead to the legitimisation of violence.[38] Understanding why a person starts potentially supporting, recognising or consenting to an extremist organisation can be deduced from radicalisation studies, which in turn allows us to address the justification through which an adversary becomes legitimised. Multiple models have emerged that explain why people become extremists, but some common elements can be found at the macro-, meso- and micro-level, which are represented in Table 11.3.[39]

Understanding the dynamics of radicalisation and the different types of justifications that lead to support, consent and recognition are important because it allows us to develop different strategies of influence to monitor, deter and compel (potential) recruits.[40] For example, persons craving for adventure will need to be offered different alternatives (e.g. a career as a travel agent, investigator, etc.)

[38] See for example Spaaij 2012, p. 47; Lowe 2014, p. 30; Alimi et al. 2015, p. 11. Note that not all radicals become terrorist or violent extremists—it is a cognitive process that can lead to the use of violence.

[39] See for example Schmid 2013; Dzhekova et al. 2016; Veldhuis and Staun 2009; Borum 2011; Christmann 2012; Young et al 2013; Kundnani 2015.

[40] Rahimullah et al. 2013; Kruglanski et al. 2014; Dzhekova et al. 2016, p. 22.

Table 11.3 Elements of radicalisation

Level	Explanation	Factors	Examples
Micro	Individual beliefs and motivations	Personal experiences	Identity seeking, quest for significance, need for adventure, need for belonging, perceived marginalisation
		Psychological characteristics	
		Motivations	
Meso	Social surrounding/group dynamics	Social identification	Identity crisis, threat to ingroup
		Network dynamics	Social rules, friendship
		Internet	Propaganda, meeting others
Macro	Societal and political environment	Political	Discrimination, marginalisation
		Economic	Poverty, deprivation
		Cultural	Globalisation, modernisation

whereas persons who perceive that they are being marginalised should be subjected to a strategy of inclusion.[41] AAA can identify micro-, meso-, and macro-level justifications and which persons, groups or populations are particularly vulnerable to radicalisation; and which relevant factors at each level have to be addressed in order to develop an effective influence strategy.

11.4.3 Compelling Active Members: Learning from de-Radicalisation

De-radicalisation is the cognitive process by which persons actively supporting, consenting or recognising an extremist group is reversed, altering the belief system and the behaviour that justify the use of violence.[42] Most authors use the concepts of push and pull factors (see Table 11.4), where push factors are negative circumstances and social forces related to an individual's experience to turn away from terrorism, whereas pull factors are positive opportunities or outside influences that attract individuals to an alternative lifestyle.[43] The interaction between these

[41] For examples on how measures are tailored to the specific audience, see Veenkamp and Zeiger 2015.

[42] Note that authors tend to differentiate between de-radicalisation (changing beliefs) and disengagement (changing behaviour), since some persons may disengage from violence but maintain radical thoughts and beliefs. See for example Metz and Millen 2004; Gompert 2007, p. 44; Horgan 2009; Fink and El-Said 2011; Clubb 2014; Jasparro 2015, p. 95.

[43] Bjorgo 2002, pp. 11–13; Bjorgo 2009, pp. 30–48; Noricks 2009, pp. 301–304; Rabasa et al. 2010, pp. 11–12; Dalfgaard-Nielsen 2013; Altier et al. 2014; Bertram 2015.

Table 11.4 Push and pull factors

Push factors	Pull factors
Criminal prosecution	New more compelling ideology or belief structure
Loss of faith in ideology or politics of the group	Financial incentives
Discomfort with group's violent activities	New employment or educational opportunities
Disillusionment with group leadership and members	Positive interactions with moderates
Loss of confidence, status or position in the group	Competing loyalties
Ejection from group	Desire for a normal life
Increased activity in a competing role	Desire to establish a family
Parental or social disapproval	New role model or social group
Exhaustion from tension and uncertainty as a member of the group	

factors is a carrot-and-stick method, noteworthy to mention since it is also used in coercion.[44]

One interesting framework that differs from the push/pull factors is Barrelle's theory, who interviewed extremists around the world from different groups and connected their motivations for joining and leaving violence behind.[45] She identified fifteen recurring themes and categorised these into five domains, as indicated in Table 11.5.

Interestingly, every interviewed person mentioned the emergence of an '*alternate social identity*' (100%, family, work or interest groups before radicalising or created after leaving radicalism behind) and '*modification of own ideas*' (100%, finding own beliefs), the most cited other themes being '*resilience, skills and coping*' (91%, ability to continue with the de-radicalisation process and access to resources to do this), '*emergence of personal identity*' (91%, loss of dominance of the group's identity), '*physical and psychological issues*' (86%, physical and/or psychological pressure from the group, stress), '*social support*' (86%), and '*prosocial engagement in society*' (86%).[46]

Two points must be made about Barrelle's framework. First, Barrelle demonstrated that identity plays an important role in leaving extremism behind, and understanding the personal and social identity of the target through AAA allows us to tailor influence in a correct manner. It is important to consider this issue, since collective identity becomes highly activated in conflict and extremist, especially terrorist, organisations are particularly capable of 'replacing' the personal identity with the collective identity.[47] Moreover, studies concerning future warfare

---

[44] Drezner 1999, p. 188; Byman and Waxman 2002, pp. 231, 227; Jakobsen 2011, p. 156; Motwani 2012, p. 14; Bjorgo 2013, p. 90.

[45] Barrelle 2014.

[46] Barrelle 2014, p. 83.

[47] Boer and Boehnke 2016, p. 131.

Table 11.5 Barrelle's framework

Domain	Elements
Social relations	Disillusionment with leaders
	Disillusionment with group members
	Relations with 'others'
Coping	Physical and psychological issues
	Social support
	Resilience, skills, and coping
Identity	Reduction in group identity
	Emergence of personal identity
	Alternate social identity
Ideology	Disillusionment with radical ideas
	Modify own ideas
	Plurality and acceptance of difference
Action orientation	Disillusionment with radical methods
	Stop or reduce radical methods
	Prosocial engagement in society

emphasise the use of identity by charismatic leaders in conflict.[48] Second, what can be considered pull factors (positive opportunities and outside influences) played a more important role than push factors (such as disillusionment with leaders and group members). This coincides with what Lieshout and Beeres state in this book, namely that positivity is stronger than negativity.[49] Moreover, outside influences can play an important factor in leaving terrorism behind.

Once the objective is determined (delegitimisation by taking away the support, consent or recognition of the adversary) and the audience we want to influence is identified (potential supporters or active recruits), relevant factors for influence have to be analysed, which on the one hand consists of the role of the communicator and on the other basic techniques of influence and persuasion.

11.5 Principles of Social Influence and Persuasion

Social influence is the process by which individuals change their beliefs, feelings and behaviours when interacting with others, be that in actual, imagined or implied presence of others.[50] Perloff differentiates between social influence and persuasion by stating that the former is a broader process of change based on other persons,

[48] See for example Jenkins et al. 2015; Mazarr 2015; Voelz 2015; Bunker 2016.

[49] More on this from a counter- and de-radicalisation perspective can be read in El-Said 2015, pp. 1–11: 'soft measures (…) have yielded more benefits than purely kinetic approaches (…) particularly in reducing the rates of terrorist incidences'.

[50] Rashotte 2007, p. 4426; Forsyth 2013, p. 305.

whereas persuasion involves sending intentional messages in order to influence others.[51] One of the most influential authors on theories of influence, persuasion and dissuasion is Cialdini, whose framework is composed of liking, social proof, authority, commitment and consistency, scarcity, and reciprocation.[52] Though there are many theories on influence and persuasion, his framework will be used as a basis to simplify relevant elements of deterring or compelling support, consent or recognition of adversaries.[53]

11.5.1 The Role of the Communicator

The communicator is the source who tries to convey a message, regardless of content. The most cited characteristic of a source is credibility (believability), which consists of expertise (credentials and knowledge the source possesses) and trustworthiness (motivation to share truthful information).[54] Other factors that can play a role in the source are authority, liking, and similarity. Authority refers to the source's position in a social structure.[55] Liking makes persuasion more effective, for example by giving compliments, showing contact and cooperation and by being physically attractive.[56] The most important factor for the purposes of this chapter is similarity, which is the basis for differentiating members of the own group (in-group), opposing groups (out-groups) and irrelevant groups (neutral groups).[57]

As stated above, the collective identity becomes more salient in the context of war, therefore the communicator must find a way to make it clear that he or she belongs to the same social category as the target.[58] People have many identities, such as a parent, profession, sport, or even specific characteristics such as intelligent or creative, a common ground can be found.[59] This is connected to the principle of social proof, which implies that behaviours and beliefs are guided by others of the in-group that considers it as correct, especially in uncertain situations.[60] Depending on the results of AAA concerning the *degree* of salient identities other than the

---

[51] Perloff 2010, p. 13.

[52] Cialdini 2007, 2011.

[53] For an extensive overview of theories of influence and persuasion, see Paul et al. 2015, pp. 317–348.

[54] Bordens and Horowitz 2002a, b, pp. 219–220; Rucker et al. 2015, pp. 31–32.

[55] Borum 2004, pp. 49–50.

[56] Petty et al. 1983; Dal Cin et al. 2004, p. 180; Cialdini 2007, pp. 171–204.

[57] Howard 2000, p. 372; Stroink 2007, p. 298; Janoff-Bulman and Werther 2008, p. 165; Ledgerwood et al. 2014, p. 549.

[58] Mackie and Queller 2000; Bodenhausen and Morales 2013, p. 239; Ledgerwood et al. 2014, p. 549.

[59] Rogers 2011, pp. 37–38; Stets and Serpe 2013, p. 31; Goerzig and Al-Hashimi 2015, pp. 19–20, 32, 140–144; Rabasa and Benard 2015, pp. 29, 58; Roberts 2015, pp. 209–212.

[60] Cialdini 2007, pp. 129–133; Wells 2015, pp. 161–162.

collective one and the contextual factors, the *most* salient identities need to be activated. Methods to accomplish this is through de-categorisation (forming new subgroups), personalisation (finding common information of the communicator that is relevant to the self of the target) and recategorisation (integrating the target in a more inclusive group that overlaps the identified salient identities).[61]

Effective communicators, depending on the results of AAA, can be family members, ex-members of the terrorist organisation, scholars, experts, mentors, friends, and religious leaders. For example, a former right-wing extremist stated that his change was a result of interaction with a prison chef who became a mentor for him: the chef sat and talked with him about his former life, family and friends. The kindness shown by the prison guard, as well as positive experiences with a hated outgroup, led to another need for belonging than the original one (which in turn led to grievances against outgroups): belong again with his family who didn't know about his activities until he ended up in prison.[62] The same happened to an Islamist extremist who was treated well by the police officers who arrested him and surprised him by treating him with respect. It showed that part of his grievances was incorrect, since the perception portrayed by the terrorist group of the police was different than what he experienced.[63] These accounts show that liking, similarity and credibility played important roles. As for social proof, Euskadi Ta Askatasuna (ETA) can form a good example. Iñaki Rekarte, an ex-ETA member, who initially joined due to political grievances, left the organisation because of he was afraid what his son and wife would think of him.[64] It is important to remember that the ceasefire of ETA was partially caused by the disapproval of the majority of the Basque population towards them and the total rejection of employed violence.[65] Finally, former combatants played an authoritative and credible role, both in Northern Ireland and in Indonesia, based on their former role in the group (high status), knowledge, motivation and independence.[66]

The military can learn from this by using the same techniques, for example by portraying themselves differently compared to what the adversary tells the target, or reminding the person of their family and friends. One important example in military context was Colombia's Operation Christmas, where military forces decorated trees with Christmas lights in rebel-held territory and a message saying 'If Christmas can

[61] Dovidio et al. 2013.

[62] Barrelle 2014, pp. 60, 101–102.

[63] O'Brien 2008.

[64] Espí 2015.

[65] Data was found in the Euskobarómetro, an analysis made every year by the University of Basque Country which measures, amongst others, support for ETA. Whereas 64% of the Basque population totally rejected ETA, whereas the remaining 36% either thought that violence was justified in the past, but not in the present, or presented a total rejection of the violence. The last group still supports the goals, but not the manner in which ETA tried to achieve them. See Universidad del País Vasco 2011. For ETA's statement, see Europa Press (2011) ETA anuncia el cese definitivo de su actividad armada (Video File). Youtube, 20 October 2011. Accessed 16 December 2016.

[66] Clubb 2014, pp. 856–857; Sim 2013.

come to the jungle, you can come home', causing the demobilisation of 331 guerrillas in a matter of days.[67] In this case, relying on the alternate social identity and demonstrating social support by using the principle of social proof served the military's purpose in compelling the targets to leave the adversary.

11.5.2 The Message: Techniques of Persuasion

The message must be tailored to the specific audience we want to influence. In general, three principles of social influence play a role in the message to persuade the relevant target: (1) reciprocation, (2) scarcity, and (3) commitment and consistency.[68] Again, the use of the various techniques is audience-specific. First, reciprocation is perhaps one of the most common and oldest forms of influence. It entails repaying what others have provided for us, which creates a sense of obligation, is cross-cultural, and includes concessions as payment.[69] In influencing extremist adversaries, one important measure is providing alternatives. First, because providing alternatives, such as participation and engagement in group processes—based on what the target needs and wants—as a reward for not supporting, consenting or recognising the adversary, can create a positive sense of belonging.[70] This can be seen in Colombia's rehabilitation programme that offered amnesty, education and healthcare if terrorists left Fuerzas Armadas Revolucionarias de Colombia (FARC), but was unsuccessful (leading to recidivism) since the programme did not provide socio-economic opportunities.[71] It shows that AAA was done correctly, but the influence technique was not sufficient because the government could not address one core grievance—the ability to address socio-economic problems. Second, active extremists fear to end up in a social vacuum, which can be solved by providing an alternative social network based on salient identities.[72] For example, Damian Clairmont (a deceased fighter of Jabhat al Nusra) joined the terrorist organisation because he was troubled, felt rejected at school and work, and wanted to find a place in society by helping others, and Jabhat al Nusra was able to give him a quest for significance, friendship and a sense of belonging.[73]

Second, scarcity builds on various facts: when something is rare or difficult to obtain, it becomes more attractive; people become more motivated when there are potential losses; and opportunities become more valuable when they are less

[67] Logan 2016.

[68] Cialdini 2007.

[69] Cialdini 2007, p. 17; Hogh-Olesen 2013, p. 213; Wells 2015, pp. 159–160.

[70] Paul 2009, p. 128; Haslam et al. 2010, p. 50; Adelman and Chapman 2011, p. 16. Kamolnick 2012, p. 14.

[71] Ribetti 2009, pp. 166–168; Villanueva O'Driscoll et al. 2013, pp. 117–140; Marsden 2015, pp. 204–205; McColl 2016.

[72] Dandurand 2015, pp. 23–37; Goerzig and Al-Hashimi 2015, p. 16; Bjorgo 2006, p. 14.

[73] Bell 2013; Boudreau 2015.

available, especially in the case of freedoms that are threatened.[74] Imposing a deadline for compliance is also part of scarcity, and of coercion, therefore using, for example, time-limits can be a useful tool.[75] If the principle of scarcity is used, it must be tailored to what the target values. For example, former ISIS-members claim they left the group because freedom of movement and speech were impaired, amongst others, something that remained important for them and was derived from their Western 'previous life'.[76] As for valuable but scarce opportunities, right-winged extremists in Germany were especially interested and committed in a specialised programme that had the highest rates of success due to its provision of new socio-economic opportunities and support, but had a strict admittance procedure.[77]

Third, commitment and consistency are interrelated concepts (once the target has committed to a certain belief or behaviour, it will act or think accordingly due to internal and social pressure) relate to prior beliefs and behaviour that need to be consistent with new situations.[78] Consistency in identity plays also an important role, but how can we influence targets that have decided to commit to the adversary and therefore act consistently?[79] The answer lies in the motivation: a distinction must be made between targets that want to enhance their self-view and targets that need to receive feedback that confirms their self-view.[80] For the first group showing inconsistency between the adversary's goals and actions can diminish the self-view, as we saw with al-Mauritani that left Al-Qaeda because of the attacks prohibited by Islamic law.[81] For the latter group social proof will become more important. Social psychology found that favourable views of the self by others are more important than consistency.[82] For example, the visit of a cardinal to a former IRA member made him leave the group and made him realise that the portrayal of the enemy by the terrorist organisation was inconsistent with what the government and the Church actually did.[83] However, commitment and consistency are more emphasised in independent societies (in which individualism prevails over collectivism) rather than interdependent ones (which value kindness and accommodation more, so reciprocation should be the better option), its applicability depending on the results of AAA.[84]

---

[74] Cialdini 2007, p. 238; Wells 2015, pp. 162–163.

[75] See for example Jakobsen 1998; Cialdini 2007; Wells 2015.

[76] Dearden 2016; O'Neill 2016.

[77] Bjorgo et al. 2009, pp. 144–146.

[78] Jacobson 2010, pp. 15–16; Perloff 2010, pp. 238–239; Gill et al. 2015, p. 252.

[79] Shapiro 2013, p. 201.

[80] Franzese 2013, p. 286.

[81] Al Jazeera 2015.

[82] For an overview, see Baumeister et al. 2013, p. 261.

[83] Evans and O'Doherty 2014.

[84] Varnum et al. 2010; Baumeister et al. 2013, p. 266.

11.6 Conclusion

This chapter has set out some of the social psychological approaches to influence and persuasion in the context of coercion by using delegitimisation as an example. Specifically, it must be taken into account that the perspective of 'one message fits all' is ineffective, but rather that each audience has a specific frame of reference to which influence and persuasion must be tailored. Influence must serve a purpose, in the case of delegitimisation it is to diminish support, consent and recognition of the adversary so it will become weakened (Table 11.6).

To do this, friendly and neutral audiences must be prevented from supporting, consenting or recognising the adversary (deterrence), and hostile audiences should be influenced to leave the adversary's organisation (compellence). A useful tool to answer the question why people support, consent and recognise the adversary and how to influence them is Actor and Audience Analysis, which steps away from traditional social psychological approaches of beliefs, attitudes and behaviour by taking various factors into account to understand the target audience, including contextual and environmental factors.

This chapter has used Cialdini's principles as the basis to explain some basic elements of influence and persuasion. The communicator must possess certain characteristics in order to effectively influence the relevant target, such as credibility, authority, liking, social proof and similarity. Furthermore, the message must be tailored to the audience. Three basic principles of social influence can be applied, which are reciprocation (repaying what others have provided for us); scarcity (making something difficult to obtain so it becomes more attractive) and commitment and consistency (once a target has committed to a certain belief or behaviour,

Table 11.6 Influence and persuasion: a delegitimisation framework using AAA

Steps AAA	Elements	
Identify the objective	*What?* Delegitimisation: take away the support, consent or recognition of the adversary	
	Deterrence: Maintain the status quo by stopping the start of support of the adversary by friendly or neutral audiences	*Compellence*: Stop or undoing active and current support by hostile audiences
Descriptive phase	*Who?* Analysis of the adversary and the relevant frame of reference of the audience	
Prognostic phase	*Why?* Analysis of why the relevant audience (potentially) supports, consents or recognises the adversary	
Transformative phase	*How?* Determining which factors are applicable to effectively influence the audience	
	Communicator: Credibility, likeability, similarity, social proof, authority	*Techniques/Message*: Reciprocation, scarcity, commitment and consistency

it will act or think in a consistent manner). It must be noted that depending on the outcome of AAA one principle or another will be applicable.

Understanding that there is not a one-fits-all approach and that context remains crucial is important to effectively influence the relevant target. There are numerous approaches to influence and persuasion, but it must always be remembered that a comprehensive account of the audience must be made in order to influence and persuade, making different kinds of communicators and techniques of influence applicable, depending on the relevant objective and context.

References

Abrams D, Levine JM (2012) Norm Formation: Revisiting Sherif's autokinetic illusion study. In: Smith JR, Haslam A (eds) Social Psychology: Revisiting the Classical Studies. Sage, London, pp 57–75

Adelman JR, Chapman A (2011) Social Identification, Influence and Why People Join Al-Qaeda in the Arabian Peninsula (AQAP). In: Chapman A, Adelman J (eds) Influencing Violent Extremist Organizations: Focus on Al Qaeda in the Arabian Peninsula (AQAP). Department of Defense, Washington, pp 13–44

Al Jazeera (2015) Former al-Qaeda mufti: I condemn ISIL attacks (21 November 2015) http://www.aljazeera.com/programmes/talktojazeera/2015/11/abu-hafs-al-mauritani-al-qaeda-mufti-reject-islamic-state-isil-baghdadi-caliphate-151121133941210.html

Alimi EY, Demetriou C, Bosi L (2015) The Dynamics of Radicalization. A Relational and Comparative Perspective. Oxford University Press, Oxford

Altier MB, Thoroughgood CN, Horgan JG (2014) Turning away from terrorism: Lessons from psychology, sociology, and criminology. Journal of Peace Research 51(5):647–661

Andersen ML, Taylor HF (2008) Sociology - Understanding a Diverse Society. Thomson Wadsworth, Belmont

Archetti C (2013) Understanding Terrorism in the Age of Global Media - A Communication Approach. Palgrave Macmillan, London

Barrelle K (2014) Pro-Integration: Disengagement and life after extremism. Monash University, Melbourne

Baumeister RF, Masicampo EJ, Twenge JM (2013) The Social Self. In: Suls JM, Tennen HA (eds) Handbook of Psychology. John Wiley & Sons Inc, Hoboken NJ, pp 247–274

Beetham D (2012) Political Legitimacy. In: Amenta E, Nash K, Scott A (eds) The Wiley-Blackwell Companion to Political Sociology. Blackwell Publishing, West Sussex, pp 120–129

Bell S (2013) 'It's between me and God': How a Calgary high school dropout joined Syria's civil war. National Post (12 July 2013) http://www.news.nationalpost.com/news/its-between-me-and-god-how-a-calgary-high-school-dropout-joined-syrias-civil-war Accessed 29 November 2016

Benard C (2011) The Mechanics of De-Legitimization. In: Canna S, Popp G (eds) Countering Violent Extremism: Scientific Methods & Strategies. NSI, Washington DC, pp 106–111

Benard C (2015) The Mechanics of De-Legitimization. In: Canna S, Popp G (eds) Countering Violent Extremism: Scientific Methods & Strategies. NSI, Washington DC, pp 95–103

Bertram L (2015) How Could a Terrorist be De-Radicalised? Journal for Deradicalization 5:120–149

Bjorgo T (2002) Exit Neo-Nazism: Reducing Recruitment and Promoting Disengagement from Racist Groups. NUPI Working Paper 627. Norsk Utenrikspolitisk Institutt, Oslo

Bjorgo T (2006) Reducing Recruitment and Promoting Disengagement for Extremist Groups: The Case of Racist Sub-Cultures. In: Benard C (ed) A Future for the Young: Options for Helping Middle Eastern Youth Escape the Trap of Radicalization. RAND, Santa Monica CA, pp 1–30

Bjorgo T (2009) Processes of disengagement from violent groups of the extreme right. In: Bjorgo T, Horgan J (eds) Leaving terrorism behind: Individual and collective disengagement. Routledge, New York, pp 30–48

Bjorgo T (2013) Strategies for Preventing Terrorism. Palgrave Macmillan, London

Bjorgo T, Van Donselaar J, Grunenberg S (2009) Exit from right-wing extremist groups - Lessons from disengagement programmes in Norway, Sweden and Germany. In: Bjorgo T, Horgan J (eds) Leaving terrorism behind: Individual and collective disengagement. Routledge, New York, pp 135–151

Bodenhausen GV, Morales JR (2013) Social Cognition and Perception. In: Suls JM, Tennen HA (eds) Handbook of Psychology. John Wiley & Sons Inc, Hoboken NJ, pp 225–246

Boer D, Boehnke K (2016) What are values? Where do they come from? A developmental perspective. In: Brosch T, Sander D (eds) Handbook of Value - Perspectives from economics, neuroscience, philosophy, psychology, and sociology. Oxford University Press, Oxford, pp 129–151

Booth K (1981) Strategy and Ethnocentrism. Holmes and Meier, New York

Bordens KS, Horowitz IA (2002) Social Psychology. Lawrence Erlbaum Associates, Mahwah NJ

Borum R (2004) Psychology of Terrorism. University of South Florida, Tampa

Borum R (2011) Radicalization into Violent Extremism II: A Review of Conceptual Models and Empirical Research. Journal of Strategic Security 4(4):37–62

Bosi L, Della Porta D (2015) Processes of disengagement from political violence - A multi-level relational approach. In: Tellidis I, Toros H (eds) Researching Terrorism, Peace and Conflict Studies. Routledge, New York, pp 81–99

Boslaugh SE, Kreuter MW, Nicholson RA, Naleid K (2005) Comparing demographic, health status and psychosocial strategies of audience segmentation to promote physical activity. Health Educ. Res. 20(4):430–438

Boudreau C (2015) My son was born in Nova Scotia. He died an IS extremist in Syria. This can be stopped. The Globe and Mail (17 February 2015) http://www.theglobeandmail.com/opinion/my-son-died-fighting-in-syria-we-need-a-dialogue-to-stop-others-joining-him/article23023630/ Accessed 29 November 2016

Bowen W (2004) Deterrence and Asymmetry: Non-State Actors and Mass Casualty Terrorism. Contemporary Security Policy 25(1):54–70

Braddock K (2014) The talking cure? Communication and psychological impact in prison de-radicalisation programmes. In: Silke A (ed) Prisons, terrorism and extremism. Routledge, New York, pp 60–74

Bratton PC (2005) When is Coercion Successful? And Why Can't We Agree on It? Naval War College Review 58(3):99–120

Bunker RJ (2016) Old and New Insurgency Forms. Strategic Studies Institute, Carlisle PA

Bunn ME (2007) Can Deterrence Be Tailored? Strategic Forum 225:1–8

Byman D (1998) The logic of ethnic terrorism. Studies in Conflict & Terrorism 21(2):149–169

Byman D, Waxman M (2002) The Dynamics of Coercion: American Foreign Policy and the Limits of Military Might. Cambridge University Press, New York

Christmann K (2012) Preventing Religious Radicalisation and Violent Extremism - A Systematic Review of the Research Evidence. Youth Justice Board, London

Cialdini RB (2007) Influence: The psychology of persuasion. Harper Collins Publishers, New York

Cialdini RB (2011) Roots, Shoots, and Fruits of Persuasion in Military Affairs. Analyses of Social Issues and Public Policy, 11(1):27–30

Clubb G (2014) "From Terrorists to Peacekeepers": The IRA's Disengagement and the Role of Community Networks. Studies in Conflict & Terrorism 37(10):842–861

Cohen E (2016) From terrorist to university expert: GW hires former Islamic extremist. CNN (30 August 2016) http://www.edition.cnn.com/2016/08/30/health/gw-hires-former-islamic-extremist/index.html Accessed 28 November 2016
Coskun BB (2007) Does "Strategic Culture" Matter? Old Europe, New Europe and the Transatlantic Security. Perceptions 12(2):71–90
Cronin AK (2009) How terrorist campaigns end. In: Bjorgo T, Horgan J (eds) Leaving terrorism behind: Individual and collective disengagement. Routledge, New York, pp 49–65
Cwalina W, Falkowski A, Newman BI (2015) Persuasion in the Political Context: Opportunities and Threats. In: Stewart DW (ed) The Handbook of Persuasion and Social Marketing, Volume 1: Historical and Social Foundations. Praeger, Santa Barbara CA, pp 61–128
Dal Cin S, Zanna MP, Fong GT (2004) Narrative Persuasion and Overcoming Resistance. In: Knowles ES, Linn JA (eds) Resistance and Persuasion. Lawrence Erlbaum, Mahwah NJ, pp 175–192
Dalfgaard-Nielsen A (2013) Promoting Exit from Violent Extremism: Themes and Approaches. Studies in Conflict & Terrorism 36(3):99–115
Dandurand Y (2015) Social Inclusion Programmes for Youth and the Prevention of Violent Extremism. In: Lombardi M, Ragab E, Chin V et al (eds) Countering Radicalisation and Violent Extremism Among Youth to Prevent Terrorism. IOS Press, Amsterdam, pp 23–37
Davis PK (2010) Simple Models to Explore Deterrence and More General Influence in the War with al-Qaeda. RAND, Santa Monica CA
Davis PK (2012) Toward an Analytic Basis for Influence Strategy in Counterterrorism. In: Wenger A, Wilner A (eds) Deterring Terrorism: Theory and Practice. Stanford University Press, Stanford, pp 67–94
Davis PK (2014) Toward Theory for Dissuasion (or Deterrence) by Denial: Using Simple Cognitive Models of the Adversary to Inform Strategy (RAND Working Paper WR-1027). RAND, Santa Monica CA
Davis PK, Jenkins BM (2004) A System Approach to Deterring and Influencing Terrorists. Conflict Management and Peace Science 21(1):3–15
Dearden L (2016) Former ISIS militant who grew up in the UK says coalition bombing will drive more jihadists to launch attacks. Mirror (29 April 2016) http://www.mirror.co.uk/news/world-news/isis-fighter-who-escaped-terrorist-7584183 Accessed 29 November 2016
Della Porta D (2009) Leaving underground organizations: A sociological analysis of the Italian case. In: Bjorgo T, Horgan J (eds) Leaving terrorism behind: Individual and collective disengagement. Routledge, New York, pp 66–87
Dovidio JF, Gaertner SL, Wright Mayville E, Perry S (2013) Social Conflict, Harmony, and Integration. In: Suls JM, Tennen HA (eds) Handbook of Psychology. John Wiley & Sons Inc, Hoboken NJ, pp 429–448
Drezner DW (1999) The Trouble with Carrots: Transaction Costs, Conflict Expectations and Economic Inducements. Security Studies 9(1–2):188–218
Dzhekova R, Stoynova N, Kojouharov A, Mancheva M, Anagnostou D, Tsenkov E (2016) Understanding Radicalisation. Review of Literature. Center for the Study of Democracy, Sofia
Eagly AH, Chaiken S (1998) Attitude Structure and Function. In: Gilbert T, Fiske ST, Lindzey G (eds) Handbook of Social Psychology. Mc-Graw-Hill, Boston MA, pp 269–322
Echevarria A (2011) American Strategic Culture: Problems and Prospects. In: Strachan H, Scheipers S (eds) The Changing Character of War. Oxford University Press, Oxford, pp 431–445
El-Said H (2015) New Approaches to Countering Terrorism - Designing and Evaluating Counter Radicalization and De-Radicalization Programs. Palgrave Macmillan, London
Espí S (2015) Entrevista al Etarra Iñaki Rekarte. Escalofriante 'Salvados': "Ser de ETA te convertía en un héroe, celebraban atentados y nos protegían en casas". Periodista Digital (11 May 2015) http://www.periodistadigital.com/3segundos/periodismo/2015/05/11/jordi-evole-salvados-tremenda-entrevista-inaki-rekarte-etarra-arrepentido-redes-sociales-podridos-por-dentro-irene-villa-twitter.shtml Accessed 29 November 2016

Evans R, O'Doherty SP (2014) 8 Terrifying Life Lessons from a Former Terrorist. Cracked (28 July 2014) http://www.cracked.com/personal-experiences-1420-8-terrifying-life-lessons-from-former-terrorist.html Accessed 29 November 2016

Ferguson N (2016) Disengaging from Terrorism: A Northern Irish Experience. Journal for Deradicalization 6(1):1–23

Fink NC, El-Said H (2011) Transforming Terrorists: Examining International Efforts to Address Violent Extremism. International Peace Institute, New York

Forsyth DR (2013) Social Influence and Group Behavior. In: Suls JM, Tennen HA (eds) Handbook of Psychology Volume 5: Personality and Social Psychology. John Wiley & Sons Inc, Hoboken NJ, pp 305–328

Franzese AT (2013) Motivation, Motives, and Individual Agency. In: Delamater J, Ward A (eds) Handbook of Social Psychology. Springer, Dordrecht, pp 191–224.

Fussey P (2011) Deterring Terrorism? Target-hardening, surveillance and the prevention of terrorism. In: Silke A (ed) The Psychology of Counter-Terrorism. Routledge, New York, pp 164–185

Gavac S, Murrar S, Brauer M (2016) Group Perception and Social Norms. In: Summers RW (ed) Social Psychology: How Other People Influence Our Thoughts and Actions. ABC-CLIO, Santa Barbara CA, pp 333–361

George AL, Smoke R (1974) Deterrence in American Foreign Policy: Theory and Practice. Columbia University Press, New York

Gill P, Bouhana N, Morrisson J (2015) Individual Disengagement from Terrorist Groups. In: Kennedy-Pipe C, Clubb G, Mabon S (eds) Terrorism and Political Violence. SAGE, London, pp 243–257

Goerzig C, Al-Hashimi K (2015) Radicalization in Western Europe: Integration, public discourse, and loss of identity among Muslim communities. Routledge, New York

Gompert DC (2007) Heads We Win - The Cognitive Side of Counterinsurgency (COIN). RAND, Santa Monica CA

Gross Stein J (2012) Deterring Terrorism, Not Terrorists. In: Wenger A, Wilner A (eds) Deterring Terrorism: Theory and Practice. Stanford University Press, Stanford, pp 46–66

Haniff Bin Hassan M (2012) Key considerations in counterideological work against terrorist ideology. In: Horgan J, Braddock K (eds) Terrorism Studies: A Reader. Routledge, New York, pp 358–384

Haslam SA, Reicher SD, Platow MJ (2010) The New Psychology of Leadership: Identity, Influence and Power. Psychology Press, East Sussex

Heuser B (2010) The Evolution of Strategy - Thinking War from Antiquity to the Present. Cambridge University Press, Cambridge

Hitlin S, Pinkston K (2013) Values, Attitudes, and Ideologies: Explicit and Implicit Constructs Shaping Perception and Action. In: Delamater J, Ward A (eds) Handbook of Social Psychology. Springer, Dordrecht, pp 319–340

Hogh-Olesen H (2013) The will to sacrifice: Sharing and sociality in humans, apes and monkeys. In: Geertz AW (ed) Origins of Religion, Cognition and Culture. Routledge, New York, pp 203–218

Horgan J (2009) Walking Away from Terrorism: Accounts of Disengagement from Radical and Extremist Movements. Routledge, New York

Hovland CI, Janis IL, Kelley HH (1953) Persuasion and communication. Yale University Press, New Haven CT

Howard JA (2000) Social Psychology of Identities. Ann Rev of Soc 26:367–393

Jackson BA (2009) Organizational Decisionmaking by Terrorist Groups. In: Davis PK, Cragin K (eds) Social Science for Counterterrorism - Putting the Pieces Together. RAND, Santa Monica CA, pp 209–256

Jacobson M (2010) Terrorist Dropouts: Learning from Those Who Have Left (Policy Focus No. 101). The Washington Institute for Near East Policy, Washington DC

Jakobsen PV (1998) Western Use of Coercive Diplomacy after the Cold War: A Challenge for Theory and Practice. Macmillan Press Ltd, Basingstoke

Jakobsen PV (2010) Coercive Diplomacy. In: Collins A (ed) Contemporary Security Studies. Oxford University Press, Oxford, pp 277–298

Jakobsen PV (2011) Pushing the Limits of Military Coercion Theory. International Studies Perspectives 12(2):151–170

Janoff-Bulman R, Werther A (2008) The social psychology of respect: Implications for Delegitimization and Reconciliation. In: Nadler A, Malloy T, Fisher JD (eds) Social Psychology of inter-group reconciliation: From violent conflict to peaceful co-existence. Cambridge University Press, Cambridge, pp 145–171

Jasparro C (2015) The Case for Cultural Heritage Protection as an Element of COIN. In: Cline LW, Shemella P (eds) The future of counterinsurgency: Contemporary debates in internal security strategy. ABC-CLIO, Santa Barbara CA, pp 91–120

Jenkins BM (2006) Unconquerable Nation: Knowing Our Enemies, Strengthening Ourselves. RAND, Santa Monica CA

Jenkins DM, Endersby L, Guthrie KL (2015) Leadership Education 2050: Changing the Spaces and Faces of Experience. In: Sowcik M, Andenoro AC, McNutt M, Murphy SE (eds) Leadership 2050: Critical Challenges, Key Contexts, and Emerging Trends. Emerald Group Publishing, Bingley, pp 127–144

Johnson DE, Mueller KP, Tafter WH (2002) Across the Spectrum of Operations - The Utility of U.S. Military Forces in the Emerging Security Environment. RAND, Santa Monica CA

Johnston AI (1995) Thinking about Strategic Culture. International Security 19(4):36–43

Kalmonick P (2012) Delegitimizing Al-Qaeda: A Jihadist-Realist Approach. Strategic Studies Institute, Carlisle PA

Kelman HC (2001) Reflection on Social and Psychological Processes of Legitimization and Delegitimization. In: Jost JT, Major B (eds) The Psychology of Legitimacy. Cambridge University Press, Cambridge, pp 54–73

Knopf JW (2010) The Fourth Wave in Deterrence Research. Contemporary Security Policy 31 (1):1–33

Knopf JW (2012) Terrorism and the Fourth Wave in Deterrence Research. In: Wenger A, Wilner A (eds) Deterring Terrorism. Theory and Practice. Stanford University Press, Stanford, pp 21–45

Koehler D (2015) De-radicalization and Disengagement Programs as Counter-Terrorism and Prevention Tools - Insights From Field Experiences Regarding German Right-Wing Extremism and Jihadism. In: Lombardi M, Ragab E, Chin V et al (eds) Countering Radicalisation and Violent Extremism Among Youth to Prevent Terrorism. IOS Press, Amsterdam, pp 120–150

Kroenig M, Pavel B (2012) How to Deter Terrorism. The Washington Quarterly 35(2):21–36

Krugslanski AW, Gelfand MJ et al (2014) The Psychology of Radicalization and Deradicalization: How Significance Quest Impacts Violent Extremism. Advances in Political Psychology 35 (1):69–93

Kundnani A (2015) A Decade Lost: Rethinking Radicalisation and Extremism. Claystone, London

Lamb RD (2014) Rethinking Legitimacy and Illegitimacy - A New Approach in Assessing Support and Opposition across Disciplines (CSIS Report). Rowman & Littlefield, Lanham MD

Lantis JS, Howlett D (2013) Strategic Culture. In: Baylis J, Wirtz JJ, Gray CS (eds) Strategy in the Contemporary World - An Introduction to Strategic Studies. Oxford University Press, Oxford, pp 76–95

Lebow RN, Gross Stein J (1989) Rational Deterrence Theory: I Think, Therefore I Deter. World Politics 41(2):208–224

Ledgerwood A, Callahan SP, Chaiken S (2014) Changing Minds: Persuasion in Negotiation and Conflict Resolution. In: Coleman PT, Deutsch M, Marcus EC (eds) The Handbook of Conflict Resolution: Theory and Practice. Jossey-Bass, San Francisco CA, pp 533–557

Logan L (2016) How Unconventional Thinking Transformed a War-Torn Colombia. CBS News (11 December 2016) http://www.cbsnews.com/news/60-minutes-colombia-after-civil-war-lara-logan/ Accessed 30 December 2016

Long JM (2009) Strategic Culture, Al-Qaeda, and Weapons of Mass Destruction. In: Johnson JL, Kartchner KM, Larsen JA (eds) Does Al Qaeda have a Strategic Culture? Palgrave Macmillan, London, pp 201–220

Long JM, Wilner A (2014) Delegitimizing al-Qaida. Defeating an "Army Whose Men Love Death". International Security 39(1):126–164

Lowe D (2014) Radicalization of Terrorist Causes: The 32CSM/IRA Threat to U.K. Security. In: Lowe D, Turk A, Das DK (eds) Examining Political Violence - Studies of Terrorism, Counterterrorism and Internal War. CRC Press, Boca Raton FL, pp 25–56

Mac Ginty R (2010) Social network analysis and counterinsurgency: a counterproductive strategy? Critical Studies on Terrorism 3(2):209–226

Mackie DM, Queller S (2000) The impact of group membership on persuasion: Revisiting "Who says what to whom with what effect?" In: Terry DJ, Hogg MA (eds) Attitudes, behavior, and social context: the role of norms and group membership. Erlbaum, Mahwah NJ, pp 135–155

Maio GR, Haddock G (2007) Attitude Change. In: Kruglanksi AW, Higgins ET (eds) Social Psychology. The Guildford Press, New York, pp 565–586

Marsden S (2015) How Terrorism Ends. In: Kennedy-Pipe C, Clubb G, Mabon S (eds) Terrorism and Political Violence. SAGE, London, pp 199–225

Mazarr MJ (2015) Mastering the Gray Zone: Understanding a Changing Era of Conflict. Strategic Studies Institute, Carlisle PA

McAllister B, Schmid AP (2011) Theories of Terrorism. In: Schmid AP (ed) The Routledge Handbook of Terrorism Research. Routledge, New York, pp 201–271

McBride MK (2011) The Logic of Terrorism: Existential Anxiety, the Search for Meaning and Terrorist Ideologies. Terrorism and Political Violence 23:560–581

McColl R (2016) Who Will Hire Colombia's Ex-Terrorists? The Daily Beast (21 May 2016) http://www.thedailybeast.com/articles/2016/05/21/who-will-employ-colombia-s-future-ex-guerrillas.html Accessed 29 November 2016

Medina RM, Hepner GF (2013) The Geography of International Terrorism - An Introduction to Spaces and Places of Violent Non-State Groups. CRC Press, Boca Raton FL

Metz S, Millen R (2004) Insurgency and Counterinsurgency in the 21st Century: Reconceptualizing Threat and Response. Strategic Studies Institute, Carlisle PA

Miller GD (2013) Terrorist Decision Making and the Deterrence Problem. Studies in Conflict & Terrorism 36(2):132–151

Motwani N (2012) The Relevance of Coercive Diplomacy despite its Mixed Results. Burgmann Journal 1:13–20.

NATO (2007) AJP-3.10.1(A): Allied Joint Doctrine for Psychological Operations, October 2007

NATO (2010) AJP-01(D): Allied Joined Doctrine, December 2010

NATO (2011), AJP 3–4.4: Allied Joint Doctrine for Counterinsurgency (COIN), February 2011

NATO (2013) AJP-5: Allied Joint Doctrine for Operational-Level Planning, June 2013

Noricks DME (2009) The Root Causes of Terrorism. In: Davis PK, Cragin K (eds) Social Science for Counterterrorism. Putting the Pieces Together. RAND, Santa Monica CA, pp 11–70

Northouse (2013) Leadership: Theory and Practice. SAGE, London

O'Neill K (2016) ISIS fighter who escaped terrorist group reveals he didn't 'support ideology' in first interview. Mirror (18 March 2016) http://www.mirror.co.uk/news/world-news/isis-fighter-who-escaped-terrorist-7584183 Accessed 29 November 2016

O'Brien N (2008) Interview with a Former Terrorist: Nasir Abbas' Deradicalization Work in Indonesia. CTC Sentinel 1:12

Paul C (2009) How do Terrorists Generate and Maintain Support? In: Davis PK, Cragin K (eds) Social Science for Counterterrorism: Putting the Pieces Together. RAND, Santa Monica CA, pp 113–150

Paul C, Yeats J, Clarke CP, Matthews M, Skrabala L (2015) Assessing and Evaluating Department of Defense Efforts to Inform, Influence and Persuade: Desk Reference. RAND, Santa Monica CA

Perloff RM (2010) The dynamics of persuasion. Communication and attitudes in the 21st century. Routledge, New York

Petty RE, Cacioppo JT, Schumann D (1983) Central and peripheral routes to advertising effectiveness: The moderating role of involvement. Journal of Consumer Research 10(2):135–146

Petty RE, Wheeler SC, Tormala ZL (2013) Persuasion and Attitude Change. In: Tennen HA, Suls JM. (eds) Handbook of Psychology, Volume 5: Personality and Social Psychology. John Wiley & Sons, Hoboken NJ, pp 369–390

Rabasa A, Benard C (2015) Eurojihad. Patterns of Islamist Radicalization and Terrorism in Europe. Cambridge University Press, Cambridge

Rabasa A, Pettyjohn SL, Ghez JJ, Boucek C (2010) Deradicalizing Islamist Extremists. RAND, Santa Monica CA

Rahimullah RH, Larmar S, Abdalla M (2013) Understanding Violent Radicalization amongst Muslims: A Review of the Literature. Journal of Psychology and Behavioral Science 1(1):19–35

Rashotte L (2007) Social Influence. In: Ritzer G (ed) The Blackwell Encyclopedia of Sociology. Blackwell Publishing, West Sussex

Ribetti M (2009) Disengagement and beyond: A case study of demobilization in Colombia. In: Bjorgo T, Horgan J (eds) Leaving terrorism behind: Individual and collective disengagement. Routledge, New York, pp 152–169

Rid T (2012) Deterrence beyond the State: The Israeli Experience. Contemporary Security Policy 33(1):124–147

Roberts K (2015) Social Psychology and the Investigation of Terrorism. In: Pearse J (ed) Investigating Terrorism. Current Political, Legal and Psychological Issues. John Wiley & Sons, West Sussex, pp 202–213

Rogers B (2011) The psychology of violent radicalisation. In: Silke A (ed) The Psychology of Counter-Terrorism. Routledge, New York, pp 34–47

Rucker DD, Petty RE, Briñol P (2015) Social Psychological Foundations of Social Marketing. In: Stewart DW (ed) The Handbook of Persuasion and Social Marketing, Volume 1: Historical and Social Foundations. Praeger, Santa Barbara CA, pp 27–60

Sauer F (2015) Atomic Anxiety. Deterrence, Taboo and the Non-Use of U.S. Nuclear Weapons. Palgrave Macmillan, London

Schaefer RT (2007) Sociology. McGraw-Hill, New York

Schelling T (1960) The Strategy of Conflict. Harvard University Press, Cambridge

Schelling T (1966) Arms and Influence. Yale University Press, New Haven

Schmid AP (2013) Radicalisation, De-Radicalisation, Counter-Radicalisation: A Conceptual Discussion and Literature Review (ICCT Research Paper). International Centre for Counter-Terrorism, The Hague

Seese G (2014) Comprehensive PSYOP Assessment and Evaluation in Counterterrorism Efforts. In: NATO Centre of Excellence Defence Against Terrorism. Strategic Communication in Counter Terrorism: Target Audience Analysis, Measures of Effect and Counter Narrative 04–05 June 2014, pp 19–23

Shapiro E (2013) Social Psychology and the Body. In: Delamater J, Ward A (eds) Handbook of Social Psychology. Springer, Dordrecht, pp 191–224

Sheehan M (2010) Military Security. In: Collins A (ed) Contemporary Security Studies. Oxford University Press, Oxford, pp 167–184

Sherif M (1937) An experimental approach to the study of attitudes. Sociometry 1:90–98

Sim S (2013) Countering Violent Extremism: Leveraging Terrorist Dropouts to Counter Violent Extremism in Southeast Asia. Qatar International Academy for Security Studies, Qatar

Spaaij R (2012) Understanding Lone Wolf Terrorism. Springer, Dordrecht

Sperandai M (2006) Bridging Deterrence and Compellence: An Alternative Approach to the Study of Coercive Diplomacy. International Studies Review 8(2):253–280

Stets JE, Serpe RT (2013) Identity Theory. In: Delamater J, Ward A (ed) Handbook of Social Psychology. Springer, Dordrecht, pp 31–60

Stroink ML (2007) Processes and Preconditions Underlying Terrorism in Second-Generation Immigrants. Peace and Conflict: Journal of Peace Psychology 13(3):293–312

Suchman MC (1995) Managing Legitimacy: Strategic and Institutional Approaches. The Academy of Management Review 20(3):571–610
Tatham S (2013) U.S. Governmental Information Operations and Strategic Communications: A Discredited Tool or User Failure? Implications for Future Conflict. Strategic Studies Institute, Carlisle PA
Tatham S (2015a) Using Target Audience Analysis to Aid Strategic Level Decisionmaking. Strategic Studies Institute, Carlisle PA
Tatham S (2015b) Target Audience Analysis. Three Swords Magazine 28:50–53
Tatham S, Giles K (2015) Training Humans for the Human Domain. Strategic Studies Institute, Carlisle PA
Thies WJ (2003) Compellence Failure of Coercive Success? The Case of NATO and Yugoslavia. Comparative Strategy 22(3):243–267
Toje A (2009) Strategic Culture as an Analytical Tool. History, Capabilities, Geopolitics and Values: The EU Example. Western Balkans Security Observer 14(4):3–23
Toros H (2008) 'We Don't Negotiate with Terrorists!': Legitimacy and Complexity in Terrorist Conflicts. Security Dialogue 39(4):407–426
United Kingdom Ministry of Defence (2014) AJP-3.10.1: Allied Joint Doctrine for Psychological Operations (Edition B version 1 with UK National Elements), September 2014 http://www.gov.uk/government/uploads/system/uploads/attachment_data/file/450521/20150223-AJP_3_10_1_PSYOPS_with_UK_Green_pages.pdf Accessed 4 March 2017
United States Department of the Army (2013) Field Manual 3-53 (FM 3-53): Military Information Support Operations, January 2013
Universidad del País Vasco (2011) Euskobarómetro Mayo 2011. University of Basque Country, Bilbao. http://www.ehu.eus/documents/1457190/1525260/EB1105.pdf Accessed 29 November 2016
Van Gorp B (2007) The Constructionist Approach to Framing: Bringing Culture Back In. Journal of Communication 57(1):60–78
Varnum MEW, Grossmann I, Kitayama S, Nisbett RE (2010) The Origin of Cultural Differences in Cognition: Evidence for the Social Orientation Hypothesis. Curr Dir Psychol Sci 19(1):9–13
Veenkamp I, Zeiger S (2015) Countering Violent Extremism: Program and policy approaches relating to youth through education, families and communities. In: Lombardi M, Ragab E, Chin V et al (eds) Countering Radicalisation and Violent Extremism Among Youth to Prevent Terrorism. IOS Press, Amsterdam, pp 151–163
Veldhuis T, Staun J (2009) Islamist Radicalisation: A Root Cause Model. Netherlands Institute of International Relations Clingendael, The Hague
Voelz GJ (2015) The Rise of iWar: Identity, Information, and the Individualization of Modern Warfare. Strategic Studies Institute, Carlisle PA
Villanueva O'Driscoll J, Loots G, Derluyn I (2013) Children Disengaged from Armed Groups in Colombia: Integration Processes in Context. Versita, London
Walton CD (2012) Weapons of Mass Destruction. In: Jordan D, Kiras JD, Lonsdale DJ et al (eds) Understanding Modern Warfare. Cambridge University Press, Cambridge, pp 292–340
Wells S (2015) Hostage Negotiation and Communication Skills. In: Pearse J (ed) Investigating Terrorism - Current Political, Legal and Psychological Issues. John Wiley & Sons, West Sussex, pp 144–166
Williams MA, Covarrubias J (2009) Rebalancing America's War on Terror: President Obama. In: Lansford T, Watson RP, Covarrubias J (eds) America's War on Terror. Ashgate Publishing Ltd, Surrey, pp 261–282
Wilner A (2010a) Delegitimising Terrorism: A better way to counter radicalization and recruitment in the West. Atlantic Institute for Market Studies, Nova Scotia
Wilner A (2010b) Targeted Killings in Afghanistan: Measuring Coercion and Deterrence in Counterterrorism and Counterinsurgency. Studies in Conflict & Terrorism 33(4):307–329
Wilner A (2011) Deterring the Undeterrable: Coercion, Denial, and Delegitimization in Counterterrorism. Journal of Strategic Studies 34(1):3–37

Wilner A (2012) Apocalypse Soon? Deterring Iran and its Terrorist Proxies. Comparative Strategy 31(1):18–40

Wilner A (2013) Fencing in Warfare: Threats, Punishment, and Intra-War Deterrence in Counterterrorism. Security Studies 22(4):740–772

Wilner A (2015a) Deterring Rational Fanatics. Pennsylvania University Press, Philadelphia

Wilner A (2015b) Contemporary Deterrence Theory and Counterterrorism: A Bridge Too Far? NYU Journal of International Law and Politics 47(2):439–462

Young HF, Zwenk F, Rooze M (2013) A review of the literature on radicalisation; and what it means for TERRA. Terra, Diemen

Author Biography

C.I.A. van Kuijck, LL.M., M.A. is a Researcher of non-kinetic influence at the Royal Netherlands Air Force's Innovation Centre AIR, co-located at the Faculty of Military Sciences, Netherlands Defence Academy. She holds master degrees in International Law and Criminal Law from Tilburg University and a master degree in Military Strategic Studies from the Netherlands Defence Academy.

Chapter 12
Profiling Terrorists—Using Statistics to Fight Terrorism

Robert Beeres, Robert Bertrand and Myriame Bollen

Some families produce baseball players. Others produce terrorists.

(Levitt and Dubner 2009, p. 62)

Abstract The financial war on terror necessitates authorities and governments to search for novel methods and tools to identify, prosecute and round up (members of) terrorist networks and, preferably, preventively thwart their activities. *Terrorist profiling,* originating in offender profiling, provides one such method. It aims to derive information from seemingly relevant and coherent sets of attributes. Grounded in *counter threat finance*, and geared towards incapacitating terrorist business models and underpinning financial resources, the toolkit used is derived from another area of misuse of financial services, namely money laundering. This chapter, based on practice and scarce research, thus far, renders an account of developments in terrorist profiling in the financial economic realm. Specifically, it is discussed whether statistical profiling based on retail banking behaviour can be considered effective in the financial fight against terrorism. We conclude that, to date, similar to the effectiveness of FATF's financial measures to fight terrorism, the effectiveness of statistical profiling in this respect remains limited.

Keywords Counter threat finance · Statistical profiling · Terrorism

R. Beeres (✉) · R. Bertrand · M. Bollen
Faculty of Military Sciences, Netherlands Defence Academy, Breda, The Netherlands
e-mail: RJM.Beeres@mindef.nl

R. Bertrand
e-mail: RMM.Bertrand@mindef.nl

M. Bollen
e-mail: MTIB.Bollen.01@mindef.nl

P.A.L. Ducheine and F.P.B. Osinga (eds.), *Netherlands Annual Review of Military Studies 2017*, NL ARMS, DOI 10.1007/978-94-6265-189-0_12

Contents

12.1 Introduction

The 9/11 terrorist attacks have caused tremendous impact on modern day society, including the financial-economic realm. Apart from *direct* economic effects (i.e., the costs of the event itself; including 3,000 people killed),[1] *first order indirect* effects (i.e., costs induced by the agents' reactions to the threat of new terrorist attacks) also, are to be taken into account. For instance, following 9/11, Blalock et al., observe a significant increase in driving fatalities, due to the fact people were prompted "to substitute air transportation for safer road transportation". From 9/11 until September 2003, Blalock et al. conclude "a total of about 2,170 driving deaths may be attributable to the attacks".[2] Last, *second order indirect* effects (i.e. costs caused by political response to potential threat and to the agent's reactions) are to be taken into account.[3] The wars in Afghanistan and Iraq, both due to 9/11, serve as one example.[4] Moreover, within the financial-economic realm, President Bush, by imposing Executive Order 13223 to "freeze assets and block financial transactions of those allegedly involved with terrorism",[5] intended to hit the global terror network another blow. According to McCulloch and Pickering, combating terrorist financing is a crucial instrument in the war against terror. Ayers argues the war on terrorism will not be fought on the battlefield, but instead "will be fought in the halls of our financial institutions" and "may be won by the destruction of check books".[6] McCulloch and Pickering add that "Under the auspices of the 'financial war on terror', 21st-centrury warriors on the neo-liberal frontier are more likely to be wearing suits than combat gear; armed with briefcases rather than weapons".[7]

Financial warfare is of course not all that novel; in fact, it dates back to at least early medieval Europe, where up until the late 19th century, citizens could be condemned to *la mort civile,* depriving them of all rights, and excluding them from

---

[1] Coleman 2011, p. 4091.

[2] Blalock et al. 2009, p. 1717.

[3] Beeres and Bollen 2012, p. 149; Brück 2005, p. 380; Krugman 2004; Müller 2011, p. 131.

[4] Levitt 2012, p. 1.

[5] McCulloch and Pickering 2005, p. 470.

[6] Ayers 2001, pp. 458–459.

[7] McCulloch and Pickering 2005, p. 470.

marriage, possessions and contract titles. Thus, effectively barred from society, indeed, *les morts civiles* can be considered dead while still alive. A true example of winning without fighting, or rather, killing without fighting.

Returning to current financial measures imposed by governments partaking in the financial war on terror, both preventive and repressive measures can be distinguished. Preventive measures aim to establish "a regulatory regime for financial institutions, intended to reduce the scope for using the financing systems to collect and transfer funds for terrorist purposes".[8] Repressive measures originate from agreements between countries and aim to alter national laws to such extents, that certain acts become illegal. Examples are establishing blacklists of suspicious persons or organizations, freezing bank accounts, and excluding certain individuals or organizations from insurances and financial services.[9]

Meanwhile, the use of repressive measures in the financial war on terror is contested. Napoleoni, in 2010, claims, "Thousands of innocent individuals ended up on blacklists, trapped in a financial Guantanamo as illegal as the Cuban one is real". "Up until ... late 2009, only four people mentioned on the lists have been found guilty. Dozens of individuals and companies have been removed from the lists due to lack of evidence, but hundreds of thousands still are waiting, pending investigations".[10] Donohue adds: "The political and humanitarian costs of the ... antiterrorist finance provisions go well beyond the security of freedom rubric. Exercise of the powers—particularly in the absence of direct evidence of a person's or business's participation in terrorist activity—may alienate international partners. And domestic and international populations ... needed to help respond to terrorist claims may find ethnic targeting and the state's refusal to respond to humanitarian issues unacceptable. Such behaviour underscores terrorist assertions that, since September 11, have painted Western states as going after individuals—and states—on the basis of race and religion".[11]

One preventive measure against (potential) terrorists focuses on the identification of individuals "likely to engage in terror acts".[12] Contra-terrorist efforts in this domain lean on information provided by informants, surveillance of communications (e.g., tapping telephone, email and internet communications) and unveiling the financial resources feeding into terrorist business models. This measure is grounded in assumptions about a worldwide financial system holding extensive information that can be used to identify and, hence, incapacitate members of terrorist organizations. In 2007, then US Treasury Secretary, Henry Paulson, voiced the workings of the system as follows.

---

[8] Raphaeli 2003, p. 80.

[9] Beeres and Bollen 2011, p. 99.

[10] Napoleoni 2010, p. 29.

[11] Donohue 2008, pp. 175–176.

[12] Levitt 2012, p. 1.

> Illicit actors who otherwise try to avoid detection often use the formal financial system because there is no good alternative and, in many cases no alternative at all … These transactions typically leave a trail of detailed information which we can follow to identify key actors and their networks. Opening an account or initiating a funds transfer requires a name, an address, a phone number; identification information that does not lie. Unlike a phone call or conversation that essentially disappears if it's not captured at the moment it occurs, the financial system produces records that tend to survive.[13]

Following from the Secretary's thoughts, statistical data generated by daily retail banking can be used to trace potential terrorists and, therefore, to profile them. Recently, attention is geared towards data-driven analytics, as a new pathway for knowledge creation, in this case, aiming to identify, prosecute and punish terrorists and other criminal actors.[14] However, to date, scientific research as to its effectiveness remains scarce. In one study, Levitt and Dubner show the use of retail banking data for identifying and profiling potential terrorists using statistics.[15] To this end, based on personal information of the terrorists that committed the 2007 London attacks, the authors browsed through the detailed administrative records of a large British bank and demonstrated a number of variables in retail data to be strongly correlated with terrorism-related activities. Moreover, a particular pattern of banking behaviour decisive to trace terrorist is put forward.[16]

Levitt's and Dubner's research will be elaborated on in Sect. 12.4 of this chapter. First, however, a prerequisite for any meaningful profiling, is understanding—perpetrators of—terrorism. Therefore, in the next section, we take a closer look at terrorists, terrorist acts and terrorist organizational networks. Subsequently, based on (semi)structured data, derived, amongst others, from retail banking, Sect. 12.3 discusses some problems regarding statistical profiling, exposing and prosecuting terrorists (or criminals). Next, in Sect. 12.4, we return to Levitt's and Dubner's research. All sections in this chapter serve the purpose of finding provisional answers as to whether intelligent data analyses are useful to monitor and fight potential terrorists and, whether, consequentially, terrorist attacks may thus be prevented (Sect. 12.5).

12.2 Terrorism

Meaningful terrorist profiling starts by defining what constitutes terrorists, terrorist acts and terrorist organizations. According to Keene, "terrorism has served as one of the most common expressions of man's inhumanity to man since time immemorial".[17] The concept of terrorism appears elusive and definitions appear either highly

[13] Paulson 2007.

[14] van Fenema et al. 2015; Leese 2014; Stielkowski and Lisin 2013.

[15] Levitt and Dubner 2009, pp. 87–96; Levitt 2012; Levitt and Dubner 2009, pp. 161–165.

[16] Levitt and Dubner 2014, p.164.

[17] Keene 2014, p. 4.

Table 12.1 Al Qaeda and Islamic State inspired terrorist attacks (Coleman 2012 and Lister et al. 2017)

Date	Location	Deaths	Injuries	Description	Attributed to
11-9-2011	New York and Washington	3,000	6000+	Hijacked planes crash into the World Trade Centre and Pentagon. The fourth plane crashes into a field in Pennsylvania	Al Qaeda
12-10-2002	Bali	202	240	Attacks against US Consulate, Sari Club and Paddy's Bar	Al Qaeda
11-3-2004	Madrid	192	2050	Bomb attacks against four trains	Al Qaeda
7-7-2005	London	53	700+	Four bombs aboard tube trains and a bus	Al Qaeda
13-11-2015	Paris	130	368	Shootings, suicide bombings, grenade, hostage taking	Islamic State
22-3-2016	Brussels	32	340	Suicide bombers attack a metro station and Brussels airport	Islamic State
28-6-2016	Istanbul	44	239	Three men open fire on Ataturk Airport before blowing themselves up	Islamic State

impressionistic, reflecting personal opinions, or else they are of an all-embracing nature to cover a broad variety of acts and actors.

Perhaps the most comprehensive and authorative definition of terrorism is put forward by Schmid and Jongman. "Terrorism is an anxiety-inspiring method of repeated violent action, employed by a (semi-)clandestine individual, group or state actors, for idiosyncratic, criminal or political reasons, whereby –in contrast to assassination—the direct targets of violence are not the main targets. The immediate human victims of violence are generally chosen randomly (targets of opportunity) or selectively (representative or symbolic targets) from a target population, and serve as message generators. Threat- and violence based communication processes between terrorist (organization), (imperilled) victims, and main targets are used to manipulate the main target audience(s), turning it into a target of terror, a target of demands, or a target of attention, depending on whether intimidation, coercion, or propaganda is primarily sought".[18] According to these authors, states, also, can be labelled terrorist organizations.

Avoiding the ongoing debate on defining terrorism, while acknowledging the variety of types of actors that may be engaged in terrorist activities, this chapter adheres to an induction based perspective.[19] By means of *induction*, based on the assumption that certain acts of violence will not be condoned in certain societies, there may be a degree of consensus to be reached on what constitutes an *act of*

[18] Schmid and Jongman 1988, p. 28.

[19] See for instance Sinai 2008, p. 11; Fitzgerald 2004, p. 387.

terrorism. In other words, in the eye of the beholder, it may be agreed certain groups and acts are considered terrorist. Proceeding along this inductive pathway, in Table 12.1 we provide a number of enactments we regard to be terrorist attacks.

By labelling specific acts as terrorist attacks, it becomes possible to search for common traits amongst the perpetrators. At the start, such potential search attributes will be ranging widely from, amongst others, countries of origin, male or female actors, religion, age, and (financial) behaviour. At its core, profiling attempts to derive information from combining apparently relevant and coherent sets of attributes to discover, trace and put a halt to potential terrorist before they commit an attack. The next section elaborates on various approaches to profiling and their effectiveness.

12.3 Profiling Terrorists

Presuming the *raison d' être* of counter terrorism is to identify (members of) terrorist organizations and to prevent them from executing attacks, it is essential to be able to draw distinctions between terrorists and non-terrorists.[20] To do so, pre-supposes an ability to construct terrorist profiles that, offer in-depth and accurate knowledge on, for instance, what terrorists look like, where they live, any typical behavioural characteristics, life styles and working habits, or even whether some families indeed raise terrorist offspring.[21]

In practice as well as in literature, from the 1970s, much attention is devoted to developing so-called offender profiles. Offender profiling, traditionally, "involves the construction of a profile of the offender from the characteristics of the crime or the crime scene".[22] Meanwhile, to the public at large, the concept has gained almost mythical proportions,[23] not in the least due to Anthony Hopkins, starring as the legendary profiler Dr. Hannibal "The Cannibal" Lecter in Silence of the Lambs. Literature distinguishes three approaches to offender profiling: (1) the criminal investigative approach, (2) the clinical practitioner approach and (3) the statistical approach.[24] The criminal investigative approach, as developed by the FBI in the 1970s, according to Alison et al. presents "the first attempt to utilize all information on a violent offence in combination with considerable investigative knowledge to make inferences about the type of unknown offender".[25] Both the clinical practitioner approach and the criminal investigative approach have "traditionally been a

[20] Rae 2012, p. 64.

[21] Attran 2006; Charlesworth 2003; Clarke 2015; Davis et al. 2004; Krueger 2007; Rae 2012; Sageman 2004.

[22] Philips and Pohl 2012, p. 151.

[23] Alison et al. 2010, p. 115.

[24] Alison et al. 2010, pp. 116–117; Philips and Pohl 2012, p. 155.

[25] Alison et al. 2010, p. 117.

mixture of experience and tacit knowledge but from a clinical practitioner's knowledge experience base rather than a law enforcement knowledge-experience base".[26] Finally, the statistical approach "is primarily based on the multivariate analysis of behavioural and other information found at the crime scene to infer an offender's characteristics and psychological processes".[27]

Within the scope of this chapter, we do not aim to elaborate extensively on either of these approaches and neither on their advantages or limitations. All three approaches endeavour to generate explanations as to why people turn into offenders. Although the actual offenses and offenders differ, there is some consensus that, principles underpinning offender profiling can be attributed to terrorist profiling as well.[28]

Both within the first and second approach to profiling, the theoretical construction of profiles is grounded in professional knowledge and experience, that, according to Leese, may tend to skew analyses towards groups that become framed as more 'risky' than others.[29] A number of authors consider these approaches therefore scientifically unsubstantiated. The statistical approach to profiling, however, aims to generate knowledge based on scientifically derived conjectures and refutations.

This means rigorous testing of hypotheses regarding racial or religious characteristics (e.g., are Arabs and Muslims more likely to commit acts of terrorism?), gender (e.g., are males more likely to engage in terrorist attacks than females?), age (e.g., is there a typical age for a person to commit a terrorist attack?) clinical assessments (e.g., is there such a thing as a terrorist personality?), or even genetics-based questions (e.g., is there such a thing as a terrorist gene?) should be tested to find out whether it is possible to construct more sophisticated terrorist profiles to distinguish between different types of terrorist (e.g., lone wolves) and terrorist organizations. The disclosure of variables that separate terrorists from non-terrorists and, consecutively, joining such variables into algorithms to flesh out terrorists would seem to offer a strong pre-emptive weapon in the fight against terrorism without killing.

However, let's take a closer look at what statistical profiling could mean in practice, by presenting two assumptions, as an example. First, let's assume it is possible to develop an algorithm, approximately accurate up to 99%, and second, let's assume there are 100 terrorists in The Netherlands. An accuracy rate of 99% would then mean for our algorithm to identify 99 out of 100 terrorists. However, in doing so, we also have to take into account population in The Netherlands amounts to 17 million and, therefore, by using this specific algorithm, 1% of the Dutch population, about 17,000 inhabitants (minus 1), as a result of false positives, would be wrongly identified as terrorists (of course presuming the aforementioned 100

[26] Phillips and Pohl 2012, p. 155.

[27] Alison et al. 2010, p. 118.

[28] Also see Dean 2007, p. 176; Phillips and Pohl 2012, p. 151.

[29] Leese 2014, p. 496.

terrorists can indeed be considered legitimate Dutch inhabitants). This would be even more aggravating to the people wrongly suspected, that, pending closer investigations, de facto would be declared *morts civils*. Moreover, based on the above, applying such a relatively accurate algorithm to identify potential terrorists could be expected to result in staggering workloads amongst personnel in security and intelligence services, the Police, the Royal Netherlands Marechaussee, and fiscal investigation departments. On account of legal constraints to information sharing between these institutions, this problem would be compounded even more.[30]

12.4 Identifying Terrorists by Means of Retail Banking Data

The origins of identifying terrorists by using banking data stems from a specific area of misuse of financial services, namely money laundering.[31] Over the last four decades, financial institutions fighting money laundering, amongst others, scrutinize banking account transactions (e.g., electronic transfers, cash deposits and cash withdrawals). To this effect, financial institutions (which in most developed countries include deposit taking institutions, security firms, broker-dealers, insurance firms and money transfer agents), identify both client characteristics and the types of transactions that indicate money laundering.[32] The legal obligation to do so is codified by governments all over the world, for example, by the 1970 United States Bank Secrecy Act. Sustained global interest in anti-money laundering policies emerged in the 1980s, primarily regarding international drug trafficking. It gained momentum upon the creation of the Financial Action Task Force (FATF) in 1989, originally consisting of sixteen OECD members. Subsequently, in less than a year, FATF published its first set of 40 recommendations, designed to provide a comprehensive plan of action for fighting money laundering.[33]

During the post 9/11 era, governments gradually extended their financial institutions' legal obligations into the area of counter threat finance. Consequentially, in October 2001 FATF issued eight (later nine) new recommendations on terrorist financing, supplementing their former 40 recommendations.[34] Although this did not come as a surprise, as money launderers often use the same modes operandi as individuals or organizations involved in terrorism funding, the two differ in that money launderers, foremost, try to hide any criminal origins and actual ownership, whereas terrorism financiers are mainly concerned with hiding the criminal

[30] Levitt and Dubner 2009, pp. 91–92 and pp. 162–163.

[31] Clarke 2015, p. 7; Johnson 2008, p. 7.

[32] Zarate 2013, p. 8 ff.

[33] Beeres and Bollen 2011, p. 99.

[34] Bogers and Beeres 2013, p. 2993; Johnson 2008, p. 8.

destination of the funding they provide. Therefore monitoring both client and transaction characteristics have become an important tool to fight both financial crime and terrorist funding.[35] In total, the original (money laundering) and the additional (terrorism funding) FATF recommendations require financial institutions to profile clients and monitor their transactions to see whether (1) proceeds of crime or money laundering and/or (2) financing of terrorism are involved. Most of this monitoring is done using methods we would classify as traditional profiling, much as described before, as client and transaction characteristics are typically compared to benchmarks derived from professional knowledge and experience that sometimes will be denominated as typologies. Typologies describe typical tactics used by launderers or terrorism financiers that indicate a higher risk. Typologies are produced and published by FATF, FATF-Style Regional Bodies and national competent authorities, such as, Financial Intelligence Units.[36]

As discussed in the former section, obviously, such profiling does not come easy. Collecting and analysing data on clients and funds (transfers) does not automatically lead to revealing terrorism. Profiling based on low distinctive characteristics can result in large numbers of false positives while the use of high distinctive characteristics carries the risk of false negatives, missing out on real launderers or terrorism financiers. Evaluating possible effects, irrevocably leads to the question whether by applying FATF's financial measures to fight terrorism would indeed have resulted in a timely apprehension of the 9/11 airplane hijackers.[37] In this respect, Donohue, based on extensive research, is negative, arguing, "Al Qaeda moved the money to fund September 11 in three ways: $130,000 was wired to hijackers in the United States from the United Arab Emirates and Germany; members physically carried cash/travellers checks to the United States; and some established overseas accounts, which they drew on via ATM or credit cards in the United States. When they arrived in the United States, they opened bank accounts under their real names in both large national banks and smaller regional ones. While they lived in the United States, they made wire transfers of between $5,000 and $70,000, making the transactions virtually invisible in comparison to the billions of dollars moving daily through the international financial system. Their banking pattern –depositing a significant amount of money and then making smaller withdrawals- fits their student profiles. They did not use false Social Security numbers, and their grasp of the US banking system was not particularly sophisticated".[38] In short, by applying the financial measures, comprising an analysis of banking behaviour, chances seem exceedingly slim the 9/11 hijackers would have been profiled as terrorists.

To date, there is only one scientific study claiming to have successfully identified terrorists based on statistics derived from their banking behaviour and, as promised

[35] Zarate 2013, p. 21.

[36] See FATF 2014.

[37] Beeres and Bollen 2011, p. 100.

[38] Donohue 2008, p. 347.

before, we will devote the remaining part of this section to a more detailed discussion, including ensuing publications by the same author(s). The research, conducted by Levitt and Dubner, holds that, although the 9/11 hijackers could not be profiled as terrorists, the banking patterns of all nineteen terrorists did reveal some behaviour that "in the aggregate, distinguished them from the typical banking customer". To clarify, the authors note the hijackers "tended to make a large initial deposit and then steadily withdrew cash over time, with no steady replenishment. Their banking didn't reflect normal living expenses like rent, utilities, insurance and so on. Some of them routinely sent or received wire transfers, but the amount invariably fell below the reporting limit".[39] Although it goes without saying, such aggregated banking behaviour per se does not suffice to accuse anyone of (planned) terrorism, let alone to persecute one, it may be considered a starting point to be added to by specific demographic characteristics and behaviours to obtain more in-depth terrorist profiles.[40]

As far as the authors of this chapter could retrieve, database sources to underpin the makings of Levitt and Dubner's terrorist profile, include (1) banking data of four Islamic suicide bombers executing the London bombing 7/7/2005 (see Table 12.1), (2) depersonalized administrative records of a large UK bank, (3) a list of people who have been arrested or investigated on terrorist charges in the UK, and, finally (4) a list of predominantly Islamic names. Based on their database, Levitt and Dubner found one of the strongest demographic markers for their algorithm to consist of a Muslim name. More specifically, according to the authors, "A person with neither a first nor a last Muslim name stood only a 1 in 500,000 chance of being a suspected terrorist. The likelihood for a person with a first *or* a last Muslim name was 1 in 30,000. For a person carrying both first *and* last Muslim names, however, the likelihood jumped to 1 in 2,000".[41] Likely terrorists, moreover, turned out to be "men, most commonly between the ages of twenty-six and thirty-five". Also, would-be terrorists disproportionately were likely to own a mobile phone, be a student, and rent rather than own, a home. Levitt and Dubner continue that although these characteristics seem to fit most students, "even these common traits began to add power to the algorithm".[42] Meanwhile, it appeared characteristics such as employment, marital status or living in close proximity to a mosque were to little avail as markers to distinguish terrorists from non-terrorists. Last, Levitt and Dubner indicate three characteristics to be extremely unlikely to pinpoint a would-be terrorist: (1) having a savings account, (2) withdrawing money from an ATM on a Friday afternoon, and last, buying life insurance.

As to the latter, Levitt and Dubner remark: "Let's say you're a twenty-six-year-old man, married with two young children. It probably makes sense

[39] Levitt and Dubner 2014, pp. 161–162; Levitt and Dubner 2009, pp. 90–91 note eight of such banking behaviours.

[40] Levitt and Dubner 2009, p. 91; Levitt and Dubner 2014, p. 163.

[41] Levitt and Dubner 2009, p. 93.

[42] Levitt and Dubner 2009, p. 93.

to buy some life insurance so your family can survive if you happen to die young. But insurance companies don't pay out for if the policyholder commits suicide. So a twenty-six-year-old family man who suspects he may blow himself up probably isn't going to waste money on life insurance. This all suggests that if a budding terrorist wants to cover his tracks, he should go down to the bank and change the name on his account to something very un-Muslim … It also wouldn't hurt to buy some life insurance".[43]

Levitt and Dubner conclude regarding the aforementioned metrics "once combined, they did a pretty good job of creating an algorithm that could distil the bank's entire customer base into a relatively small group of potential terrorists." To top it all, however, the authors shroud themselves in mystery by adding "What finally made it work was one last metric that dramatically sharpened the algorithm. In the interest of national security, we have been asked not to disclose the particulars; we call it *Variable X*." According to Levitt and Dubner, variable X provided their algorithm great predictive power, to the extent of enabling them to derive a list of about 30 highly suspicious individuals that was then handed to the proper authorities.

In another publication, Levitt elaborates on statistical techniques used to achieve these findings. By now, Variable X is dubbed Variable Z, and, again, from this paper it becomes clear that based on the algorithm it has been possible to successfully distinguish terrorists from non-terrorists. Now, the database produces a list of 90 customers not previously investigated on terrorist charges. After one year and a half, from this list, 2 out of 90 turn out 'positives', that is to say, suspected of terrorist activities.[44] This paper, also, refrains from disclosing the nature of variable Z, "in the interest of national security".[45]

In *Think like a Freak*, finally, Levitt and Dubner unveil their mysterious and particularly powerful variable in the algorithm to be *buying life insurance*. The authors comment: "From the outset of the project, we recognized that finding a few bad apples out of millions would be difficult. Our odds would improve if we would somehow trick the bad apples into revealing themselves … Do you know anyone who buys life insurance through their bank? No, we don't either. Many banks do offer it, but most customers use banks for straight banking and, if they want insurance they buy it thought a broker or directly from an insurer".[46]

The studies performed by Levitt and Dubner, to us, leave an impression of an academic ambush, set deliberately for people with terrorist intentions to trap themselves. But, if so, to what avail and at who's request? Although this research may offer an interesting and relevant view on possible gains to be yielded from statistics- based terrorist profiling, Levitt's and Dubner's data bases, however, include people *suspected* of terrorist activities, something altogether different from

[43] Levitt and Dubner 2009, pp. 94–95.

[44] Levitt 2012, p. 17.

[45] Levitt 2012, p. 2.

[46] Levitt and Dubner 2014, pp. 164–165.

being a terrorist. Besides, it took one and a half year of further investigations to be able to come up with 2 positives out of 90 and it remains unclear whether the proffered 'ambush' has paid off. In other words: did some twenty-six-year-old-man, bearing two Muslim names, married with two young children, without savings, and while opening an account at a local bank indeed ask to buy life insurance? And if so, what happened next? In short, according to us, the claim to be able to successfully identify terrorists based on statistics derived from their banking behaviour seems premature.

12.5 Conclusion

Over the last decades, in the slipstream of the financial war on terror, both research and practical applications in the field of law enforcement and intelligence have provided insights and methods to be put to use to fight potential terrorists. Methods and techniques used for profiling have become increasingly sophisticated, mainly due to the sharp increase of data collected and filed on individuals as well as to developments in science and technology. Statistical profiling grounded in data sciences and analytics provides one such method. This chapter asks whether statistical terrorist profiling based on retail banking behaviour can indeed be considered effective in the financial fight against terrorism. More in particular, the questions put forward are whether intelligent data analyses on banking behaviour are useful to monitor and fight potential terrorists and consequently, can be used to prevent terrorist attacks?

To shed some light on these questions, we have discussed the prior introduction of FATF's financial measures in the financial war on terror, comprising, amongst others, analyses of banking behaviour. However, research on the effectiveness of these measures has made clear chances are regarded exceedingly slim that, by applying FATF's financial measures, the 9/11 hijackers would have been timely profiled as terrorists. We continued our quest to find answers to our questions by envisaging what would happen if and when we should be able to provide algorithms accurate up to 99 percent to statistically distinguish terrorists from non-terrorists. As it turns out, applying such rather accurate algorithms would result in subsets of selected individuals containing far too many false positives. Not only would the investigations, even when based on such accurate algorithms, result in human tragedies to the people who would later turn out to be a false positive and information overloads too big to handle, but, moreover, even if meaningful information, in the end would be available, legal constraints to information sharing between various (inter)national security and intelligence organisations and departments would provide another obstacle to concerted and meaningful profiling. Finally, we discussed Levitt and Dubner, the authors of the only research that claims to have successfully identified terrorists based on statistics derived from their banking behaviour. As it turns out, we can find no substantiation as to their claim. Although from these authors' findings it may be deduced that research may be used to lure terrorists into trapping themselves by adjusting their modus operandi, other studies as to the effects of such reverse signalling are not yet available.

We therefore conclude, that, at this day and age, statistical profiling based on banking behaviour, as a method, cannot be regarded suited to distinguish terrorists from non-terrorists and consequently, cannot be expected to be effective in preventing terrorist attacks. Despite technological innovations, organizational and legal constraints in information sharing and information handling as well as considerations regarding the re-creation of *morts civils* in today's society stand in the way of effectiveness. Perhaps at some point, it may be possible to apply data analytics in real-time scenarios, to detect potential terrorists and consequently prevent terrorist attacks. Financial data, such as retail banking data, may play a large part therein, but they will need to be supplemented with other sources and approaches.

References

Alison L, Goodwill A, Almond L, van den Heuvel C, Winter J (2010) Pragmatic solutions to offender profiling and behavioural investigative advice. Legal and Criminological Psychology 15(1):115–132

Atran S (2006) The moral logic and growth of suicide terrorism. The Washington Quarterly 29 (2):127–147

Ayers A (2001) Financial Action Task Force: The war on terrorism will not be fought on the battlefield. The New York Law School Journal of Human Rights 18:449–459

Beeres R, Bollen M (2011) 'The global financial war on terror': analyse en cijfers. In: Osinga F, Soeters J, van Rossum W (eds) Nine eleven tien jaar later. Boom, Amsterdam, pp 92–106

Beeres R, Bollen M (2012) An economic analysis of cyber attacks. In: Ducheine P, Osinga F, Soeters J (eds) Cyber warfare critical perspectives. TMC Asser Press, The Hague, pp 147–164

Blalock G, Kadiyali V, Simon DH (2009) Driving fatalities after 9/11: a hidden cost of terrorism. Applied Economics 41(14):1717–1729

Bogers M, Beeres R (2013) Burden Sharing in Combating Terrorist Financing. World Academy of Science, Engineering and Technology, International Journal of Social, Behavioral, Educational, Economic, Business and Industrial Engineering 7(12):2992–2998

Brück T (2005) An economic analysis of security policies. Defence and peace economics 16 (5):375–389

Charlesworth W (2003) Profiling terrorists: A taxonomy of evolutionary, developmental and situational causes of a terrorist act. Defense & Security Analysis 19(3):241–264

Clarke CP (2015) Terrorism, Inc. - The financing of Terrorism, Insurgency, and Irregular Warfare. Praeger Security International, Santa Barbara CA

Coleman L (2012) Testing equity market efficiency around terrorist attacks. Applied Economics 44 (31):4087–4099

Davis LM, Riley KJ, Ridgeway G, Pace J, Cotton SK, Steinberg PS, Damphousse K, Smith B (2004) When terrorism hits home. How prepared are state and local law enforcement? RAND Corporation, Santa Monica

Dean G (2007) Criminal profiling in a terrorism context. In: Kocsis RN (ed) Criminal Profiling. International Theory, Research and Practice. Humana Press, Totowa NJ, pp 169–188

Donohue LK (2008) The cost of counterterrorism: Power, politics, and liberty. Cambridge University Press, Cambridge

FATF (2014) Guidance for a risk-based approach the banking sector. FATF, Paris (October 2014) http://www.fatf-gafi.org/media/fatf/documents/reports/Risk-Based-Approach-Banking-Sector.pdf Accessed 2 March 2017

Fitzgerald V (2004) Global financial information, compliance incentives and terrorist funding. European Journal of Political Economy 20(2):387–401

Johnson J (2008) Is the global financial system AML/CFT prepared? Journal of Financial Crime 15(1):7–21

Keene SD (2014) Operationalizing Counter Threat Finance Strategies. Strategic Studies Institute and US Army War College Press, Carlisle PA

Krueger AB (2007) What makes a terrorist - Economics and the roots of terrorism. Princeton University Press, Princeton NJ

Krugman P (2004) The costs of terrorism: What do we know? Nexus of terrorism and WMDs: Developing a consensus, Princeton University. http://www.uvic.ca/research/centres/globalstudies/assets/docs/publications/l20/the-costs-of-terrorist.pdf Accessed 26 Feb 2017

Leese M (2014) The new profiling: Algorithms, black boxes, and the failure of anti-discriminatory safeguards in the European Union. Security Dialogue 45(5):494–511

Levitt SD (2012) Identifying terrorists using banking data. The B.E. Journal of Economic Analysis & Policy 12(3) DOI:10.1515/1935-1682.3282

Levitt SD, Dubner SJ (2009) Superfreakonomics - Global cooling, patriotic prostitutes, and why suicide bombers should buy life insurance. William Morrow, New York

Levitt SD, Dubner SJ (2014) Think like a freak. William Morrow, New York

Lister T, Sanchez R, Bixler M, O'Key S, Hogenmiller M, Tawfeeq M (2017) ISIS goes global: 143 attacks in 29 countries have killed 2,043. CNN (updated 17 Feb 2017) edition.cnn.com/2015/12/17/world/mapping-isis-attacks-around-the-world/ Accessed 26 Feb 2017

McCulloch J, Pickering S (2005) Suppressing the financing of terrorism: Proliferating state crime, eroding censure and extending neo-colonialism. British Journal of Criminology 45(4):470–486

Müller C (2011) The economics of terrorism from a policy-maker's perspective. Defence and Peace Economics 22(2):125–134

Napoleoni L (2010) Terrorism and the Economy. Seven Stories Press, New York

Paulson H (2007) Remarks by treasury secretary Paulson on targeted financial measures to protect our national security (14 June 2007) http://www.treasury.gov/press-center/press-releases/Pages/hp457.aspx. Accessed 26 Feb 2017

Phillips PJ, Pohl G (2012) Economic profiling of the lone wolf terrorist: can economics provide behavioral investigative advice? Journal of Applied Security Research 7(2):151–177

Rae J (2012) Will it ever be possible to profile the terrorist? Journal of Terrorism Research 3 (2):64–74

Raphaeli N (2003) Financing of terrorism: sources, methods, and channels. Terrorism and Political Violence 15(4):59–82

Sageman M (2004) Understanding terror networks. University of Pennsylvania Press, Pennsylvania

Schmid AP, Jongman AJ (1988) Political terrorism: A new guide to actors, authors, concepts, data bases, theories, and literature. North Holland Publishing, Amsterdam

Sinai J (2008) How to define terrorism. Perspectives on Terrorism 2(4):9–11

Strielkowski W, Lisin E (2013) International migration, remittances and intelligent analysis of semi-structured data from the open sources. International Economics Letters 2(2):6–11

van Fenema PC, Rietjens SJH, Kalden J, van der Sluis W, Schimmel R (2015) Big data analytics en Defensie: visie en aanpak. Militaire Spectator 184(9):347–387

Zarate JC (2013) Treasury's war. The Unleashing of a new era of financial warfare. Public Affairs, New York

Author Biographies

Professor Dr. R.J.M. Beeres is Professor of Defence Economics, Department of Military Management Studies at the Faculty of Military Sciences, Netherlands Defence Academy and an Associate Professor at Nyenrode Business University. He obtained a Ph.D in administrative sciences at Radboud University Nijmegen.

Colonel Dr. R.M.M. Bertrand RA RC RO is Associate Professor of Accounting, Control and Defence Economics, Department of Military Management Studies at the Faculty of Military Sciences, Netherlands Defence Academy. He defended his Ph.D. in 2016 at Maastricht University.

Professor Dr. M.T.I.B. Bollen is Professor of Civil-Military Interaction and Chair of the Department of Military Management Studies and a member of the Board of the Faculty of Military Sciences, Netherlands Defence Academy. She defended her Ph.D. in 2002 at Wageningen University. Professor Bollen is a Visiting Professor at the Baltic Defence College.

Chapter 13
Reframing Lawfare

Joop Voetelink

Abstract Law, in particular international law, plays an increasingly significant role in today's military operations. In accordance with the principle of the rule of law, States can deploy their armed forces abroad based on a clear legal foundation while the deployed forces must operate within the applicable rules. The growing emphasis on this legal framework for military operation entails that today, law is not just a factor to be reckoned with when preparing for and conducting military operations, but can be a possible means to reach a military objective as well. Recently, to help understand this development the term lawfare has been introduced. Lawfare it is not generally accepted as a term of art yet. As it is a catchy phrase the term is also used for politically motivated reasons, criticizing the application of law to certain aspects of military operations, as it is felt to endanger national security or to curb armed forces unnecessarily, especially when dealing with less law abiding opponents. From a military operational law perspective, manipulating lawfare in this way threatens to turn the concept in an empty shell devoid of any meaningful content. As lawfare is a useful term to describe and understand the dynamics of the instrumental role law can play in the context of contemporary complex military operations, the sometime random use of it blurs the scope of its application. This article will, therefore, explore and clarify the meaning and scope of the concept.

Keywords Anti-terrorism legislation · Evidence based operations · International law · Lawfare · Law of armed conflict · Military operational law · National security · Proportionality · State immunity · Status of forces

J. Voetelink (✉)
Faculty of Military Sciences, Netherlands Defence Academy, Breda & Amsterdam Centre of International Law (ACIL), Breda, The Netherlands
e-mail: jed.voetelink@mindef.nl

P.A.L. Ducheine and F.P.B. Osinga (eds.), *Netherlands Annual Review of Military Studies 2017*, NL ARMS, DOI 10.1007/978-94-6265-189-0_13

Contents

13.1 Introduction

The character of military operations has changed dramatically over recent decades. Traditionally, these operations took place within the framework of inter-State armed conflicts. Today, this type of conflict, or in terms of the laws of armed conflict: international armed conflicts, continues to exist, but the majority of armed conflicts now take place within the boundaries of particular States and are referred to as non-international armed conflicts. In addition, international organisations, such as the United Nations (UN), the North-Atlantic Treaty Organisation (NATO), the European Union (EU), and the African Union (AU) have developed the practice of establishing international forces to conduct crisis management operations. Within the framework of these operations, the participating military forces may be tasked to perform a wide variety of tasks, ranging from delivering humanitarian assistance to conducting full-scale combat operations. In addition, States increasingly call upon their armed forces to fulfil duties that can generally be described as law enforcement type of functions, such as anti-piracy operations,[1] counter-terrorism activities, and anti-migrant smuggling operations.[2]

Some feel that of late military operations are governed by a growing body of law. However, at a closer look it appears not necessarily to be the case. As the character of military action changes continuously, so does the legal framework within which the armed forces must operate. It is not so much an increase in rules the military are facing as that a different set of rules becomes applicable when the character of operations shifts. In modern society based on the rule of law, it is only natural that in such a dynamic environment people become more conscious of the relevant legal framework. In the military the increasing emphasis on law led to the maturation of

[1] E.g. Operation Ocean Shield (NATO) and Operation Atalanta (EU).
[2] E.g. Operation Sophia (EU) and NATO activities in the Aegean Sea.

military operational law (Oplaw) as a specialized field of law covering all legal aspects pertaining to the preparation and execution of operations. As such, law can become a non-kinetic capability, creating opportunities for a military commander to help him to achieve his operational goals. In this context the phrase 'lawfare' was introduced,[3] eventually defined as "the strategy of using—or misusing—law as a substitute for traditional military means to achieve an operational objective".[4]

Despite this often quoted definition, lawfare is still not a well-established term of art and, as it is a catchy phrase as well, some use the term rather loosely to serve their own, sometimes politically motivated goals. Adding to the confusion, lawfare has become subject of research in other disciplines, such as geography, offering yet another view on the subject from a completely different perspective.[5] As lawfare reflects the importance of law, in particular Oplaw, in present-day operations,[6] it is a useful term to describe and understand the dynamics of the constructive, instrumental role law can play in the context of contemporary military operations. In order to continue to fulfil this role and prevent erosion of the concept, lawfare must not be interpreted and applied too broadly. By exploring and reframing the concept of law as a means of reaching a military goal, this article promotes a better understanding of lawfare as a non-kinetic capability. As achieving military goals is at the heart of the lawfare concept, this chapter takes Oplaw as the vantage point to explore the concept of lawfare and refine its definition. Therefore, it starts by introducing and analysing Oplaw against the backdrop of the concept of the rule of law before exploring the origin and development of lawfare. Continuing from the vantage point of Oplaw, the lawfare definition mentioned above is analysed and on details further refined to ensure that the concept continues to be of relevance to help explain the role of law in contemporary military operations.

13.2 Rule of Law and Military Operations

Many sovereign States pride themselves on being societies built on the principles of the rule of law and refer to the concept routinely. Such statements are generally accepted without any arguments. Apparently, 'rule of law' is a concept that is

[3] Some prefer another term, such as 'juridical warfare', Rishikof 2008.

[4] Dunlap 2008, p. 146. Note that Dunlap uses 'operational objective' as a generic term to describe the results to be achieved by law. He does not use the term 'military objective', which has a specific, well-defined meaning under the law of armed conflict; article 52(2), Additional Protocol I to the Geneva Conventions ("military objectives are limited to those objects which by their nature, location, purpose or use make an effective contribution to military action and whose total or partial destruction, capture or neutralization, in the circumstances ruling at the time, offers a definite military advantage"). In order to prevent possible confusion this chapter will use the term operational goal instead of operational objective when discussing the results to be achieved by lawfare.

[5] E.g. Morrissey 2011; Jones 2015.

[6] Scharf and Andersen 2010, p. 13.

widely understood and does not seem to be in need of a further introduction. Nevertheless, rule of law has a very broad and multifaceted meaning[7] that encompasses all the State powers (legislation, enforcement, and adjudication) and prevents a State from abusing its authority while at the same time safeguarding its citizens and their property from violation of their rights.[8] In essence, under the rule of law all persons within a State, including its authorities and officials, are bound by and entitled to the benefits of laws.[9]

The concept of the rule of law is equally crucial to the military. The armed forces are key State organs and are, therefore, part and parcel of States. As a consequence, all military action is governed by the law. In addition, mandates for contemporary international forces often include an explicit or implicit rule of law component, tasking armed forces to contribute to rule of law efforts in the States where they are deployed. Rule of law applied to armed forces activities abroad entails that all military action needs to have a sound legal foundation in international law (the legal basis) and, subsequently, that all military activities are consistent with international and domestic law (the legal regime).[10]

Taking all these legal aspects together, the "various bodies of national and international law which are applicable to and regulate the planning and conduct of military operations",[11] "also informed by political-strategic interests"[12] are today described as Oplaw. This specialized field of law, balancing between operations and law, is still young and developing, covering a wide array of themes. However, many of these themes, such as status of forces, international humanitarian law, and the rules on resort to force by States (the 'jus ad bellum'), have been around much longer. In that respect Oplaw does not offer something completely new. What is different now is "the way and extent to which these various areas of law interact with each other and influence and regulate and shape the way in which contemporary operations are planned and conducted".[13] Law, Oplaw, is the legal cornerstone of military operations and sets out the limits within which military action must take place. Nevertheless, it is not a one-way street. Military actions and military operations can, to a certain extent, help to shape the legal framework as well. For example, a status of forces agreement geared to the operational tasks the deployed forces are mandated to carry out, can actually expedite the entry of the

---

[7] E.g. the much-quoted definition by the Secretary-General of the United Nations in: UN Doc. S/2004/616, Report of the Secretary-General on the Rule of Law and Transitional Justice in Conflict and Post-Conflict Society.

[8] Strikwerda-Verbeek 2012, p. 17.

[9] Bingham 2011, p. 8.

[10] *Dutch Parliamentary Papers* 2006/07, 29 521, nr. 41 [Letter of the Ministers of Foreign Affairs and Defence, 22 June 2007, Note on the legal basis and mandate of missions in which Netherlands armed forces units participate]. Which specific rules come into play depend on the legal basis of the operation and the circumstances at hand.

[11] Gill and Fleck 2015, p. 3.

[12] Jones 2015, p. 21.

[13] Gill and Fleck 2015, p. 5.

forces into host State territory and facilitate day-to-day operations of the troops during their presence abroad.[14] In this way, law can be constructive to mission accomplishment.[15] The contributory part law thus plays in military operations is captured in the concept of lawfare.

13.3 A Primer on Lawfare from an Oplaw Perspective[16]

Lawfare is a fairly new term that has a completely different meaning today than it had at its introduction in the 1970s.[17] At that time it was used to criticize what was regarded by some the "individualistic and accusatorial aspects of law in Western societies."[18] In the 1990s, it had yet another, less profound meaning as it was used in advertisements by travel agencies to offer lawyers airfares at reduced prices.[19] In 1999, two Chinese colonels introduced the term 'international law warfare'[20] as one of the many forms of unconventional warfare to be used combined against a superior opponent.[21] Their ideas touched upon the instrumental role of law in military operation as is reflected in lawfare.

13.3.1 Definition

The Chinese colonels did not use the word lawfare in their book 'Unrestricted Warfare', though. That term was introduced a couple of years later by the then US Air Force Colonel Charles J. Dunlap, Jr. Initially, he saw lawfare just as a catchy phrase, a resonating 'bumper sticker', he could use to inform a broad audience of the role of law in the conduct of hostilities[22] and to help the armed forces realize why law should be part of their thinking and planning.[23] In 2001, he introduced the term to emphasize the growing influence of international law on military operations, defining the concept as "a method of warfare where law is used as a means of

[14] Voetelink 2015, p. 230; Morrissey 2011, pp. 293–294.

[15] Dunlap 2010, p. 128.

[16] Based on Voetelink 2013a, pp. 71–73.

[17] See for details on its development Werner 2010.

[18] Werner 2010, p. 63.

[19] Sadat and Geng 2010, p. 157.

[20] 'International law warfare' means joining international or multinational organizations in order to subvert their policies and the interpretation of legal ruling, Liang and Xiangsui 2002, xii.

[21] Liang and Xiangsui 2002, Chap. 7.

[22] Dunlap 2008, p. 146.

[23] Scharf and Andersen 2010, p. 12.

realizing a military objective".[24] He pointed out that enemies of the US could abuse law in an armed conflict making it appear that the US, instead of strictly following the laws of armed conflict, conducted military operations in gross violation of these rules.

Over time, Dunlap's ideas matured and in 2008 he published a modified, more dynamic definition of lawfare: "The strategy of using—or misusing—law as a substitute for traditional military means to achieve an operational objective".[25] He built his thoughts on, *inter alia*, the notion that law had become an inherent part of contemporary armed conflicts, just as it had penetrated modern society in general.[26] The instrumental nature of lawfare is reflected in either the use or misuse of law by warring parties instead of utilization of traditional military means to prevent further physical violence.[27] Dunlap stresses that military commanders should face this reality and start to use law in an efficient manner instead of viewing law as an impediment to the conduct of military operations.[28] As Kennedy noticed, law "... can be a strategic ally....".[29] At the same time, commanders must appreciate that an opponent can conduct lawfare as well and they must, therefore, take a proactive stance and be prepared to identify and counter these lawfare threats.

In the literature, other definitions of lawfare are discussed as well. Bachmann and Munoz Mosquera, for example, define it as "using law as a weapon with a goal of manipulating the law by changing legal paradigms".[30] They wrote their article against the backdrop of the concept of hybrid warfare and Russia's activities in Ukraine. Another example of this approach is the legal characterization of the second Palestinian uprising against Israel (*Intifada)* that started September 2000. Lawyers from the International Law Department of the Israel Defence Forces (IDF) came to the conclusion that violence used by the Palestinians was no longer a law enforcement issue, but had passed the armed conflict threshold, making the IDF operations subject to the laws of armed conflict.[31] Jones considers it lawfare as the legal designation was not just a response to the *Intifada*, but a well-prepared pre-emptive action that started months before the Intifada began.[32] Although this definition also takes the instrumental role of law as starting point, it is, from my point of view, too limited in scope to help explain the role law plays in modern military operations.

---

[24] Dunlap 2001, p. 11.

[25] Dunlap 2008, p. 146.

[26] Dunlap 2009, 35; Kennedy 2007, pp. 176 ff.

[27] Dunlap 2008, p. 147.

[28] Dunlap 2009, p. 28 citing Nathaniel Burney's vision on lawfare: http://www.burneylawfirm.com/international_law_primer#lawfare.

[29] Kennedy 2007, p. 173.

[30] Bachmann and Munoz Mosquera 2015.

[31] Jones 2015, p. 12.

[32] Jones 2015, p. 12.

To my mind, not all use of law in a military context should be called lawfare. First, the rule of law dictates that deployed armed forces base their actions on the law and conduct their operations in accordance with the law, legitimizing military action. Thus, law is the context for military action and not an instrument of military action. Therefore, operating within the legal framework as such it is not lawfare. Second, the consequence of operating in accordance with the law entails that armed forces are accountable to the law as well. This is the flip side of the rule of law coin, so to say, and, as a consequence, military action can always be tested against the law. In fact, it is mere application of the law as long as a party to a conflict is not exploiting it with a view to achieve a particular military goal. Therefore, the civil suit by the 'Mothers of Srebrenica', an association representing survivors of the 1995 Srebrenica massacre by Bosnian Serb forces in Bosnia and Herzegovina, demanding compensation from the UN and the Netherlands because of their alleged responsibility for the failure to prevent the massacre, is a legal action that can be used to hold the armed forces involved accountable and should not be called lawfare. Neither should the prosecution of a service member for committing a war crime or another criminal offence during military operations, even if some feel that the proceedings are politically motivated and damaging to the armed forces. In order to exclude these legitimate legal proceedings from the lawfare definition it may be considered to emphasize that just using/misusing does not yet qualify as lawfare by including that use or misuse of law must be deliberately designed to achieve an operational goal.

13.3.2 Use of Law

Use of law in a military operation can take any form. A telling example of the deliberate use of law designed to achieve a specific goal is the introduction of Evidence Based Operations as part of ISAF's rule of law activities in Afghanistan. Evidence Based Operations are officially defined as: "operations where Afghan law enforcement authorities, supported by ISAF, effectively investigate, apprehend, search and seize criminal suspects and affiliated property in accordance with Afghan law".[33] In practice, ISAF and coalition forces were critical in this approach. They were to gather intelligence on insurgent activity and then convert it into evidence admissible in Afghan criminal court. Consequently, they had to apprehend the insurgents, preferably in close coordination and cooperation with Afghan criminal justice officials, to prosecute them before Afghan courts. Furthermore, in order to enable their Afghan partners to assess and use the collected evidence ISAF and coalition forces, *inter alia,* trained and assisted the Afghan justice authorities in

[33] HQ ISAF FRAGO 058-2012 Support to Afghan led Evidence Based Operations, 18 March 2012; the definition has been taken from the NATO 2014 Rule of Law Field Support Officer Deskbook. Guidelines and Lessons Learned for Military Support to Justice Sector Development in Afghanistan, Camp Phoenix, Kabul, Afghanistan, June 2012, p. 60.

making use of forensic techniques and mentored prosecutors and judges of the court and the court of appeal in Parwan (collectively known as the Justice Centre in Parwan).[34]

The Evidence Based Operations-concept was initiated when the focus of military operations in Afghanistan shifted towards Counter-Insurgency (COIN) operations. A key element of COIN is the focus on the civilian population and the need for creating a safe environment;[35] "securing the population from the insurgents".[36] Based on detention operations in Iraq, ideas were developed so as not just to fight the ongoing insurgency in Afghanistan, but to criminalize it as well. After all, most insurgent activity, such as planting bombs, attacking security forces, and trafficking drugs, qualify as criminal offences punishable under Afghan law. Consequently, military commanders had to transition from combat operations to law enforcement type of operations. Insurgents were after capture by ISAF or coalition forces to be turned over to the Afghan criminal justice system for prosecution. In this way Afghan law and court system could be leveraged as a tool to effectively remove insurgents from the battlefield while at the same time making clear to the Afghan people that insurgents were ordinary criminals and that the Afghan courts were capable of holding them accountable for their crimes committed under Afghan law.[37]

13.3.3 Misuse of Law

Misuse of law in military operations, or, as some call it: malicious employment of the law,[38] on the other hand, is not a recent phenomenon. Throughout history warring parties have disrespected rules to gain an advantage over their opponents. A more recent development is that a weak or military less advanced party to a conflict, unable to directly engage a better equipped, stronger opposing force, leverages its opponent's sensitivity to alleged violation of law, in particular the laws of armed conflict. They attempt to create the image that their opponents "cause disproportionate collateral damage and civilian casualties, thereby undermining the perception of legitimacy".[39] Tactics based on this concept are designed to have a double effect on a law-sensitive opponent.[40] Firstly, it may lead an opponent to impose additional constraints on his own actions whenever civilian casualties or damage to civilian property are to be expected, rendering its forces less effective.

[34] Voetelink 2013b.

[35] U.S. Field Manual 2006, pp. 1–23.

[36] Dunlap 2010, p. 124.

[37] Voetelink 2013b, pp. 194–201.

[38] Munoz Mosquera 2015.

[39] Osinga and Roorda 2016, p. 58.

[40] Kittrie 2016, p. 285.

Secondly, the tactic may cause a loss of popular and international support for allegedly operating in violation of the laws of armed conflict.

Reynolds claims that the Vietnam War was the first example of a party systematically exploiting its opponents', i.e. US concerns over civilian casualties.[41] Something that now seems a common tactic, especially in asymmetric warfare. Since the Vietnam conflict, many forms of this type of lawfare have been reported. In his 2001 paper Dunlap took the US attack on the 'al-Firdos bunker' in the Amiriyah neighbourhood in Baghdad as an example.[42] On 13 February 1991 US aircraft attacked the bunker, which had been in use as an air-raid shelter before, but had been turned into a military command and control facility according to US intelligence, making it a legitimate target. At the moment of attack, however, hundreds of Iraqi civilians, including family members of high-ranking State officials, had found refuge in the bunker against the ongoing coalition attacks on the capital of Iraq.

One of the basic rules of the laws of armed conflict is that parties to an armed conflict "shall direct their operations only against military objectives".[43] This rule does not fully protect the civilian population from an attack, however, as they can become victims of attacks directly aimed at military targets. This can be justifiable under the laws of armed conflict if the attack is proportionate.[44] In other words, attacks are allowed if expected civilian casualties or damage to civilian property (collateral damage) is not excessive in relation to the concrete and direct military advantage anticipated.[45] It is not fully clear whether the 'al-Firdos' bunker was indeed in use with the Iraq military and as such qualified as a legitimate military target. Fact is that many civilians lost their lives.

Pictures of the bodies of the victims of the bombing that were being excavated from the remains of the bunker were exploited by the Iraqi regime to discredit coalition air operations. The message was clear: coalition forces operate in total disregard of the safety of the civilian population and in violation of the laws of armed conflict. The news was picked up by news outlets and published worldwide. As a result, coalition attacks on military targets in Baghdad were stopped and attacks on bridges over the Tigris river were cancelled,[46] "rendering downtown Baghdad immune from attack", in Dunlap words.[47] Moreover, the targeting process was reviewed and from then on all selected targets had to be reviewed at the highest military command level in Iraq.[48] Iraqi exploitation of the 'al-Firdos' attack and

---

[41] Reynolds 2005, p. 22.

[42] Dunlap 2001, p. 12.

[43] Article 48 of the Additional Protocol I to the Geneva Conventions.

[44] Article 51(5)(b) of the Additional Protocol I to the Geneva Conventions.

[45] Principle of proportionality.

[46] Reynolds 2005, p. 34.

[47] Dunlap 2001, p. 12.

[48] Reynolds 2005, p. 33.

other incidents led to an increased coalition sensitivity to loss of civilian life, resulting in more stringent targeting procedures.

Other armed conflicts generated similar examples. Especially in the conflicts between Israel and Hamas in 2009 and 2014 lawfare played a significant role. Hamas frequently employed "tactics designed to gain advantage from the greater leverage which the law of armed conflict and its processes exert over Israel than Hamas".[49] On many occasions Hamas operated from residential areas using the presence of civilians as protection against Israeli attacks. Hamas was fully aware that the IDF was poised to reduce collateral damage as much as possible, knowing that Hamas would document collateral damage incidents and exploit them in the media. As the number of civilian casualties had risen, so would have international pressure on Israel to end the hostilities which would typically have entailed making concessions to Hamas.[50]

13.3.4 Counter Lawfare

Thus, misuse of law can be an effective means of conducting military operations and parties to a conflict should be prepared to identity the challenge it may pose to them and to design appropriate tactics to counter the threats in due time. As Hamas was operating from populated areas any IDF attack on their positions would most probably have led to civilian casualties. Even if such an attack had passed the proportionality test, pictures of the victims would nevertheless have shown up in the media suggesting the IDF was deliberately attacking civilians in violation of the laws of armed conflict.

In Operation Cast Lead (Gaza 2008–2009) IDF had developed the practice of warning civilians in or near the object of attack by calling them by telephone and warning them of the imminent attack. This practice is in accordance with the laws of war that require an attacker to give an effective advance warning of attacks which may affect the civilian population.[51] However, the warning did not always have the desired effect. Hamas encouraged civilians to get to the roof of the targeted objects in full view of the unmanned IDF aircraft that kept the objects under close surveillance.[52] This, of course, would raise the stakes for the IDF as an attack on the object would definitely lead to civilian casualties and could even be interpreted as a deliberate and thus forbidden attack on civilians. With this approach, deliberately bringing local citizens in harm's way, Hamas deviously explored the laws of armed conflict to compel the IDF to refrain from attacking the objects.

[49] Kittrie 2016, p. 283.

[50] Kittrie 2016, p. 287.

[51] Article 57(2) of the Additional Protocol I to the Geneva Conventions.

[52] Kittrie 2016, p. 286.

IDF countered this lawfare threat by initiating the 'roof knocking' method.[53] IDF-aircraft deployed a small calibre bomb targeted to an empty area of the roof, making it clear to the people gathered there that their presence would not deter IDF from attacking the building. After the people had indeed fled the building the aircraft would then attack it with precision-guided missiles. Israel employed this counter-lawfare method later on in the 2014 conflict with Hamas (Operation Protective Edge) as well and reports indicate that the US has used similar tactics in the fight against ISIS.[54] Other counter-lawfare tactics include attaching camera teams to combat units. Their footage of ongoing operations can be released to the media in situations where an opponent is likely to exploit the incidents that occurred in an operation.[55] In addition, other information could be released to substantiate the footage, such as data showing that an opponent is operating from residential areas and is setting up and launching weapon systems from civilian areas.

13.4 Refining the Lawfare Definition

All examples discussed in the previous paragraphs are, what Kittrie calls, 'battlefield lawfare'[56] and qualify as lawfare under Dunlap's definition. A wide range of other types of legal actions have been categorized as lawfare as well, although these actions were taken sometimes far from an actual area of operation, such as China's international law warfare, Guantanamo Bay court cases and, US anti-terrorism legislation and litigation. What they all have in common is that they include a legal element that bears some kind of relation with military operations, State security or terrorism. Thus, it seems the term lawfare is sometimes used rather loosely, entailing the danger of lawfare becoming an empty shell; a slogan devoid of any significant content. From an Oplaw perspective, lawfare highlights the changing role of law in military operations and emphasizes that law can be a means to reach a military goal. It may also help identify legal challenges as well as opportunities and as such needs to be clearly defined. Dunlap's definition ("the strategy of using—or misusing- law as a substitute for traditional military means to achieve an operational objective") largely serves that purpose, but as lawfare continues to evolve some details may be in need for specification or refinement. As a starting point, military operations can be conducted at a strategic, operational and tactical level. The strategic level is the highest political-military level where the grand design for operations are developed. The word 'strategy' in the lawfare definition does not

[53] Boogaard 2011, pp. 264–265.

[54] E.g. Popovich 2016.

[55] U.S. Joint Publication 3-05, pp. 3-05, IV-10.

[56] Kittrie 2016, pp. 293–309.

refer to this military concept and must be understood to have a general meaning, such as a concept or an idea.[57]

13.4.1 Operational Goal

Key to the lawfare concept is achieving an operational goal, using or misusing law. Use or misuse of law can take many forms as is already addressed above. From the context of Dunlap's ideas, it is clear that military goals are meant to be achieved in the course of armed conflicts.[58] Contemporary military operations take place in a dynamic and complex environment where the character of developing conflicts is difficult to establish yet or a crisis management operation can develop from a peacekeeping mission to participation in an armed conflict and vice versa. Therefore, the element 'operational objective' in the lawfare definition should not be limited to armed conflict situations, but should be broadly interpreted to include goals to be achieved in the course of military operations in general.

Legal actions designed to help shape a possible future battle space do not qualify as lawfare then. China's legal warfare, introduced in 2003 and building on the ideas of 'international law warfare' mentioned above, is part of the concept of the Three Warfares[59] and is partly designed to "condition the operational environment in China's favour".[60] Typically, it is employed before the actual outbreak of physical hostilities and may include selective use or interpretation of treaty provisions for unintended purposes to substantiate territorial claims.[61] An example are the arguments China raised with respect to the rights it claimed to resources in the South China Sea.[62] These actions may indeed have an effect on a possible future conflict, but should be regarded as part of the policy and diplomatic instruments of a State when advancing its international, political goals. As such, it cannot be called lawfare from an Oplaw perspective. An option would be to introduce a broader category for actions that are aimed at having an effect on a political level under the header political lawfare.

---

[57] E.g. Dunlap 2015, p. 823.

[58] Dunlap 2010, p. 121.

[59] The Three Warfares encompasses the following forms of interrelated and mutually reinforcing forms of warfare: public opinion warfare, psycho-logical warfare, and legal warfare; Cheng 2012, 1.

[60] Halper 2013, p. 31.

[61] Halper 2013, p. 29.

[62] 12 July 2016 PCA Case N° 2013 in the matter of the South China Sea Arbitration before an arbitral tribunal constituted under Annex vii to the 1982 UN Convention on the Law of the Sea between: The Republic of the Philippines v. The People's Republic of China http://www.pca-cpa.org/wp-content/uploads/sites/175/2016/07/PH-CN-20160712-Award.pdf.

13.4.2 Object

The definition is silent on the question who can actually conduct warfare (actor) and what is the target audience (object). The concept of lawfare was developed against the backdrop of armed conflicts. Therefore, actors and objects must, in any case, be understood to include armed forces and armed groups involved in the conflict. As armed forces operate under State authority, the State itself and its organs can either conduct lawfare or be targeted by lawfare. More important, however, is the question how civilians fit in the concept. The laws of armed conflict prevent civilians from being deliberately targeted by a party to the conflict. Legal action directed at civilians does usually not qualify as an attack under the laws of armed conflict, though. Consequently, it is conceivable that lawfare can be directed at others than the armed forces and armed groups participating in hostilities.

In 2010, IDF naval forces stopped a flotilla on its way to bring relief supplies to the Hamas controlled Gaza strip, which was largely cut off from the outside world because of a blockade imposed by Egypt and Israel in 2007. As marines boarded one of the ships fighting broke out leaving nine activists dead. Israel was widely criticized because of the action and when in 2011 another flotilla was assembling in Greece, Israel decided on a different approach. The Shurat HaDin Law Center in cooperation with Israeli government officials contacted several companies providing services, such as insurance and satellite communications, to the ships participating in the flotilla.[63] They were informed that their clients were preparing to break a naval blockade in support of a foreign terrorist organisation (Hamas). If the companies would knowingly continue to provide their service they were open to both criminal charges and civil liability under US law.[64] Also, Greek authorities were informed that several ships were not insured and lacked proper communications which was in breach of Greek rules on seaworthiness. As a result, authorities prohibited several ships from leaving port. Because departure of the flotilla was delayed again and again, the volunteers who were to accompany the ships had to return home to their work as their period of leave came to an end.[65] The flotilla then fell apart and the remaining ships that managed to set sail to Gaza were stopped by Israeli naval vessels without any problems.

The legal actions, although taken well outside the area of operations, had a direct effect on an ongoing military operation (enforcing a naval blockade) and helped achieve a specific military goal, i.e., prevention of an actual breach of the blockade. Taking action against the civilian actors was in this example critical in achieving the military goal and threatening them with legal actions was not in violation of the law. Viewed from that perspective, it can be said that the status or position of the target audience is not relevant for establishing if an action qualifies as lawfare.

[63] Kittrie 2016, p. 313.

[64] Kittrie 2016, p. 314 and p. 316.

[65] Heijne 2011.

13.4.3 Actor

In this particular case civilians, e.g. the Shurat HaDin Law Center, were also actively engaged in lawfare activities as actor. They operated, however, with consent of and in coordination with Israeli authorities. In that respect, it can be argued that their action can be attributed to the State involved. In other examples of legal actions marked as lawfare involvement of a State either is not so obvious or is completely absent. Do these actions nevertheless qualify as lawfare? A great number of the examples pertain to civil litigation in US courts against foreign States. Traditionally, States are immune from the jurisdiction of other States' courts. Foreign State immunity is not absolute, however, and exceptions are well established under international law today. This restrictive theory of State (or: Sovereign) immunity is in the US set out in the 1976 Foreign Sovereign Immunities Act (FSIA).

Over the course of time, the FSIA was amended on several occasions, broadening the range of exceptions to State immunity under US law. In 1996 the Anti-terrorism and Effective Death Penalty Act introduced the terrorism exception, which was later further amended to facilitate legal proceeding before US courts against foreign States designated as State sponsors of terrorism.[66] Ensuing civil litigation (Bankrupting of Terrorism[67]) primarily focused on Iran, which had supported organizations such as Hezbollah, which was held responsible for a number of terrorist attacks claiming American victims. The total of damages US courts awarded were enormous. In the Peterson-case, for example, the judge held Iran responsible for the 1983 bombing of a Marine barracks in Beirut and ordered Iran to pay $2.657 billion in compensation.[68] In other cases related to the Marine barracks bombing the judge awarded some $7 billion in additional damages.

Although litigation is based on US laws enacted by Congress, official US involvement in the court cases is limited to the general designation of Iran as sponsor of terrorism. Claimants do not need official approval for their actions and litigation is not coordinated by US officials either. Furthermore, although civil litigation may have an effect, either positive or negative, on ongoing operations (if one accepts the assumption that fighting terrorism is a military operation), the outcome of a court case serves the personal goals of the claimants. Their legal actions were not and cannot be intended to achieve a military goal. Therefore, claimants cannot be regarded as lawfare actors. Only participants in an operation and entities acting under the direction of, with the consent of or in coordination with the participants involved are lawfare actors.

---

[66] At the date of writing, three states are officially listed as State sponsors of terrorism: Iran, Sudan and Syria http://www.state.gov/j/ct/list/c14151.htm, accessed 10 March 2017.

[67] Bachmann 2012, p. 93.

[68] Peterson v. Islamic Republic of Iran; discussed in Kittrie 2016, pp. 76 ff.

13.5 Synthesis and Conclusion

This chapter took Oplaw as the vantage point to explore the concept of lawfare and refine its definition. First, the concept of Oplaw was discussed against the backdrop of the rule of law. The rule of law concept, on which modern societies are built, basically means that all persons and authorities within a State are subject to the law. It follows from this principle that armed forces have to base their actions on the law and conduct their operations in accordance with the law. This increasingly significant legal framework consisting of domestic and international rules and regulations applicable to military operations is today referred to as Oplaw. Military operations, in their turn, can have an influence on the legal framework as well, facilitating military action. Thus law can become instrumental to a military commander. This instrumental character of law is reflected in the concept of lawfare.

The chapter then explored the origin and development of lawfare, which is indeed a very broad concept, covering a wide variety of actions. It is continuing to develop, making it sometimes hard to establish whether an action can be labelled as lawfare. Some legal actions although taken in a military context do not qualify as such. When the military is conducting its operations in accordance with the law, law is merely the context for military action and not the instrument. Furthermore, the flip side of the rule of law coin entails that the armed forces and their service members can be held accountable for their actions.

When law is either used or misused to achieve an operational goal, an action can be qualified lawfare. An example of the use of law to reach a military goal was the development of EvBO in Afghanistan. ISAF and coalition forces went to great lengths to capture insurgents in order to bring them to trial in Afghan courts. Intelligence gathered by ISAF and the coalition had to be converted into evidence admissible in Afghan criminal courts and Afghan partners were trained and mentored to enable them to asses and use that evidence in criminal proceedings. All efforts were aimed at removing insurgents from the battlefield while at the same time bolstering the Afghan authorities through the use of the Afghan law and court system. Conducting lawfare by misuse of the law is a common tactic used by less advanced parties to an armed conflict. They try to exploit their opponents' sensitivity to alleged violations of the law by creating the image that they operate in violation of these laws. Both the use and misuse of law can be effective instruments to a party to a conflict. Therefore, its adversary must be conscious of the challenges it poses and also be prepared to counter the threats, *inter alia,* by taking counter lawfare actions.

In the literature the term lawfare is sometimes used rather loosely, often encompassing any legal action that has some kind of bearing on military operations or national security. From an Oplaw perspective lawfare is a useful concept to help identify and understand the challenges and opportunities of law in contemporary operations. In order to continue this role the term lawfare must not be interpreted too broadly. Dunlap's definition still captures the essentials of lawfare. Some elements, however, may need to be further specified or defined.

Lawfare was a general concept or idea that developed over time. The term 'strategy' in the definition should be interpreted accordingly and not be literally understood as an (political-military) strategy. Furthermore, lawfare is more than simply using or misusing law. It requires an action deliberately designed to achieve an operational goal. The ideas about lawfare matured in the context of military operations as the actions were aimed at achieving an operational goal. On the one hand, these military operations should not be restricted just to armed conflict situations governed by the laws of armed conflict, but include crisis management and other operations by the military as well. On the other hand, from an Oplaw perspective political and legal actions that are not intended to have a direct effect on ongoing military operations should be left out of the interpretation as it would bring lawfare outside its actual intended playing field.

Because lawfare is intended to achieve an operational goal, all participants in an operation, such as armed forces and armed groups, can be lawfare actors. Other entities not directly involved in the operation may have their own reasons for initiating certain legal activities that may impact on the ongoing operation, but these actions can only be aimed at achieving an operational goal when conducted under control of, with the consent of or in coordination with one of the participants of the operation.

Taking all remarks into account it is proposed Dunlap's definition be read as follows:

The concept of using or misusing law by a participant in a military operation, or an entity acting under its control, with its consent or in coordination with it, as a substitute for traditional military means deliberately designed to achieve an operational military goal in an ongoing military operation. A wide variety of activities that are now referred to as lawfare, do not qualify as such under this definition. It must be stressed that the proposed interpretation results from an analysis carried out from an Oplaw perspective. A political analysis, for example, would undoubtedly yield other results. As lawfare is an attractive term, it cannot be claimed for exclusive use in the field of Oplaw. Therefore, it is suggested to refer to other legal activities by taking the field of research as starting point, for example political lawfare or financial lawfare.

References

Bachmann S (2012) Bankrupting terrorism: the role of US anti-terrorism litigation in the prevention of terrorism and other hybrid threats: a legal assessment and outlook. Liverpool Law Rev 33(2):91–109

Bachmann S, Munoz Mosquera AB (2015) Hybrid warfare and lawfare. The Operational Law Quarterly 16(1):2–4

Bingham T (2011) The rule of law. Penguin Books, London

Boogaard JC van den (2011) Voorzorgsmaatregelen bij aanvallen en de legaliteit van de knock on the roof method. Militair Rechtelijk Tijdschrift 104(5):262–275

Cheng D (2012) Winning Without Fighting: Chinese Legal Warfare. Backgrounder no. 2692. The Heritage Foundation (21 May 2012) http://www.heritage.org/research/reports/2012/05/winning-without-fighting-chinese-legal-warfare Accessed 1 March 2017
Dunlap CJ (2001) Law and military interventions: preserving humanitarian values in 21st century conflicts. Working paper "Humanitarian challenges in military intervention", Harvard's Carr Center for Human Rights Policy, Boston. http://people.duke.edu/~pfeaver/dunlap.pdf Accessed 1 March 2017
Dunlap CJ (2008) Lawfare today. Yale Journal of International Affairs 3(1):146–154
Dunlap CJ (2009) Lawfare: a decisive element of 21st–century conflicts? Joint Forces Quarterly 54:34–39
Dunlap CJ (2010) Does lawfare need an apologia? Case Western Reserve Journal of International Law 43(1):121–143
Dunlap CJ (2015) Lawfare. In: Moore JN, Roberts GB, Turner RF (eds) National security law & policy, 3rd edn. Carolina University Press, Durham, pp 823–838
Gill TD, Fleck D (2015) Concept and sources of the international law of military operation. In: Gill TD, Fleck D (eds) The Handbook of International Law of Military Operations. Oxford University Press, Oxford, pp 3–13
Halper S (2013) China: the Three Warfares, Study prepared for the Office of Net Assessment. http://images.smh.com.au/file/2014/04/11/5343124/China_%2520The%2520three%2520warfares.pdf?rand=1397212645609 Accessed 1 March 2017
Heijne S (2011) Onrust groeit bij Gazavloot: de vakantiedagen raken op. De Volkskrant (1 July 2011)
Jones C (2015) Frames of law. Targeting advice and operational law in the Israeli Defense Force. Environment and Planning D: Society and Space 33(4):1–21
Kennedy D (2007) Modern War and Modern Law. Baltimore Law Review 36(1)176–194
Kittrie OF (2016) Lawfare - Law as a weapon of war. Oxford University Press, Oxford
Liang Q, Xiangsui W (2002) Unrestricted Warfare: China's Master Plan to Destroy America (introduced by Santoli A). Panamerican Publishing Company, Panama City
Morrissey J (2011) Liberal lawfare and biopolitics: US juridical warfare in the war on terror. Geopolitics 16(2):280–305
Munoz Mosquera AB (2015) Hybrid warfare - When lawfare is employed maliciously. University of Exeter, Workshop on the legal framework of hybrid warfare and influence operations
NATO (2014) Rule of Law Field Support Mission and Rule of law Field Force - Afghanistan, NROLFSM/ROLFF-A, 2010–2014, Camp Phoenix, Kabul, Afghanistan, 13 March 2014
Osinga FPB, Roorda MP (2016) From Douhet to drones, air warfare, and the evolution of targeting. In: Ducheine PAL, Schmitt MN, Osinga FPB (eds) Targeting: the challenges of modern warfare. TMC Asser Press, The Hague, pp 27–76
Popovich E (2016) Knock-on-the-roof: the U.S. Air Force's new tactic. War on the rocks (9 May 2016) http://www.warontherocks.com/2016/05/knock-on-the-roof-the-u-s-air-forces-new-tactic/ Accessed 1 March 2017
Rishikof H (2008) Juridical warfare. Joint Forces Quarterly 48(1):11–13
Sadat LN, Geng J (2010) On legal subterfuge and the so-called "lawfare" debate. Case Western Reserve Journal of International Law 43(1):153–161
Scharf M, Andersen E (2010) Is lawfare worth defining? Report of the Cleveland Experts Meeting September 11, 2010. Case Western Reserve Journal of International Law 43(1):11–27
Strikwerda-Verbeek A (2012) Rule of law makes sense - A way to improve your mission. Civil-Military Co-operation Centre of Excellence. http://www.cimic-coe.org/wp-content/uploads/2014/06/Rule-of-Law-Makes-Sense.pdf Accessed 1 March 2017
U.S. Field Manual (2006) Counterinsurgency. U.S. Field Manual 3–24 (December 2006)
U.S. Joint Publication 3-05, Special Operations (16 July 2014). http://www.dtic.mil/doctrine/new_pubs/jp3_05.pdf Accessed 1 March 2017
Voetelink J (2013a) Lawfare. Militair Rechtelijk Tijdschrift 106(3):69–79

Voetelink J (2013b) EvBO: Evidence-Based Operations - How to remove the bad guys from the battlefield. Journal of the International Law of Peace and Armed Conflict 26(4):194–201 http://www.papers.ssrn.com/sol3/papers.cfm?abstract_id=2400965

Voetelink J (2015) Status of Forces: Criminal Jurisdiction over Military Personnel Abroad. TMC Asser Press, The Hague

Werner WG (2010) The curious career of lawfare. Case Western Reserve Journal of International Law 43(1):61–72

Author Biography

Colonel (RNLAF) Dr. J.E.D. Voetelink, LL.M. is Associate Professor of Military Law at the Department of War Studies, Netherlands Defence Academy and a senior researcher at the Amsterdam Centre of International Law (ACIL). He holds a degree in law from Tilburg University. He defended his Ph.D. thesis Status of Forces Agreements as an Impetus for Military Operational Law at the University of Amsterdam.

Chapter 14
Non-Kinetic Capabilities and the Threshold of Attack in the Law of Armed Conflict

Bart van den Bosch

Abstract In this chapter it is argued that within the framework of the Law of Armed Conflict (LOAC), attack and kinetic operations can be used as equivalents where military operations that do not constitute attacks have a large overlap with non-kinetic operations. These topics present more than just a linguistic exercise because many (restrictive) rules in LOAC are constructed around the definition of 'attack', placing fewer restrictions on other military operations. Traditionally, the qualification of attack is determined by the (intended) physical consequences of a military operation. If it does not (intend to) result in physical consequences, it is not defined as attack. If this approach is applied to non-kinetic military capabilities, especially in cyberspace with its non-physical components, the question is raised whether the traditional physical consequences are still valid criteria for determining whether or not a military operation qualifies as attack.

Keywords Act of violence · Attack · Cyberspace · Cyber operation · Damage · Data · LOAC · Non-kinetic operations

Contents

B. van den Bosch (✉)
Faculty of Military Sciences, Netherlands Defence Academy, Breda, The Netherlands
e-mail: glc.vd.bosch@mindef.nl

P.A.L. Ducheine and F.P.B. Osinga (eds.), *Netherlands Annual Review of Military Studies 2017*, NL ARMS, DOI 10.1007/978-94-6265-189-0_14

14.1 Introduction

If it would be possible to delete the digital bank accounts of Islamic State (IS) with the use of software only, generating the same effects as the physical destruction of a cash depot,[1] would that be considered a kinetic or a non-kinetic operation? Chances are, the answer would be: non-kinetic. Almost everybody has a certain conception of the kinetic capabilities of the military, associated with explosives, bombs and shooting that fits within the idea of wounding and killing individuals or physically destructing objects. A relevant question remains unanswered: what separates a kinetic from a non-kinetic capability?[2]

From a legal perspective, the issue resembles the question: what is the threshold of attack in the Law of Armed Conflict (LOAC)?[3] For the answer to both questions, one has to look at the goals, the means and the results of an operation. Is it only the use of kinetic energy to generate the result, or is it the result itself, that defines the kinetic (attack) or non-kinetic (not an attack) nature of a capability? Is it necessary that the result is physical injury or damage to qualify as attack or kinetic operation and if so, to what extent should psychological injury and immaterial damage be considered (i.e. virtual money in the example mentioned above)?

Although the divide between non-kinetic and kinetic capabilities, at least linguistically, seems rather clear, Western armed forces have resorted to "a variety of descriptions and terms to explain what non-kinetic targeting means",[4] sometimes referring to the lethal/non-lethal dichotomy. This may seem only a definitional issue as "the common denominator is that kinetic and non-kinetic targeting (whether or not the lethal/non-lethal dichotomy is used) jointly refer to a spectrum of means and methods employed to influence targets throughout the physical and non-physical dimensions by generating kinetic and non-kinetic effects",[5] it becomes important from a legal perspective when answering the related question about the threshold of attack. The latter can have serious implications since the rules of LOAC regulating attacks are not necessarily the same as those regulating other military operations

---

[1] As actually happened in January 2016 when a declassified video was released by the US DoD. See: CNN 2016.

[2] Capability is used in the sense of having the will, the knowledge and the means (capacity) to perform a certain action. Once a capability is actually used, it is referred to as an operation.

[3] LOAC is also known as Law of War, International Humanitarian Law (IHL) or *ius in bello*.

[4] Ducheine 2015, p. 208.

[5] Ducheine 2015, p. 210.

that are not attacks,[6] possibly providing opportunities in situations where attacks are restricted or even forbidden.

Given the basic assumption that "legitimacy is a principle of democratic societies vested in the rule of law, especially applicable to armed forces"[7] and the goal of this publication, influencing a party by the use and utility of non-kinetic capabilities on the strategic and operational level to generate political objectives,[8] it becomes obvious that in order to discuss the legitimacy of non-kinetic capabilities, a basic understanding of the applicable rules is required.

This chapter will focus on the M of the DIME acronym,[9] bearing in mind that this covers only a part of the options available to a state to bring about a change in a party's behaviour by non-kinetic capabilities. With that limitation in mind, the purpose of this chapter can be described as: establishing the delineation between kinetic and non-kinetic capabilities from a legal perspective in order to get a better understanding of the applicable rules for the use military non-kinetic capabilities without endangering the legitimacy of an operation.

Because neither of the terms 'kinetic' nor 'non-kinetic' are delineated in LOAC, where 'attack' is defined,[10] the focus will first be on the threshold of attack and then indicate how this knowledge can be used in relation to non-kinetic capabilities. In line with the title of this publication, this chapter will concentrate on situations where the military use of non-kinetic capabilities is used as an alternative for (or supplementary to) fighting, which means the focus will be on situations where an armed conflict is ongoing and LOAC is applicable. With this restriction, discussions on *ius at bellum* and the threshold of armed conflict in *ius in bello* can be skipped. This limitation, however, does not necessarily mean that the observations and conclusions arrived at this chapter are restricted to situations of armed conflict. First, the knowledge of the demarcation between kinetic and non-kinetic capabilities in situations of armed conflict can help us understand the possibilities and risks of using non-kinetic capabilities outside situations of armed conflict. Second, some states also apply, at least, the principles of LOAC for policy reasons in situations that do not (yet) meet the threshold of an armed conflict.[11]

---

[6] Examples of military operations that are not attacks are jamming, psychological operations, (strategic) movements, deception (decoys) or intelligence operations. See for other examples Chap. 9 by Ducheine, Van Haaster and Van Harskamp in this volume.

[7] Ducheine and Pouw 2012, p. 33.

[8] The overall aim of this volume.

[9] DIME is the acronym for Diplomacy, Information, Military and Economy as sources of power of a state.

[10] Additional Protocol I, Article 49(1): "'Attacks' mean acts of violence against the adversary, whether in offence or in defence".

[11] The Netherlands is one of these states, see for example *Parliamentary Papers (Kamerstukken) II 2003–2004, 29 200 X, C,* p. 3 "Within NATO, including the Netherlands, it is standing policy that also in operations to which the law of armed conflict formally is not applicable, the restrictions of this body of law (like the prohibition on attacks on the civilian population as such) will be observed" (translation: BvdB).

This chapter will start with the threshold of attack in LOAC in a traditional sense by answering the question what distinguishes an attack from other military operations, not qualifying as attacks. After this, some of the new non-kinetic capabilities will be introduced briefly before moving to the specific area of cyberspace where the subjects of attacks, damage in cyberspace and the status of data will be discussed.

14.2 What Constitutes Attack?

In LOAC, the term 'attack' is defined as 'acts of violence against the adversary, whether in offence or in defence'.[12] This definition contains three elements, 'acts of violence', 'against the adversary' and 'in offence or in defence'. The second and third element can be dealt with quite briefly for the purpose of this chapter. The 'act of violence' deserves a more detailed discussion as this is, for the discussion on non-kinetic operations, the determining element.

The element 'against the adversary'. A naive reader could take the position that acts of violence against civilians would not fall within the definition of an attack because civilians are not part of the adversary. This reasoning however is contrary to one of the main principles of LOAC, the principle of distinction. This principle is considered to be part of customary law,[13] and can be found in treaty law as a basic rule that dictates that "all Parties to the conflict shall at all times distinguish between civilian population and combatants and between civilian objects and military objectives and accordingly shall direct their operations only against military objectives".[14] 'Attacks' being a species of the more generic term 'operations' can therefore (legally) only be directed against combatants and military objectives, in other words only 'against the adversary'.

The second element 'in offence or in defence', has been subject to objections because "the use of 'attacks' to refer to acts of defence is etymologically inconsistent with its customary use".[15] Indeed, attack in ordinary military vocabulary is often associated with offensive operations.[16] Limitation to offensive operations, however, was not intended when the definition of 'attack' was drafted. The authors chose the wider scope "since it—justifiably—covers defensive acts (particularly 'counter-attacks') as well as offensive acts, as both can affect the civilian population".[17] The definition of attack is widely accepted as authoritative. The reference to

---

[12] Additional Protocol I, Article 49(1).

[13] Henckaerts and Doswald-Beck 2005, black letter rule 1 and 7.

[14] Additional Protocol I, Article 48.

[15] Hays Parks 1990, p. 114.

[16] The Collins English Dictionary 2007 defines attack as: to launch a physical assault (against): to begin hostilities.

[17] Sandoz et al. 1987, p. 603.

defensive actions is understandable from a LOAC point of view as it emphasizes that the defensive character of a specific act of violence does not exempt it from the provisions of LOAC.[18]

Finally, the third element of attack, the determining element for this chapter, 'acts of violence' needs to be discussed in reference to LOAC. Until recent years there was no doubt that acts of violence referred to physical force.[19] The term 'attack', as defined in LOAC, is sometimes equated with 'combat action',[20] clearly referring to the application of physical force of combat as opposed to other military operations such as the relocation of troops or intelligence operations. The drafting history of the definition of 'attack' supports this conclusion as the draft version referred to a hostile act, which was later transformed to attacks, as "acts of war that by their nature or purpose struck at the personnel and materiel of enemy armed forces".[21]

Where the definition of 'attack' is clearly constructed around physical activities (acts of violence), other rules of LOAC are also phrased in terms of physical consequences. Article 50 of Additional Protocol I, for instance, states that attacks that "may be expected to cause incidental *loss of civilian life, injury to civilians, damage to civilian objects*, or a combination thereof, which would be excessive in relation to the concrete and direct military advantage anticipated" are to be considered indiscriminate and therefore forbidden.[22] This wording clearly refers to the expected physical consequences of an attack independent of the way these consequences are triggered. The logical conclusion is that within LOAC, to determine whether an act qualifies as an 'act of violence', an assessment of *the physical consequences rather than the action that caused it* is required. The latter is the reason why the release of a chemical agent spread by the wind in order to kill enemy forces, falls within the definition of an act of violence and therefore qualifies as 'attack' for the purpose of LOAC.[23]

This latter example can be used to indicate the relationship between attack and (non-) kinetic capabilities. One could argue that the release of a chemical agent does not fall within the literal meaning of 'kinetic' because it does not rely on the release of kinetic energy and should therefore qualify as 'non-kinetic'. Even if the agent is delivered by an exploding shell, it is most likely not the explosion that causes the harm but the subsequently released agent. If the physical consequences are the same (death of the enemy forces), the argument can be made that, in analogy with the term 'attack', the physical consequences make that the release of a chemical agent

[18] Oeter 2013, p. 167.

[19] Bothe et al. 1982, p. 289.

[20] Sandoz et al. 1987, p. 603; Oeter 2013, p. 167.

[21] Levie 1980, p. 61.

[22] Additional Protocol I, Article 50(5)(b), emphasis added.

[23] Schmitt 2013, p. 106. Although such an attack is forbidden in LOAC amongst others based upon the 1899 Hague Declaration on Asphyxiating gases, it is still an attack.

should be considered as kinetic.[24] This leads to the first sub-conclusion, from a LOAC perspective, that terms 'attack' and 'kinetic operation' can be used as equivalents.

Does this mean that non-kinetic operations are then all other military operations that do not qualify as 'attack'? Logic reasoning tells us that this cannot be the case. Kinetic/non-kinetic is, at least linguistically, a dichotomy where 'attack' is a subset of the bigger concept 'military operation'. In LOAC the term 'attack' is defined, but 'military operation' is a wider concept lacking an authoritative definition. It is sometimes described as: "the process of carrying on combat, including movement, supply, attack, defense and maneuver needed to gain the objective of any battle or campaign",[25] in other words, all 'attacks' are 'military operations' but not all 'military operations' are 'attacks'. Looking at how the two pair, kinetic/non-kinetic and 'attack'/'military operation', relate to each other, the term 'non-kinetic operation' cannot be an equivalent of the term 'military operation that is not an attack'.

Beside this linguistic impossibility there is another argument, more relating to the content. The description of 'military operation' implies that at least a *nexus* with the ongoing armed conflict is required. Looking at this implication it might be hard, especially for the use of non-kinetic capabilities, to determine whether or not a non-kinetic operation would qualify as a military operation. Two examples may explain this. The first example is intelligence gathering through personnel of an embassy during an ongoing armed conflict. Would that be considered to be related to the armed conflict closely enough to qualify as a military operation (and would it make a difference by whom or with what means, military or non-military, the intelligence is gathered)?

The second example is the deprivation of liberty of individuals or security detention.[26] When a person is detained for security reasons, there can be a clear *nexus* with the armed conflict and as a result, the detention could be qualified as a military operation. Security detention has physical consequences, however it does not fit within the definition of 'attack', not even when it involves the use of force to detain a person.[27]

The above does not imply that the equation between 'attack' and kinetic operation becomes useless. As already mentioned in the introduction, the consequence of the equation is bigger than the linguistic determination only. "The notion of 'attack' is a concept that serves as the basis for a number of specific limitations and prohibitions in the law of armed conflict".[28] As an example, it is a basic principle within LOAC that civilians and civilian objects cannot be attacked, or, using the

---

[24] Other examples could be given such as the opening of a dam, resulting in a flood.

[25] Bothe et al. 1982, p. 286.

[26] Ducheine 2015, p. 221.

[27] Kalshoven and Zegveld 2011, p. 100.

[28] Schmitt 2013, p. 106.

concepts of 'attacks' and 'kinetic operations' interchangeably, they cannot legally be the object of a kinetic operation.

But how about a non-kinetic operation? This question will be dealt with later in this chapter where 'non-kinetic operations' and 'military operations that are not attacks' will be used interchangeable. The reason for this is that, despite the fact that these terms are not equivalents, they have a big overlap. For the understanding of the applicable rules in LOAC this suffices.

Before turning to the next paragraph in which some new non-kinetic capabilities will be addressed, two additional remarks on the element of 'act of violence' and their relationship with physical consequences are required. The first remark is about intent. The very intent to generate physical consequences suffices. If the attack is unsuccessful and no harm is done to the enemy, it still qualifies as an attack. As argued by Schmitt, "if an attack is successfully intercepted and does not result in actual harm it is still an attack under the law of armed conflict".[29]

The mirror image is that the physical consequence must be intentional, or at least could be reasonably foreseen, and is not the result of, for instance, an accident. A military airplane that crashes unintentionally, i.e. due to engine failure, and causes casualties or damage, does not qualify as an attack, not even if it crashes on enemy forces. As long as the crash is unintentional, it cannot be considered a military operation and it therefore fails to qualify as 'attack'.

Although in most cases it is clear what the intent is, sometimes the (lack of) intent of physical harm can lead to arguable qualifications. Take the example of laying mines and in the same time clearly marking it as a minefield. The purpose is probably to hinder or delay the enemy but does it qualify as 'attack'? The general feeling with the authors of Additional Protocol I was that whenever a person was directly endangered by a mine laid, it was an attack.[30]

The second remark on physical consequences is that the severity of the consequences, or the question of whether the physical consequences can be reversed, does not matter for the qualification of an operation as 'attack'. As far as its qualification is concerned, there is no difference between a military operation that causes a single wounded soldier and one that kills a hundred enemies. Similarly, both an operation that results in the puncture of one tire of a military transport truck which prevents the truck from carrying out its mission and an operation that blows up a battle tank qualify as attack as defined under LOAC.

The above discussion leads to the conclusion that for LOAC 'attack', and by analogy a 'kinetic operation', means any combat action that is aimed at, or results in, physical consequences.

---

[29] Schmitt 2013, p. 110.

[30] Sandoz et al. 1987, p. 603.

14.3 New Non-Kinetic Capabilities

The traditional understanding of the term 'attack', as explained above, has not been the subject of academic discussion for quite some time. This has changed with the commencement of the so-called cyber age.[31] With the identification of the cyber domain[32] as the fifth domain for military operations,[33] new possibilities emerged for influencing actors, and perhaps even changing the outcome of an armed conflict with military operations that do not necessarily qualify as 'attacks' as defined under LOAC. In other words, the potential relevance of stand-alone non-kinetic operations has risen, resulting in an increased attention for these operations. This does not mean they have replaced the existing means of non-kinetic capacities, but they have raised discussions regarding some interesting LOAC principles. For that reason, the following part will focus on the question of how 'attack' must be interpreted in cyberspace, with relating questions on the status of data and the interpretation of damage in cyberspace.

Two introductory remarks are required. First, this author, with others,[34] does not believe that, in the near future, we can expect an armed conflict exclusively be fought in cyberspace. Traditional military force (kinetic) will most likely be used in any armed conflict in the foreseeable future. Having established this, military operations in, or with the use of, cyberspace are a reality and therefore attention for the legal regime that regulates them, in armed conflict that would be LOAC, is opportune.

Secondly, no attention will be paid to cyber operations that are a part of any kinetic attack because when the rules regulating 'attacks' are applied, an attack has to be seen as a whole, and not by its constituent elements. As part of a kinetic attack, the cyber operations fall within the scope of that attack which means the full spectrum of LOAC rules on 'attacks' applies.[35] From a LOAC point of view there is no reason to treat a kinetic attack with support in or through cyberspace differently from attacks that do not have this support. However, if a cyber operation is conducted as a stand-alone operation, it should be regarded separately. For the remainder of this chapter cyber operations will be considered as stand-alone operations.

---

[31] See for instance Schmitt 2013, part II on the law of cyber armed conflict.

[32] Until now there is no authorative definition of the cyber domain, or as it sometimes referred to as cyberspace. For this contribution, the definition as it is given in the Tallinn Manual will be used: The environment formed by physical and non-physical components, characterized by the use of computers and the electromagnetic spectrum, to store, modify, and exchange data using computer networks. Schmitt 2013, p. 258.

[33] The cyber domain is regularly referred to as the fifth domain following land, sea, air and space. See for example Defensie Cyber Strategie (Defence Cyber Strategy), Kamerstukken (Parliamentary Papers) II 2011–2012, 33 321 nr 1. Declaration NATO Summit 2016 Warsaw, see the Chapter by Theo Brinkel in this Volume.

[34] Gill and Ducheine 2013, p. 438. ILA Report 2014, p. 5.

[35] Schmitt 2013, p. 110.

14.4 What Constitutes Attack in Cyberspace?

To answer this question in LOAC terms, we have to take a step backwards and realize that in common usage, attack in cyberspace, or cyber attack, is used in a much broader sense then the term 'attack' in LOAC. Websites that are temporarily not available as a result of a (D)DOS operation are regularly described in the media as having been under attack.[36] In LOAC, however, the term 'attack' in used to indicate injury, death, damage or a combination thereof to the adversary. Cyber attacks that do not meet the LOAC definition criteria of 'attack' present unchartered, but heavily discussed, territory. The question is whether or not the traditional focus on physical consequences is still appropriate as the distinguishing criterion for 'attack' versus non-kinetic operations in cyberspace?

If a cyber operation has physical consequences comparable to that of kinetic operations,[37] there is little doubt that this cyber operation will constitute an attack in the sense of LOAC.[38] It becomes less clear if these physical consequences do not emanate directly. This problem will be commented in two ways, firstly in relation to humans with attention for the psychological component of injury and secondly in relation to objects and the question of what constitutes damage, or even destruction, in a virtual world.[39]

14.4.1 Cyber Operation in Relation to Persons

As indicated above, it is clear that any military operation that results in physical injury is an attack as defined under LOAC. This is not different when the operation is executed through or by cyber means. As an example, imagine that, during an armed conflict, someone is able to hack a mobile phone in a way that it manipulates the transmission of a signal disrupting any insulin pump in the immediate surroundings of the telephone so that it would change the dosage causing serious health damage or even death. This technique constitutes an attack as defined under LOAC. That would mean that in certain circumstances, this attack would be

[36] (D)DOS: (Distributed) Denial of Service, a situation in which people intentionally prevent a website from operating by sending too many requests to use it (Cambridge Business English Dictionary).

[37] Just as with cyberspace there is no authorative definition of cyber operations. For this chapter, again the definition of the Tallinn Manual will be used: The employment of cyber capabilities with the primary purpose of achieving objectives in or by the use of cyberspace, See: Schmitt 2013, p. 258.

[38] See for instance back letter rule 30 "definition of cyber attack" in the Tallinn manual and its following commentary (Schmitt 2013, pp. 106–110) that defining cyber attack as: a cyber attack is a cyber operation, whether offensive or defensive, that is reasonably expected to cause injury or death to persons or damage or destruction to objects.

[39] Following the widely accepted dichotomy between humans and objects in LOAC.

permitted against a combatant but would be prohibited when directed against a civilian.

But what about psychological injury? In traditional kinetic operations, it is without a doubt that psychological injury cannot be ignored. One of the rules for protecting civilians is the prohibition of "acts or threats of violence the primary purpose of which is to spread terror among the civilian population".[40] The fact that not only acts of violence but also threats of violence are forbidden, clearly indicates the concern for the mental health of civilians. This rule clearly prohibits operations that are intended to cause physical injury and psychological injury alike.[41] Another indication that psychological injury cannot be ignored can be found in one of the basic rules mentioned in Additional Protocol I, "it is prohibited to employ weapons, projectiles and Materials and methods of warfare of a nature to cause superfluous injury or unnecessary suffering".[42] The use of the words 'injury' and 'suffering' clearly indicate both physical and psychological injury.[43]

The practical implementation of these provisions is however problematic. The first rule, prohibiting terror attacks, focuses not on the effects of attacks, as any act of warfare will be frightening to the civilian population to some degree, but on its intent (the primary purpose of). And even in a clear case of intent, how serious must this be to qualify as terror and not merely as an intention to frighten the civilian population? A similar problem occurs with the term 'unnecessary suffering' indicating that there is also necessary suffering, namely just enough to put an enemy combatant 'hors de combat'. Because it is almost impossible, if possible at all, to establish an instrument to measure the suffering, it will be even harder to establish the line between necessary and unnecessary suffering.

As demonstrated, many variables have to be taken into account and these will, besides some real serious and obvious exceptions,[44] stand in the way for a practical implementation of the psychological injury criterion. If the practical implementation of causing mental injury is nearly impossible to measure in the classical kinetic sense, it will be even more complex in a non-kinetic environment where there is no (fear for) physical injury or damage. A logical conclusion therefore is that the psychological component of non-kinetic operations must be taken into account in theory, however, in practice, it is safe to state that it can be ignored.

The sub-conclusion for cyber operations in relation to persons is that, as long as there is no intent to harm any person physically or actually to cause any physical injury, a cyber operation does not qualify as 'attack' in LOAC.

---

[40] Additional Protocol I, Article 51(2).

[41] Sandoz et al. 1987, p. 618. In the Commentary to Additional Protocol I, Article 51(2) they state that breaching this provision is defined a grave breach (Article 85 Additional Protocol I) when it results in death or serious injury to body or health.

[42] Additional Protocol I, Article 35(2).

[43] Bothe et al. 1982, p. 196. See also Schmitt 2013, p. 108. Since terror is a psychological condition resulting in mental suffering, inclusion of such suffering in this Rule is supportable through analogy.

[44] Like the announcement of terror attacks.

14.4.2 Cyber Operation in Relation to Objects in Cyberspace

The discussion on what constitutes 'attack' on objects in cyberspace is fuelled by the text of Article 52 of Additional Protocol I. This article starts with a negative definition of civilian objects stating that "Civilian objects are all objects which are not military objectives as defined [below]",[45] followed by the definition of military objectives that are "in so far as objects are concerned, military objectives are limited to those objects which by their nature, location, purpose or use make an effective contribution to military action and whose total or partial destruction, capture or neutralization, in the circumstances ruling at the time, offers a definite military advantage".[46] The ICRC Commentary to this article, further describes the word 'object' as "something that is visible and tangible"[47] and is clearly referring to a material object.[48]

This wording raises interesting issues with reference to cyberspace. Cyberspace is comprised of physical (i.a. hardware) and non-physical components like data and software.[49] The non-physical components do not fulfil the requirements of the meaning of objects of Article 52 Additional Protocol I as given in the ICRC Commentary. What does this mean? One interpretation is that because they are not 'objects', they cannot meet the criteria for military objectives and therefore, they would, by definition, never qualify as a 'military objective'. Because "attacks shall be limited strictly to military objectives",[50] data and software may not be attacked under LOAC. This reasoning would lead to the unrealistic and paradoxical conclusion that, according to LOAC, bombing and destroying computers running enemy radar systems would be permitted, though it would not be allowed to infiltrate the same computers by cyber means and delete or alter software or data, hampering their functionality.

Another interpretation is that the reference to visible and tangible, in relation to 'objects' in the definition of military objective, is meant only to clearly exclude the "general objective (in the sense of aim or purpose) of a military operation",[51] rather than "to specifically exclude intangible objects from the definition".[52] With this interpretation, the non-physical components could become a military objective if

[45] Additional Protocol I, Article 52(1).

[46] Additional Protocol I, Article 52(2).

[47] Sandoz et al. 1987, p. 634.

[48] Sandoz et al. 1987, p. 633.

[49] For this chapter the definitions of the Tallinn Manual will be used which are: Data: The basic elements that can be processed or produced by a computer (Schmitt 2013, p. 258) and Software: The non-Physical components of a computer system and of cyber infrastructure. These components include programs, applications, and related data (Schmitt 2013, p. 261). The difference will be addressed later in this chapter.

[50] Additional Protocol I, Article 52(2).

[51] Sandoz et al. 1987, p. 634.

[52] Harrison Dinniss 2015, p. 43.

they fulfil, by analogy, the two-pronged test of Article 52(2) of Additional Protocol I. They have to "make an effective contribution to military action",[53] and "the total or partial destruction, capture or neutralization offers a definite military advantage".[54]

This reasoning bears the disadvantage that it uses the definition of the term, 'military objective', to clarify the meaning of another term, 'attacks', for which there is no indication that this was ever meant by the authors of Additional Protocol I. Another disadvantage is that it "opens the floodgates to prohibiting many kinds of cyber operations that state practices and the international community has thus far not been willing to accept as prohibited attacks"[55] such as DDOS attacks or alteration of websites. The same disadvantages can be seen in an alternative way of looking at Article 52(2) of Additional Protocol I as Mučak proposed, which is to treat data as objects.[56]

A third school of thinking takes a different approach on the text of Article 52 Additional Protocol I, stating that the definition given in Article 52(2) is given only "in the sense of targets for attack".[57] If a military operation does not fulfil the criteria of 'attack', in the sense of having no (intent to generate) physical effects, the definition of military objectives is not applicable.[58] This third line of reasoning is preferred by the present author, because of the system and sequence of Additional Protocol I. This interpretation requires revisiting the traditional kinetic interpretation of 'attack'[59] and whether or not a re-interpretation is required in the light of the non-physical components of cyberspace and the issue of how 'acts of violence' should be interpreted in that case and how this relates to damage?

14.5 What Constitutes Damage in Cyberspace?

The traditional kinetic approach of damage, as the result of 'acts of violence', focused on physical damage in order to fulfil the definition of the term 'attack'. The question is whether this has to be expanded to non-physical components in cyberspace. The Tallinn Manual takes a balanced approach between the fixation on physical damage and the possibilities of expanding the ambit of damage to non-physical elements. In the final black letter rules, the Manual sticks to the

[53] Additional Protocol I, Article 52(2).

[54] Additional Protocol I, Article 52(2).

[55] Harrison Dinniss 2012, p. 3.

[56] Mučak 2015, p. 65.

[57] Bothe et al. 1982, p. 321.

[58] Schmitt 2002, p. 381.

[59] As mentioned before, this was 'any combat action that is aimed at, or results in, physical consequences for the adversary'.

criterion of physical consequences.[60] The Commentary, however, intended to identify differing positions as to the scope or interpretation of the rule,[61] provides clear examples of new interpretations that also include non-physical consequences. The definition of cyber attack, for example, states that "reasonably expected damage or destruction to objects is necessary"[62] to qualify as a cyber attack. Whether interference with the functionality of an object constitutes damage or destruction for the purpose of qualifying as 'attack', raised extensive debate.[63] This debate resulted in a minority of the International Group of Experts arguing this was not the case, while the majority was "of the view that interference with functionality qualifies as damage if restoration of functionality requires replacement of physical components".[64] Within this majority group, there was a split over the issue "whether the 'damage' requirement is met in situations where the functionality can be restored by reinstalling the operating system",[65] whilst a few experts went even further by suggesting "that interference with functionality that necessitates data restoration, while not requiring physical replacement of components or reinstallation of the operation system, qualifies as attack".[66]

Another example of the balanced approach can be found in the rule on civilian objects and military objectives.[67] Clear reference is made to Additional Protocol I, Article 52(2) with the addition that military objectives may include computers, computer networks and cyber infrastructure. The Commentary to this rule clearly mentions that data (as non-physical entities) "are not excluded from the ambit of the term attack. Whenever an attack results in […] damage or destruction of physical objects, those […] objects constitute the 'object of attack' and the operation therefore qualifies as an attack".[68] A minority argued that an operation against data without any physical consequences should also qualify as attack. The majority, however, characterized this position as *de lege ferenda.*[69]

What is clear from the outcome of the discussion is that the 'old' kinetic interpretation of attack is also valid for cyber operations. Any cyber operation that is aimed at, or results in physical consequences for the adversary is an attack. There is, however, no unambiguous answer to the question of whether or not this is still the demarcation line between attacks and other (non-kinetic) military operations in LOAC.

[60] Which is logical because "The Rules were adopted employing the principle of consensus", see Schmitt 2013, p. 6.

[61] Schmitt 2013, p. 6.

[62] Schmitt 2013, black letter rule 30, p. 106.

[63] Schmitt 2013, p. 108.

[64] Schmitt 2013, p. 108.

[65] Schmitt 2013, p. 109.

[66] Schmitt 2013, p. 109.

[67] Schmitt 2013, black letter rule 38, p. 125.

[68] Schmitt 2013, p. 108.

[69] Schmitt 2013, p. 127.

The examples given above from the Tallinn Manual demonstrate that the loss of, or damage to, non-physical elements do not automatically lead to the same conclusion. There seem to be different categories of non-physical elements, each with their own status, at least for the legal consequences when they are lost or altered. This raises the final question of this chapter, the question of the status of data.

14.6 The Status of Data

Before discussing the status of data it is necessary to define data. The definition of cyberspace gives a dichotomy of physical and non-physical components.[70] It is possible to make a further classification in the non-physical components with, depending on the purpose of the further classification, a number of layers ranging from two[71] up to six.[72] To determine the legal status of data, it suffices to distinguish two separate layers. Layer one is the logical network layer, some prefer the term operation-level data,[73] that consists of protocols, operating systems software, applications etc. that "gives hardware its functionality and ability to perform the tasks we require".[74] Layer two consists of the cyber persona layer, or content-level data,[75] consisting of elements such as e-mail accounts and social-media accounts[76] but also the contents of databases such as text files, medical files, library files etc.[77]

An interesting question, associated with this classification, is based on "the traditional interpretation of the dichotomy of targets which constitute legitimate military objectives in IHL, namely persons and objects".[78] The question is, for the application of LOAC, whether a difference exists between the definition of data relating to persons and data relating to objects. The traditional dichotomy in LOAC offers more protection to persons then to objects. A typical example of this can be found in Additional Protocol I where "the civilian population and individual civilians shall enjoy general protection against dangers arising from *military operations*",[79] where "civilian objects shall not be the object of *attack or of*

[70] The environment formed by physical and non-physical components, characterized by the use of computers and the electromagnetic spectrum, to store, modify, and exchange data using computer networks, see: Schmitt 2013, p. 258.

[71] US Department of Defense (2013), JP 3-12: Logical Network Layer and Cyber persona Layer.

[72] Ducheine and Van Haaster 2014, p. 310. The Open System Interconnection model: datalink, network, transport, session, presentation and application layers.

[73] Harrison Dinniss 2015, p. 41.

[74] Harrison Dinniss 2015, p. 41.

[75] Harrison Dinniss 2015, p. 41.

[76] Ducheine and Van Haaster 2014, p. 310.

[77] Harrison Dinniss 2015, p. 41.

[78] Mučak 2015, p. 64. This dichotomy is found in many other places e.g. Sandoz et al. 1987, p. 635 and Schmitt 2013 the structure of part 4.

[79] Additional Protocol I, Article 51(1). Emphasis added.

reprisals".[80] For persons, the wider term 'military operations' is used where the more limited terms 'attack and reprisals' is used for objects. An explanation for this difference is found in one of the basic principles of LOAC, humanity. Humanity, per definition, is related to persons so in the balance between military necessity and humanity, it is logical that persons enjoy a higher level of protection than objects. Returning to the related question on data, this author argues that the dichotomy persons and objects for *this* discussion should be persons and 'non-persons'. Data, although it might be related to persons, never reaches the status of 'person' for the protective rules of LOAC. Therefore, on the basis of simple logic, regardless of the status of data in the discussion related to objects, data never becomes a person.

Returning to the given division of the non-physical components provided at the beginning of this section, the relevance of it is in the effect of deleting or changing this data. If something is changed or deleted in the logical network layer, interference with the functionality of the system is likely. In most cases this is even the very purpose of the cyber operation. It depends on the position in relation to the question what constitutes damage or destruction of the functionality of a system, as described above, whether such a cyber operation constitutes an attack or not.[81]

The deletion or alteration of data in the cyber persona layer or content-level data is even a step further away from constituting 'attack' in the sense of LOAC. Such an operation might "leave the system intact, albeit with corrupted or missing data".[82] As stated above, if the direct foreseeable result of such a cyber operation is physical injury, death or damage, it automatically constitutes an 'attack'.[83] Although this is possible, the majority of this type of cyber operations will only result in 'mere inconvenience'. Because the system itself is still intact and running and most likely, back-ups will be available including cyber persona layer type of data, the situation can possibly be restored to the pre-cyber operation situation. In a situation like this, is it still desirable to talk about damage in the sense of LOAC and thus qualify this cyber operation as a cyber attack?

To answer this latter question, it is wise to take a step back from the individual rule to the wider *raison d'être* of LOAC and to be aware of the aim and purpose of this legal regime. It is well accepted that this purpose is twofold, regulating the conduct of hostilities and at the same time the protection of those that do not, or no longer, participate in the hostilities. This results in "a balance between military necessity and the requirements of humanity".[84] Trying to give new interpretation to a rule, because of a new situation or development, focusing on either military necessity or humanity, disregarding the balance might lead to a biased and therefor undesired, outcome.

[80] Additional Protocol I, Article 52(1). Emphasis added.

[81] See the aforementioned discussion on functional disruption.

[82] Harrison Dinniss 2015, p. 42.

[83] A possible example is changing the blood type in the health record of an enemy wounded soldier in a hospital who, as a result, gets a blood transfusion with blood not of his type.

[84] O'Connell 2013, p. 37.

It is true that strict adherence to the physical consequence criterion "does dramatically expand the possibilities for 'targeting' (but not attacking) non-military objectives".[85] If the emphasis is on the military necessity side of the balance, this is not a weakness, "it simply represents an expansion of permissible methods and means resulting from advances in technology, while existing norms remain intact".[86] Taken to its extreme, this means that shutting down the civilian water supply of a town by manipulating the pumps (without physically damaging them) would not be an attack, whilst bombing it would be. In this case, the attack would be prohibited because of the civilian status of the object. However, where the same result could be achieved by a cyber operation, this does not qualify as attack and therefore, the operation would not be prohibited. This is a rather simplistic conclusion because non-kinetic operations still have to fulfil the principles and rules of LOAC that go beyond 'attacks'.[87] It is, however, sufficient to identify the negative outcome of such an interpretation. Focusing on military necessity, and with that sticking to the physical consequences, provides a clear advantage of having a distinguishable criterion that is practical to apply, though this might lead to undesired openings for targeting (not attacking) civilian objects.

If, on the other hand, the emphasis is on humanity, one might go as far as to say that "one of the main objectives of this body of law [LOAC] is to protect the civilian population from the effects of warfare",[88] and that "the civilian population must be protected not only against attacks, but also more generally against the effects of military operations".[89] Taken this reasoning to its extreme implies that any cyber operation which has an effect on the civilian population would fall under the scope of attack, even if it would lead to mere 'inconvenience'. This would contradict the well-established opinion that every armed conflict will have negative effects on the civilian population.[90] It is for this reason that Additional Protocol I refers to the "general protection against dangers arising from military operations",[91] not to protection against the effects of military operations. Focusing on humanity, by considering loss or alteration of content-level data to constitute damage as a result of an act of violence, would automatically bring almost all cyber operations under the definition of 'attack'. The advantage would be that civilians, and civilian

[85] Schmitt 2002, p. 378, makes the analogy with psychological operations against the civilian population. As long as they do not cause physical harm (and they are not intended to terrorize), they are entirely permissible.

[86] Schmitt 2002, p. 378.

[87] Although some writers seem to suggest that if the effects of an operation are limited to cyberspace, these operations are outside the scope of LOAC, Mačak 2015, p. 59. Also see Schmitt 2013 p. 127 where some experts argued that data per se should be regarded as object because otherwise "datasets would potentially *escape the regulatory reach of the law of armed conflict*" (emphasis added).

[88] Droege 2012, p. 533.

[89] Droege 2012, p. 555.

[90] Think only of the economic effects because of the costs involved in an armed conflict.

[91] Additional Protocol I, Article 51(1).

objects, would be protected to the maximum extent. The disadvantage is that it would deprive military forces of possibilities in cyberspace that LOAC never imposed in the physical world (nor was it ever the intent to do so).

The sub-conclusion is that there is no clear status of data within LOAC yet. It is not clear whether deletion or alteration of data alone constitutes an act of violence in the sense of the definition of 'attack'. There seems to be a tendency to qualify cyber operations that lead to the interference with the functionality of a system closer to attacks than operations that merely interfere with data in the cyber persona layer or content level data.

14.7 Conclusions

In this chapter it is concluded that 'kinetic operations' and 'attacks' can be used as equivalents for the purpose of LOAC. 'Non-kinetic operations' and 'military operations that are not attacks' have a large overlap but cannot be used as equivalents. To understand the applicable rules of LOAC, it suffices to focus on the overlap.

Traditionally, the delineation between 'attacks' and 'non-kinetic operations' was determined by the physical consequences: 'attacks' have (intended) physical consequences such as damage or destruction of objects or injury or death to humans, whereas 'non-kinetic operations' did not have these physical manifestations.

With the rise of new non-kinetic capabilities, especially in cyberspace, it is unsettled whether or not this criterion is still effective. It is without a doubt that cyber operations that result in physical consequences fall within the definition of 'attack'. It is however unclear whether or not operations that cause only non-physical consequences, such as (temporary) loss of functionality or alteration or loss of data, also qualify as 'attack'.

From a legal point of view this is a somewhat unsatisfactory situation because of the importance of LOAC as a normative framework for the conduct of hostilities. A substantial number of rules from LOAC is constructed around the concept of 'attack' and it is therefore important to get a clear understanding of whether the interpretation of 'attack' should be expanded to non-physical consequences. A tendency that treats the most serious non-physical consequences in the same way as physical consequences can be observed, however, new questions on how serious the non-physical consequence must be, fuel this ongoing debate.

How to deal with this legal uncertainty in the (near) future? Sticking to the 'old' physical consequences criterion seems the best way of serving the purpose of LOAC. It recognizes the new possibilities for achieving the military necessities without compromising the protective goals. As soon as military operations have physical effects, there is little doubt that they fall within the scope of 'attack', with its subsequent rules and its protective effects. The cases of only non-physical consequences can then be considered 'mere inconvenience'. The latter has never been considered to be regulated by LOAC and neither should it.

Returning to the overall aim of this publication and this chapter, in a democratic society legitimacy is vested in the rule of law that is especially applicable to the armed forces.[92] To protect the legitimacy of any operation, kinetic and non-kinetic, it is necessary to respect the applicable legal rules. For LOAC it is clear that any operation that is aimed at, or has, physical consequences, has to follow the rules regarding 'attacks'. For non-kinetic operations with non-physical consequences, the rules are rather ambiguous.

Acknowledgements The author would like to thank Lieutenant Colonel (RNLAF) Guus de Koster and Jeroen van den Boogaard for their most welcome suggestions and assistance.

References

Additional Protocol to the Geneva Conventions of 12 August 1949, and relating to the protection of victims of international armed conflicts (API) of 8 June 1977

Bothe M, Partsch K, Solf W (1982) New rules for victims of armed conflicts. Martinus Nijhoff Publishers, The Hague

CNN (2016) US bombs ISIS cash money. cnn.com (15 January 2016) http://edition.cnn.com/2016/01/15/politics/us-bombs-isis-cash-money-iraq. Accessed 1 March 2017

Defensie Cyber Strategie (2011–2012) Kamerstukken II 2011-2012, 33 321 nr. 1

Droege C (2012) Get off my cloud: Cyber warfare, international humanitarian law and the protection of civilians. International Review of the Red Cross 94(886):533–578

Ducheine P (2015) Non-Kinetic Capabilities: Complementing the Kinetic Prevalence to Targeting. In: Ducheine P, Schmitt M, Osinga F (eds) Targeting: Challenges of Modern Warfare. TMC Asser Press, The Hague, pp 201–220

Ducheine P, Pouw E (2012) Legitimizing the use of force: legal bases for operation Enduring Freedom and ISAF. In: van den Meulen J, Vogelaar A, Beeres R, Soeters J (eds) Mission Uruzgan: collaborating in multiple coalitions for Afghanistan. Amsterdam University Press, Amsterdam, pp 33–46

Ducheine P, van Haaster J (2014) Fighting Power, Targeting and Cyber Operations. In: Brangetti P, Maybaum M, Stinissen J (eds) Proceedings of the 6th International Conference on Cyber Conflict, NATO CCD COE, Tallinn, pp 303–328

Gill T, Ducheine P (2013) Anticipatory self-defense in the cyber context. International Law Studies 89:438–471

Harrison-Dinniss H (2012) Cyber warfare and the laws of war. Cambridge University Press, Cambridge

Harrison-Dinniss H (2015) The nature of objects: Targeting networks and the challenge of defining cyber military objectives. Israel Law Review 48(1):39–54

Hays Parks W (1990) Air war and the law of war. Air Force Law Review 32:1–225

Henckaerts J, Doswald-Beck L (2005) Customary international humanitarian law - Volume 1: Rules. Cambridge University Press, Cambridge

International Law Association Study Group (2014) The conduct of hostilities and international humanitarian law - Challenges of 21st century warfare. http://www.ila-hq.org/en/committees/study_groups.cfm/cid/1040. Accessed 1 March 2017

Kalshoven F, Zegveld L (2011) Constrains on the waging of war, 4th edn. Cambridge University Press, Cambridge

[92] See Sect. 14.1 of this chapter.

Levie H (1980) Protection of war victims: Protocol I to the 1949 Geneva Conventions (Volume 3). Oceana Publications, New York
Mučák K (2015) Military objectives 2.0: The case for interpreting computer data as objects under international humanitarian law. Israel Law Review 48(1):55–80
O'Connell M (2013) Historical development and legal basis. In: Fleck D (ed) The handbook of international humanitarian law, 3rd edn. Oxford University Press, New York
Oeter S (2013) Methods and means of combat. In: Fleck D (ed) The handbook of international humanitarian law, 3rd edn. Oxford University Press, New York
Sandoz Y, Swinarzki C, Zimmermann B (eds) (1987) Commentary on the Additional Protocols of 8 June 1977 to the Geneva Conventions of 12 August 1949. Martinus Nijhoff Publishers, Geneva
Schmitt M (2002) Wired warfare: Computer network attack and jus in bello. International Review of the Red Cross. 84(846):365–399
Schmitt M (ed) (2013) Tallinn manual on the international law applicable to cyber warfare. Cambridge University Press, Cambridge UK
US Department of Defense (2013) Joint Publication Cyberspace Operations (JP 3–12(R). http://www.dtic.mil/doctrine/new_pubs/jp3_12R.pdf. Accessed 1 March 2017

Author Biography

Lieutenant-Colonel G.L.C. van den Bosch (Bart), LL.M., M.A. is Assistant Professor of Military Law, Department of War Studies, Netherlands Defence Academy. He holds master degrees in Law and in Public Administration (Dutch Open University) and is preparing his Ph.D. thesis on the Law of Armed Conflict for Cyber Operations.

Chapter 15
Jamming and Spoofing: Effective Cyber Weapons Looking for a Defence

Barend Lubbers, Eric Theunissen and Patrick Oonincx

Abstract The Global Positioning System has long been a great tool for military—and, of course, civilian—activities, but the military's heavy reliance on it also makes it a likely target. Due to the character of the satellite navigation system, security vulnerabilities exist that threaten information continuity and integrity. In this chapter possible threats for global satellite systems are discussed briefly and some actual civil and military actions of Navigation Warfare are mentioned. The chapter also gives an overview of past and current research done on these type of vulnerabilities by the Netherlands Defence Academy. Within this research different approaches to Navigation Warfare actions are explored. For this purpose mitigation concepts are designed and tested in a laboratory setting. The chapter also addresses alternative means for obtaining position and timing information.

Keywords Global Navigation Satellite Systems · Navigation Warfare · Cyber · Jamming · Spoofing · Resilient Positioning

Contents

B. Lubbers (✉) · E. Theunissen · P. Oonincx
Faculty Military Sciences, Netherlands Defence Academy, Den Helder, The Netherlands
e-mail: b.lubbers.02@mindef.nl

E. Theunissen
e-mail: e.theunissen@mindef.nl

P. Oonincx
e-mail: pj.oonincx@mindef.nl

P.A.L. Ducheine and F.P.B. Osinga (eds.), *Netherlands Annual Review of Military Studies 2017*, NL ARMS, DOI 10.1007/978-94-6265-189-0_15

15.1 Introduction

The increasing demand for unmanned systems creates new challenges for navigation functions operating at higher levels of system autonomy. Like many other systems that rely on communication links, vulnerabilities in the area of cyber are increasing. On one hand this concerns cyber vulnerabilities similar to other system. For unmanned systems two examples are the potential hijacking of the control link and injection of malicious code into one or more of the on-board systems. On the other hand, vulnerabilities exist that always have been present but until recently were not explicitly classified in the Cyber category.

When considering the navigation function of an autonomous system in terms of the Observe-Orient-Decide-Act (OODA) loop, the requirements on the Orient follow from the need to autonomously make the right decisions (Decide) without human involvement. To be able to make such decisions, the Orient phase poses data requirements in terms of availability, accuracy, continuity and integrity on the Observe phase. Part of this data has to be obtained by the on-board sensors in real-time. With jamming, an attacker denies the Observe function the information that it is looking for. With spoofing, an attacker provides falsified information to the Observe function.

15.2 Jamming and Spoofing

Jamming can be seen as the equivalent to a DDoS cyberattack and is defined as "the emission of radio frequency energy of sufficient power and with proper characteristics to prevent receivers in the target area from tracking GPS signals".[1] Jammers can produce different types of jamming signals.

A noise jammer transmits broad- or narrowband noise or just a single frequency (tone) within a GPS frequency band. In effect it lowers the signal to noise ratio of the victim receiver causing a loss of lock when the jamming signal power reaches a certain receiver dependent threshold. Since a receiver needs a higher signal to noise ratio in the acquisition phase a noise jammer can prohibit a receiver of locking onto the satellite signals using less power. When the noise or tone is not transmitted continuous but in pulses, it has the same effects as continuous noise. However, when certain characteristics of the victim receiver are known, it can have an extra

[1] Stated in US Department of Transportation 2001.

negative effect on the Automatic Gain Control (AGC) and on parts of the receiver that control the necessary processes and status (Control Laws). In case the Intermediate Frequency (IF) of a victim receiver is known, a jammer can use this knowledge to induce a direct current signal into the receiver that hampers the detection of the navigation message. Design of systems that rely on position and timing data have to take this vulnerability into account. System redundancy based on dissimilarity (i.e. a completely independent system using a different means to estimate position and measure time) can be used to mitigate the jamming threat.

With spoofing, an attacker attempts to deceive the Observe function by sending falsified information which mimics the properties of truthful information to the extent that the falsification goes undetected. Spoofing is far more difficult than jamming, but can also be far more dangerous since a properly executed spoofing attack is likely not be detected by the victim until it is too late.

The lack of encryption in the civil GPS signal makes it particularly vulnerable to meaconing and spoofing. The verb 'meacon' is a contamination of 'm(islead)' and '(b)eacon' and was first used by the British during the Second World War.[2] In the case of meaconing, a malevolent transmitter receives a GNSS signal, amplifies and retransmits it. A victim receiver now receives the original signal and the retransmitted signal. Because the latter has more power the receiver might lock onto this signal thus generating the PNT information that concurs with the position and time of the malevolent transmitter instead of the PNT information of its own position and time.[3] Instead of retransmitting an original GNSS signal, a transmitter might also create and transmit the signal itself using existing ranging codes and navigation messages. For spoofing to be effective the spoofer needs to know the position, course and speed of the victim receiver. When properly executed the receiver will in first instance just show the correct PNT information once it is locked onto the spoofing signal. From this moment on the spoofer has control of the receiver and can generate any PNT output he wishes by manipulating the spoofing signal.[4]

15.2.1 State-Based Jamming and Spoofing

Deliberate, intentional jamming is occurring with increasing regularity and should now be expected as a routine aspect of operations to disrupt and deceive. As with other aspects of electronic warfare (EW), attacks on PNT capabilities are often made before any "kinetic" fighting. Reports of jamming and spoofing activities continue to increase.

North Korea has reportedly broadcast a jamming signal on 100 occasions. A total of 962 planes have been affected by the jamming, as well as nearly 700

[2] See also Psiaki et al. 2013.

[3] See Akos 2012.

[4] For a rigorous description, see Bhatti et al. 2012.

fishing vessels. The jamming has also affected cell phone base stations. Previous jamming attacks have affected approximately 1,000 civil aircraft and unmanned aerial vehicles of the South Korean military. North Korea developed its GPS jamming capability in response to GPS-guided weapons that could be used by South Korean and U.S. forces in the event of war. The country maintains a regiment-sized GPS jamming unit near the capital of Pyongyang and battalion-sized units near the demilitarized zone. In the past, North Korea purchased truck-mounted GPS jammers from Russia with a range of 30–60 miles, and the country is developing even longer-range jammers.

Whereas the jamming performed by North Korea has been known to occur at least since 2010, the Russians are apparently spoofing GPS near the Kremlin and China has used GPS jamming to prevent U.S. aircraft from conducting surveillance missions in the Spratly Islands. Furthermore, Russia plans to modify it Cell tower network to add the capability to act as a GPS-jamming network. GPS jamming has affect operational capabilities of unmanned systems in Ukraine. In December 2016 it was reported that the 72 Raven RQ-11B UAS provided to Ukraine were effectively rendered ineffective by jamming of the Russian-backed separatists. In this particular case the main vulnerability was caused by the use of analogue video and control links. Russia and the separatist forces it supports both intercepted and jammed the video feeds and data. However, even the later digital versions can still be fooled, but it is more a Cyber challenge.

A far more advanced cyber attack on an unmanned system may already have occurred five years earlier. In 2011, GPS-spoofing may have been used by Iranian forces to capture an American Lockheed Martin RQ-170 Sentinel unmanned aerial vehicle (UAV). The Iranian government announced that their cyber warfare unit commandeered the aircraft and safely landed it. They said the aircraft was detected in Iranian airspace 225 km (140 mi) from the border with Afghanistan. The government of the United States initially claimed that its forces in Afghanistan had lost control of a UAV that there was a possibility that this is the vehicle that crashed near Kashmar. Later, the drone appeared to be largely intact, except for possible minor visible damage on its left wing. Dan Goure, an analyst at the Lexington Institute, stated the largely intact airframe ruled out the possibility of an engine or navigational malfunction. It was suggested that either this was a cyber or electronic warfare attack that brought the system down or it was a glitch in the command-and-control. A Christian Science Monitor article relates an Iranian engineer's assertion that the drone was captured by jamming both satellite and land-originated control signals to the UAV, followed up by a GPS spoofing attack that fed the UAV false GPS data to make it land in Iran at what the drone thought was its home base in Afghanistan.

15.2.2 Spoofing Going Mainstream

In it was indicated that the means to locally deny availability of the information (i.e., GPS jamming) have become commonly available. Although the involvement of Russia, China, Iran in the GPS-spoofing examples may suggest that the more complex spoofing is a weapon that can only be wielded with access to significant resources, this is no longer the case. With the rapidly increasing availability of transmitters that are programmable in terms of signal and frequency, the creation of falsified signals that 'fool' one or more of the sensors of navigation systems has become relative easy. In a previous publication[5] a 20 year timeline (1994–2014) of GPS jamming and spoofing was presented, and it was concluded that although twenty years ago GPS was deemed rather invulnerable to a range of potential jamming and spoofing threats, today this is no longer the case. Furthermore, it was pointed out how in recent years the concept of ADS-B spoofing has gained attention in the hacker community, as evidenced by presentations at DEFCON 17, 18 and 18. Both in 2015 and 2016, GPS spoofing has become a topic of interest among the hacker community and at DEFCON 23 and 24. This was to be expected (and was already predicted in 2003[6]) since with the availability of Software Defined Radios that include a transmit capability the challenge is basically reduced to generate fake messages that cannot be distinguished from real ones.

15.3 NLDA Research into Resilient Position, Navigation and Timing

To keep track of these developments, explore different approaches to jamming and spoofing and design and test mitigation concepts NLDA research address both alternative means for obtaining position and timing information.

15.3.1 GPS Augmentation for Artillery Operations, a Case Study

In 2010–2014 FMS research on Navigation vulnerabilities was dominated by investigating the use of terrestrial systems, like eLoran. This system uses fixed stations on the earth's surface transmitting high power low frequent signals, making it hard to be jammed. The user receives these signals and can determine its position by comparing travel times from station to receiver. Unfortunately the method is less

[5] See Theunissen 2016.

[6] During a lecture at ION-GNSS: Scott 2003.

Figure 15.1 PzH2000 mobile artillery of the Royal Netherlands Army

precise as GPS with best possible positioning errors of 10–20 m. The possible use of eLoran for artillery deployment, for instance the PzH200 (see Fig. 15.1), was investigated for the Dutch army in 2012. Here GPS is used for positioning of the gun. The gun location error (GLE) will always be a parameter in the final accuracy of delivery. This accuracy is called the mean point of impact (MPI) error. Positioning accuracy can also influence the Target Location Error (TLE). Together with the (mean) MPI error and round-to-round error, the aim point is shifted to the point of impact (see Fig. 15.2). When using GPS, the value of MPI is typically around 50 m for a firing range of 20 km, which is also a NATO requirement.

As a second navigation system artillery uses a strap-down inertial navigation system. As a characteristic of INS, the accuracy will now degrade with time and travelled distance. To make sure the GLE stays within limits, the artillery platform must return to a known position after it has driven 5 km. This can be a surveyed landmark or a road intersection for example. At this known position, INS can be calibrated and starts with a position error equal to the known position's accuracy. This limits the freedom of movement significantly. Moreover, in case there are no surveyed positions nearby, the GLE will increase even more. In this example study, the use of a stand-alone eLoran as a backup system for an artillery position has been investigated. Results have also been published in a scientific journal on Navigation Technology.[7]

[7] Griffioen and Ooinincx 2013.

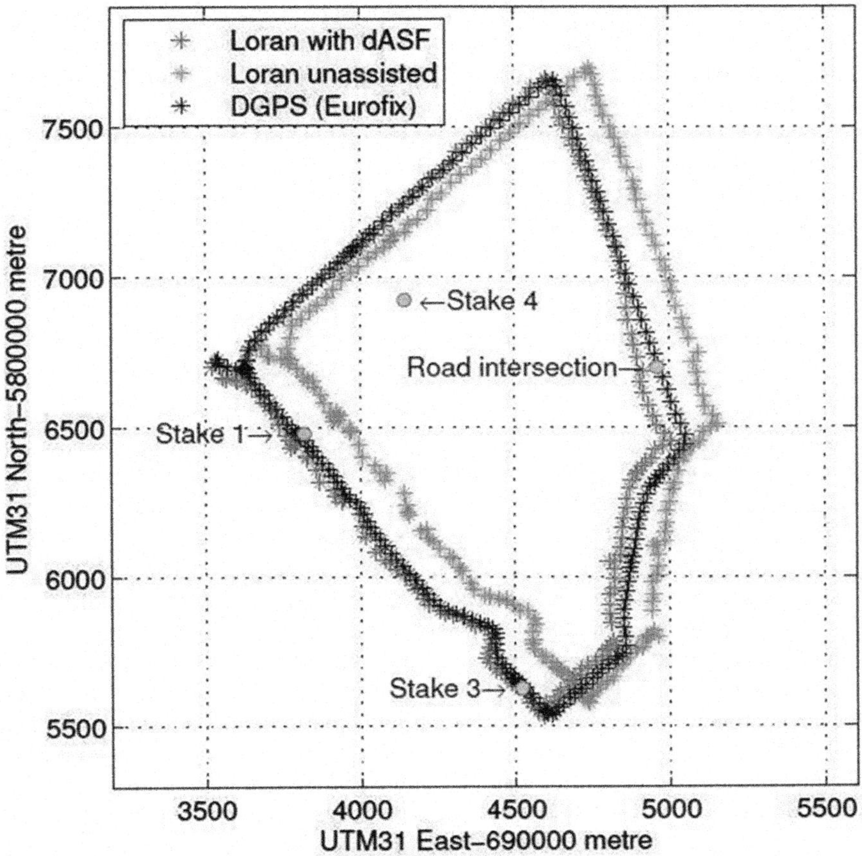

Figure 15.2 Loran positions of dynamic measurements with correction maps (dASF)

For accurate positioning using eLoran, the environment where the artillery will be deployed has to be modelled as well for correcting propagation errors. At the end this will lead to better positioning results. The environment we have chosen is the Oldebroekse Hei, a heath land area in the Mid-Eastern part of the Netherlands. It has been chosen because of its soil and scenery (sand and forest) and relatively long distance from sea to land transition and urban areas.

Here static measurements have been done using an E-field antenna setup to generate ASF maps that have been used for correcting dynamic measurements in a second experiment. To study temporal variations in eLoran positions the experiment was conducted during 24 h twice in a 17 day period. For both static and dynamic experiments an antenna pole was fixed to the back of a Mercedes-Benz 290 GD (MB) modified for Netherlands' army purposes.

Figure 15.2 shows the results of the dynamic survey. As can be denoted, using the environmental information as a correction on the eLoran signal yields positioning information that comes close to GPS positioning. As mentioned before, the value of the error in the mean point of impact with GPS positioning is typically around 50 m for a firing range of 20 km. Based on our survey results and the use of ballistic models we get an error for the MPI of 60.2 m for a firing range of 20 km. Furthermore, the mean MPI error is now reached 5.2 km earlier (50 m at 14.8 km firing range).

15.3.2 Spoofing Design and Testing

In 2015 NLDA started a project to implement jamming and spoofing devices based on Software Defined Radios (SDRs). A SDR can be programmed to transmit signals on the frequencies used by GNSS. These signals can be designed to jam GNSS, which is an easy task, or to spoof GNSS, which is more difficult.

Spoofers vary in complexity from relatively simple to extremely sophisticated. One of the easiest ways to implement a spoofer is to record the GNSS signals and resubmitting them as presented at DEFCON 23 in 2015.[8] Their setup consisted of two SDR's one for receiving the signals and another to transmit them. The total cost of such a setup would be approximately 2000 euros. This method is also known as meaconing (masking beacon). The effectiveness of meaconing, however, is limited because it is not possible to fully control the spoofed Position, Velocity and Timing (PVT) solution. It could be effectively used to confuse and increase the workload of operators or decrease the performance of integrated navigation systems. Or, in the case of ill-trained operators or badly designed navigation systems, the target may be forced to make mistakes jeopardizing their own safety. It is however very limited in a controlled attack.

A slightly more advance spoofer generates the GNSS signals instead of recording them. The generated GNSS signals are subsequently transmitted using a SDR. The demands on the hardware are lower than the record-replay spoofing, so the hardware setup could be cheaper. On the other hand, the generation of GNSS signals is rather complex. If implemented well, this type of spoofer is more capable in deceiving a target as the signals can be made coherent to every possible PVT solution. It is, however, difficult to predict when and if the target receiver is locked to the counterfeit signals. If the receiver is locked to the true signals, the tracking loops in the receiver are designed with a narrow bandwidth such that other signals are not observed, unless they have the same frequency, code and carrier phases. This prevents the receiver from locking to a counterfeit signal. To increase the chances of the target receiver locking on the counterfeit signals, the spoofer can be combined with a jammer. The jammer will cause a loss of lock, forcing the receiver

[8] DEFCON-23.

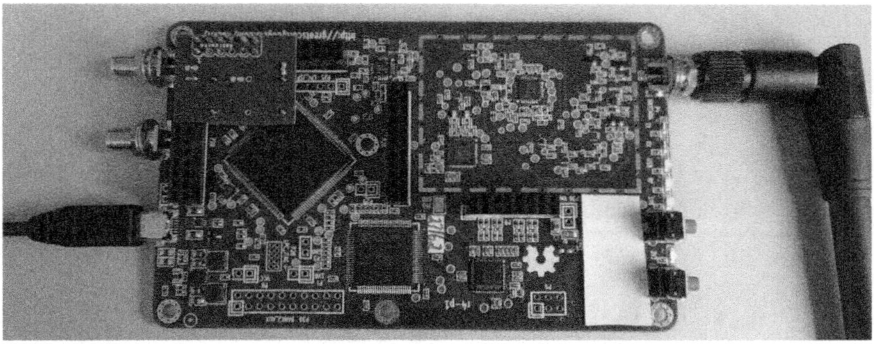

Figure 15.3 The software defined radio, used as spoofing device

to re-acquire the signals where a much larger bandwidth is covered. If the counterfeit signal is slightly stronger than the real signal, the receiver is likely to lock on the counterfeit signal. By the use of the jammer, however, the target might be alarmed. Also, because the spoofer does not have knowledge on the current true signals, a jump in the PVT solution will occur, which is also an indication of a spoofing attack. If the jump and the jamming, however, goes unnoticed, the spoofing attack is successful. Also, if the spoofing attack is detected, the attack might cause a higher workload and will deny the use of GNSS, which will reduce the navigation performance of most navigation systems.

Currently at NLDA this type of spoofing device is implemented based on a €350 widely available SDR (Fig. 15.3) using free to download software. This software generates the signals based on a user defined position, or track. The implementation will take, once the correct hardware is selected, less than an hour. It has been successfully tested on many civil GPS receivers; see Fig. 15.4 for an example of a civil GPS receiver that is commonly used by armed forces. After some changes in the source code of the software, also a military Defense Advanced GPS Receiver (DAGR) was successfully spoofed. Because the receiver was unauthorized to use the encrypted P(Y) code, it was using the unencrypted C/A code. Therefore, a successful spoofing attack was possible. A student project is currently investigating whether a crypto DAGR, using the encrypted P(Y) code, can be successfully spoofed by jamming the P(Y) signal and simultaneously spoofing the unencrypted C/A code.

The used software first generates the signals and saves them into a file. This file can then be transmitted using a SDR. On the internet, also a real-time version of the spoofing software is available, which is capable of simultaneously generating and transmitting the signals. This opens the possibility of a hijack of a receiver and potentially control the target, as presented at DEFCON-24.[9]

[9] DEFCON-24.

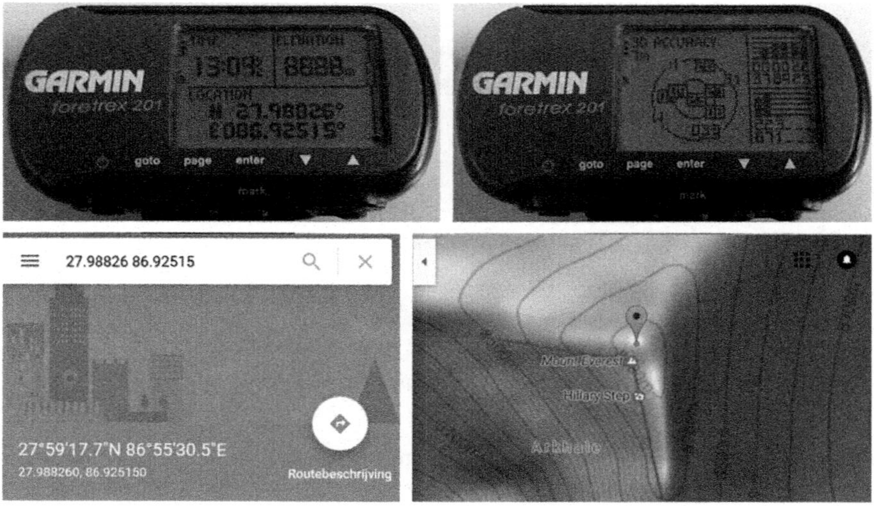

Figure 15.4 Although the authors have never been there, the GPS receiver has. Or … (map view from Google Maps)

Because of the low cost and readily available software to build these low-complexity spoofers, it presents a high risk, and it's only a matter of time before the first attacks occur, if they have not already. It could be effective against most civil receivers as it is not yet common practice to implement anti-spoofing methods into receivers. It can be used for targeted terroristic attacks but also just to cause annoyance amongst GNSS users.

A more sophisticated spoofer combines a (software) receiver and transmitter. The receiver is used to synchronize time and obtain information on the code and carrier phase of the actual signals, as well as the content of the navigation data message. Based on this information, and the current position and velocity of the target receiver, a prediction can be made on the code and carrier phase at the location of the target. The spoofer is now able to replicate accurately the true signal at the position of the target such that both the true and counterfeit signals perfectly align. If the signal strength of the counterfeit signal is then increased such that it is slightly stronger than the real signal, the receiver will follow the counterfeit signal. By slowly changing the code and carrier phase of the signal, the receiver can be slowly steered away from the true position. In 2012 Humphreys claimed[10] that this can be successfully executed over a distance of over 600 m. The take-over this way is more predictable and, because no jamming is used, it is less likely that the target is alarmed. Also, because the counterfeit signals are synchronized to the true signals, no sudden jump in position, velocity or time is induced, also reducing the probability of a detection.

[10] Humphreys 2012.

It is expected that this more sophisticated spoofer could cost around 1000 euros or less on the required hardware. The implementation however is very complicated and requires a deep knowledge on GNSS signals. Therefore, it is less likely for an attack to occur, however it is certainly not impossible and we should take into account that many countries and possibly terrorist organizations could develop these sophisticated spoofers. If implemented well, an attack by this type of spoofer is difficult to detect. Successful attacks might lead to serious accidents such as airplane crashes, grounded ships or major disruptions of infrastructures relying on precise timing. Also it will be possible to hijack (un)manned systems such as UAV's, unmanned ships or maybe even autonomous cars. One effective way of mitigation, amongst others, is to use multiple antennas, or multi-element antennas to determine the direction from which the signal arrives at the receiver. If this direction is not coherent with the expected satellite position, an alarm should be raised.

The most sophisticated spoofing system, apart from taking over the GNSS satellites themselves, consists of many transmitters, preferably attached to flying platforms. This allows the attacker to also control the direction from where the signals are transmitted, such that the angle of arrival is coherent with the satellite position. It is even possible to define a completely fake constellation base on, for example UAV's instead of satellites, transmitting their own constellation data message. This multi-transmitter spoofing, however, is very complex, but if implemented well, it will be very difficult to detect an attack. Because of the complexity of such a system, however, it is unlikely for such an attack to happen, at least for the time being.

15.4 Conclusion

Since the selective availability restriction on civilian Global Positioning System (GPS) signals has been deactivated in 2000, the use of GPS services for various applications has increased rapidly. This quick rise in GPS usage comes with ever greater dependence on the availability and integrity of the system. More often than not the presence of proper GPS signals is taken for granted. This comes with certain dangers, as civil GPS signals are unencrypted and information on signal details are openly available. Therefore, it is certainly possible to create faked signals to intentionally disrupt or misguide the signals localization and timing services. These actions are called jamming, in the case of preventing reception of the signal by the receiver, and spoofing, when signals are counterfeited to misguide the receiver. Although jamming will definitely be brought to attention, the main focus of this work will be on spoofing.

In 2001 the Volpe report brought to attention the vulnerabilities that come with relying on GPS and recommended further research. Two years later, Logan Scott posed the spoofing threat as underestimated, predicting the development and online sharing of spoofing software by "script kiddies". Since then, many studies have been conducted regarding the vulnerabilities of GPS signals, the development of spoofing devices and signals, and implementation of several countermeasures to prevent spoofing or harden receivers. Research at FMS on this topic focuses on understanding the vulnerabilities of GPS thoroughly and developing countermeasures for it.

References

Akos D (2012) Who's Afraid of the Spoofer? GPS/GNSS Spoofing Detection via Automatic Gain Control (AGC), NAVIGATION. Journal of The Institute of Navigation 59(4):281–290

Bhatti J, Shepard D, Humphreys T (2012) Drone Hack - Spoofing Attack Demonstration on a Civilian Unmanned Aerial Vehicle. GPS World (1 August 2012) http://gpsworld.com/drone-hack/ Accessed 1 March 2017

DEFCON-23 notices at https://media.defcon.org/DEF%20CON%2023/DEF%20CON%2023%20presentations/DEFCON-23-Lin-Huang-Qing-Yang-GPS-Spoofing.pdf

DEFCON-24 notices at https://media.defcon.org/DEF%20CON%2024/DEF%20CON%2024%20presentations/DEFCON-24-Aaron-Luo-Drones-Hijacking-Multi-Dimensional-Attack-Vectors-And-Countermeasures-UPDATED.pdf

Griffioen J, Oonincx P (2013) Suitability of Low-Frequency Navigation Systems for Artillery Positioning in a GNSS Denied Environment. Journal of Navigation 66(1):35–48

Humphreys T (2012) Statement on the vulnerability of civil unmanned aerial vehicles and other systems to civil GPS spoofing. Submitted to the Subcommittee on Oversight, Investigations, and Management of the House Committee on Homeland Security (18 July 2012) http://homeland.house.gov/files/Testimony-Humphreys.pdf Accessed 1 March 2017

Psiaki M, Powell SP, O'Hanlon BW (2013) GNSS Spoofing Detection. GPS World 24(6):53–58

Scott L (2003) Anti-Spoofing & Authenticated Signal Architectures for Civil Navigation Systems. Proceedings ION GGNSS (9–12 September), Portland OR

Theunissen E (2016) Cyber threats to position and timing data and their impact on safety and security. Netherlands Annual Review of Military Studies pp 309–324

US Department of Transportation (2001) Vulnerability Assessment of the Transportation Infrastructure Relying upon the Global Positioning System. http://ntl.bts.gov/lib/31000/31300/31379/17_2001_Volpe_GPS_Vulnerability_Study.pdf Accessed 1 March 2017

Author Biographies

B. Lubbers, M.Sc. is Assistant Professor Operational Processes for Navigation Technology at the Department of Military Technical Studies, Netherlands Defence Academy. His research focusses on navigation integrity and navigation warfare.

Professor Dr. Eric Theunissen is Professor of Avionics at the Department of Military Technical Studies, Netherlands Defence Academy.

Professor Dr. P.J. Oonincx is Professor for Navigation Technology, Department of Military Technical Studies and Acting Dean of the Faculty Military Sciences, Netherlands Defence Academy. He obtained his Ph.D. degree from the University of Amsterdam in 2000 on mathematical signal processing for physics.

Chapter 16
Winning Without Killing in the South China Sea

Niels Woudstra

Abstract This chapter concerns itself with the way in which China uses mainly non-kinetic means to broaden its power and influence in the South China Sea. The old writings of Sun Tzu and the modern writings of Liang and Xiangsui, together with the analysis of the tensions in the South China Sea, demonstrate that the strategies being used by China do not always concur with Western ways of thinking. China did not launch a large-scale offensive against its neighbours to control the South China Sea with its natural resources and sea lines of communication. Over the last decades it has been patiently broadening its influence in the area, often by using civilian or paramilitary means. It now occupies most of the Spratlys, Paracels and Scarborough Shoal. Because of its economic and diplomatic dominance China is persuasive behind the scenes in bilateral negotiations with neighbouring countries. The military, symmetric, deterrent response of the West, by sailing through these waters with warships, may be effective, but may not be sufficient to secure the status of international waters and freedom of navigation in the long term. In dealing with China in the disputes over the South China Sea, Westerners need to acquire knowledge about Chinese strategic thinking in order to come up with the most adequate response.

Keywords Maritime Strategy · Non-kinetic Effects · Sun Tzu · Unrestricted warfare

Contents

N. Woudstra (✉)
Faculty of Military Sciences, Netherlands Defence Academy, Breda, The Netherlands
e-mail: na.woudstra@mindef.nl

P.A.L. Ducheine and F.P.B. Osinga (eds.), *Netherlands Annual Review of Military Studies 2017*, NL ARMS, DOI 10.1007/978-94-6265-189-0_16

16.1 Ba and Wang

Samuel P. Huntington wrote his bestseller *The Clash of Civilizations and the remaking of World Order* twenty years ago. His work can be seen as a reaction to Francis Fukuyama's *The End of History and the Last Man*. Fukuyama argued that the end of the Cold War implied the ultimate victory of Western liberalism in the world. Huntington however saw a revival of civilizational divides. He predicted that the fault lines between civilizations would be the future conflict lines. International relationships would no longer be dominated by "isms" like communism, capitalism or liberalism, but by deeply rooted, sometimes more than thousand years old civilizations. A consequence of his views is the need of a paradigm shift. When Westerners consider China and the Chinese civilization, they inevitably have to refrain from looking exclusively from their own civilizational perspectives and adopt a different angle. This means that the Chinese way of achieving political, diplomatic and strategic goals may differ from the Western way of thinking. According to Paul An-hao Huang in *The Maritime Strategy of China in the Asia-Pacific Region* (2010), it becomes clear that the Western paradigms on international relations, like (neo)realism and similar schools, are insufficient for analysing Chinese strategy. Chinese thinking, influenced by Confucianism for instance, cannot be simply explained by Western power thinking. According to sinologist Henk Schulte Nordholt, Chinese President Xi has repeatedly stated that Confucianism represents the soul of the nation and the teachings of the old philosopher are enjoying a true revival. As to the nature of a state, Confucianism makes a crucial distinction between 'ba' and 'wang.' 'Ba'—hegemony—is practiced by states that subjugate others by a show of force. States that choose 'wang'—the princely way—instead, achieve their goals through moral persuasion and the superiority of their virtuous example.[1] Unsurprisingly, Chinese opinion leaders attribute 'ba' to the United States and 'wang' to China. A well-known exponent of this is "the peaceful rise of China." Even though this might sound like mere propaganda, the Chinese seem to be masters in achieving their aims without using large-scale violence. Does this mean that Chinese rulers always apply soft power to achieve their goals? No. The Chinese offensive in the Korean War (1950–1953), the Chinese seizure of the Paracel Islands from Vietnam (1974) and the state violence

[1] Schulte Nordholt 2016, p. 38.

on Tiananmen Square (1989) remind us of Chinese hard power and the will to use it. But the wisdom of the old strategist Sun Tzu prevails: achieving victory without use of force is the preferred way.

This brings us to the broader perspective of non-kinetic influencing in this book, to which this chapter aims to contribute. How do the Chinese pursue their interests? While power is shifting from the West to Asia and particularly China, which may cause conflict, it is crucial to have an understanding of the Chinese way of strategic thinking. Without this knowledge, it will be hard to respond adequately to it. This chapter will therefore begin with a short introduction of the strategist Sun Tzu, followed by Chinese views on modern warfighting, expressed in the book *Unrestricted Warfare*. Then this chapter analyses China's expansion in the South China Sea, with an emphasis on non-kinetic influencing. How do the Chinese increase their influence without the use of high end military means? We will have a look at Chinese grand strategy, followed by what is at stake in the South China Sea. The next sections deal with maritime strategy and unrestricted warfare in the South China Sea. This chapter has a final section on "all warfare is based on deception", before it wraps up with a conclusion.

16.2 Sun Tzu

Sun Tzu has been reflecting on Chinese strategic thinking since his thoughts were written in approximately 500 BC. Westerners became gradually acquainted with his work, *The Art of War*, much later, in the 20th century. Sun Tzu is well-known today: a search on the internet provides us with numerous applications of his wisdom, also for civilian use. One hundred years ago however, he was unknown. The first high-quality scientific translation was made by General Samuel B. Griffith in 1963. The quotes in this article are based on the last translation.[2] This chronology provides the explanation why great Western strategists, like von Clausewitz and Jomini, or maritime strategists like Mahan or Corbett, made no reference to Sun Tzu.[3]

Sun Tzu expressed his principles in thirteen concise chapters. No comprehensive treatises, but short eloquent statements with great profoundness. For instance: "Generally in war the best policy is take a state intact; to ruin it is inferior to this. (…) For to win one hundred victories in one hundred battles is not the acme of skill. To subdue the enemy without fighting is the acme of skill".[4] This is paradoxical. In military affairs, is it not all about fighting and winning? As a strategist, Sun Tzu however focussed on the higher purpose: *subduing the enemy*. It is preferable to do this without bloodshed, high costs, destruction and causing deep hatred. I assess

[2] Sun Tzu 1963.

[3] Woudstra 2013.

[4] Sun Tzu 1963, p. 77.

that such a strategy eventually worked in ending the Cold War. Deterrence proved to be effective. The Warsaw Pact had to give up and disintegrated without fighting. According to Sun Tzu this amounts to *the acme of skill*. In general though, the Western paradigm of warfare during the 20th century was marked by force-on-force attrition style warfare. Indeed, as Russell Weighley asserted, technological and logistical superiority and attrition strategy was the American Way of War since the Civil War. Sophisticated approaches such as advanced by Sun Tzu were generally not part of the military lexicon. War was conceptualized primarily in military terms, as a clash of armed forces and a contest in which economic and psychological instruments played decidedly subordinate roles.

What might Sun Tzu have proposed? One could think of buying and corrupting the enemy leadership, or eliminating them silently, or employing spies on a large scale to find out enemy weaknesses.[5] In addition, one could exploit the cognitive domain by presenting false or ambiguous information, confusing the leadership, or chasing commanders physically and psychologically round the clock.[6] According to Chinese tradition this should be done discretely yet effectively.

What we call irregular, asymmetric or hybrid warfare are not "new" phenomena. Seen through the Chinese lens of Sun Tzu, these types of warfare have been common practice since at least 500 BC. Indeed, creating and exploiting asymmetries is the essence of strategy. These ideas have been further developed in the book *Unrestricted Warfare*, which is the subject of the next section.

16.3 Unrestricted Warfare

A Chinese view on contemporary warfare and the use of force is reflected by the book *Unrestricted Warfare*, published in 1999 by two Chinese (senior) colonels, Qiao Liang and Wang Xiangsui. Written well before 9/11, as a reflection on Western military performance in Operation Desert Storm (1991), it provides an almost prophetic insight in the Chinese views on warfare in the 21st century. The writers coined unrestricted warfare by stating that whilst non-warlike actions may be the new factors constituting future warfare, a new name for this new form of war is needed: Warfare which transcends all boundaries and limits. This, in short, is meant with unrestricted warfare: all instruments of state power will be available, information will be omnipresent, and the battlefield will be everywhere. It means that all weapons and technology can be superimposed at will, all the boundaries separating the two worlds of war and non-war, of military and non-military, will be

[5] Sun Tzu 1963, p. 67: "Anger his general and confuse him. Pretend inferiority and encourage his arrogance".

[6] Sun Tzu 1963, p. 66: "All warfare is based on deception", and p. 68: "Keep him under strain and wear him down".

destroyed, many of the current principles of combat will be modified, and even the rules of war may need to be rewritten.[7] Indeed, two years later, on September, 11th, 2001, these words almost surrealistically came true: the non-state actor Al Qaeda launched a coordinated attack of a military scale, using ordinary civilian means, with thousands of casualties and a high psychological impact, whilst striking at the economic and military symbols of the United States: the Twin Towers and the Pentagon. The attack on the political symbol—the White House—only failed thanks to the courageous intervention of the airplane's passengers. As Liang and Xiangsui predicted, there is nothing in the world today that cannot become a weapon. Hence, our understanding of weapons should be redefined as the choice of weapons is unrestricted and only limited by human fantasy and innovation. With technological enablement the increase in the types of weapons, cognitive and creative processes will have the same effect. As we see it, a single man-made stock-market crash, a single computer virus, a single rumour or scandal affecting the enemy's currency exchange rates or exposing the leadership of an adversary on the Internet, all can be included in the ranks of new-concept weapons.[8]

According to the two Chinese colonels, the necessary new method is to create a complete military Machiavelli. Achieve objectives by fair means or foul, that is the most important spiritual legacy of this Italian political thinker of the Renaissance. In the Middle Ages, this represented a breakthrough against romantic chivalry and the declining tradition of knighthood. It meant using means, some possibly comprehensively, without restraint to achieve an objective; this holds for warfare also. Even though Machiavelli was not the earliest source of "an ideology of going beyond limits" as China's Han Feizi preceded him, he was its clearest exponent.[9] One may conclude that the influence of the principles of medieval chivalry is diminishing in modern warfare. These principles were reflected by the Geneva Conventions, aimed to set international rules to limit the dehumanization of warfare. Unrestricted warfare does not necessarily acknowledge these Conventions per se: no uniforms, no legal protection for combatants, no rules.

Does this mean that unrestricted warfare will be more violent, as it tends not to abide by any rules? Probably not. Sun Tzu argued that subduing the enemy without fighting is the acme of skill. Liang and Xiansui confirm that technological progress has given us the means to strike at the enemy's nerve centre directly without harming other things, offering numerous new options to achieve victory, and all these make people believe that the best way to achieve victory is to control, not to kill.[10]

One might argue that unrestricted warfare and hybrid warfare are synonyms. For the purpose of this chapter, unrestricted warfare is the preferred term, as it was coined by the Chinese.

[7] Liang and Xiangsui 1999, p. 12.

[8] Liang and Xiangsui 1999, p. 25.

[9] Liang and Xiangsui 1999, p. 180.

[10] Liang and Xiangsui 1999, p. 27.

16.4 Chinese Grand Strategy

Between 1405 and 1433, the Ming Dynasty (1368–1644) undertook seven trans-oceanic voyages. Emperor Yong-Le appointed Admiral Zheng He as the commander of the most powerful navy in the world, comprising 255 ships and 27,800 men. The strategic purpose of the Ming Dynasty's battle fleet was to re-establish an imperial global order with the Middle Kingdom in prime position. It pursued China's political and economic interests by controlling the strategic lines of communication in the South China Sea.[11] Since the voyages of Admiral Zheng He in the fifteenth century, China has been a mere continental power with a continental, inward looking strategy. The Chinese Wall is a symbol of its preoccupation with outside threats on land. The loss of face for China during the Opium Wars, also called the Century of Humiliation, has shaped the present Chinese defence posture in an important way. In the 20th century, the navy was a small coastal service and stayed in the shadow of the land forces. Therefore its name makes sense: People's Liberation *Army* Navy (PLAN), the navy as part of the army. The coastal role for the PLAN has gradually changed since the need to counter the increasing piracy threat in the Gulf of Aden. For the first time in centuries, the Chinese navy went out-of-area to fight pirates in the vicinity of Somalia. China also started a comprehensive building program for its navy. In August 2011, the first sea trials took place with its then recently procured aircraft carrier, the ex-Soviet *Varyag*. This raises questions about China's maritime strategy. Though it will last several years before this ship will be fully operational capable, China made a step on the way to maritime power projection.[12]

In 1890, A.T. Mahan coined Seapower in his famous book *The Influence of Seapower upon history (1660–1783)*; he argued that naval power and economic expansion reinforce each other mutually. According to Frans-Paul van der Putten and Minke Meijnders, the Chinese government is currently developing an ambitious programme of maritime infrastructure construction along the main Asia-Europe shipping route. China's initiative for a so-called '21st Century Maritime Silk Road' is aimed at port development in South-East Asia, around the Indian Ocean and in the eastern Mediterranean region. The Chinese leadership publicly presented its initiative for the Maritime Silk Road in October 2013. Earlier that year, China had already launched its Silk Road Economic Belt initiative, which is aimed at infrastructure cooperation in a zone that stretches from Xinjiang (the north-western part of China) to the Baltic Sea. The Chinese government uses the term 'One Belt, One Road' to refer to the combination of these two initiatives. Chinese investment in large infrastructure projects constitutes the basis of One Belt, One Road. These projects are financed, constructed, supplied and sometimes operated by Chinese firms that are either state-owned or that otherwise have close

[11] Jenner and Thuy 2016, p. 297.

[12] Woudstra 2011, p. 24.

relations with the Chinese government.[13] Earlier, the maritime expansion was expressed by the metaphor of the String of Pearls:

> The "String of Pearls" describes the manifestation of China's rising geopolitical influence through efforts to increase access to ports and airfields, develop special diplomatic relationships, and modernize military forces that extend from the South China Sea through the Strait of Malacca, across the Indian Ocean, and on to the Arabian Gulf.[14]

This maritime strategy is not limited to Asia. Chinese expansion also reaches Europe and Africa. The Greek harbour Piraeus, the Pakistan harbour of Gwadar and the harbour or Djibouti are now (almost) in Chinese hands. At the beginning of this decade, the Port of Rotterdam had to intervene when the Hong Kong owned Hutchison Whampoa was about to achieve a dominant position in the harbour. The conglomerate has owned Europe Container Terminals (ECT), which treats more than 60% of the containers in that port, since 2002. Together with their share in the Euromax terminal, their influence increased further. Therefore the Rotterdam port authority decided to invite other players as a counterweight to the Chinese position.[15] Taking into account that 90% of world trade goes by the seas, the Chinese maritime expansion seems to be inevitable to satisfy the needs of more than 1.3 billion inhabitants.[16]

So what is the purpose of this strategy? According to Ryan Martinson, who assessed Chinese maritime strategy from primary sources, China strongly pursues economic wealth through maritime expansion. At the 30 July 2013 study session, the Politburo discussed the topic of building China into a "maritime power." A close reading of this document reveals that the party-state's definition of "maritime power" is very broad, inter alia: maritime powers use the ocean to build wealth, their marine science and technology are advanced and they are able to exploit marine sources effectively. However, maritime powers are also able to safeguard their maritime rights and interests, and a careful reading of Xi Jinping's remarks also suggests endorsement of an assertive turn—in his words, a "transformation"—in how China handles its maritime disputes. In Xi's words: "We love peace and will continue along the path of peaceful development, but we absolutely cannot abandon our legitimate rights and interests, much less sacrifice our core interests." "Ocean defence" involves much more than preventing foreign invasion along China's coast; it refers to actions to defend all "Chinese" space from encroachment, including offshore islands and remote waters under Chinese jurisdiction.[17] In sum, in order to serve its economic interests, China's maritime expansion and power have increased in the 21st century.

[13] Van der Putten and Meijnders 2015, p. 6.

[14] Pehrson 2006, p. v.

[15] NRC Handelsblad 2010.

[16] US Department of Defense 2015, p. v.

[17] Martinson 2016, pp. 26–27.

16.5 What Is at Stake in the South China Sea?

In the previous section we have seen the purposes of the broader Chinese maritime strategy. We now zoom into Chinese interest in the South China Sea in order to understand Chinese behaviour in that sea. The South China Sea is semi-enclosed. It contains a wealth of maritime resources and some of the world's most important strategic Sea Lines of Communication (SLOCs). It is one of the most diversified tropical seas. The rich fishing stock and the high potential of hydrocarbon reserves make it an attractive pool of resources and an important means to feed the surrounding populations. The seas around the Spratly and Paracel Islands contain important fossil fuel and gas reserves. China claims sovereignty over 80% of the South China Sea based on the "nine-dash-line" which figures on a map of 1947.[18] Gradually, by occupying the Paracel Islands, Spratly Islands, Scarborough Shoal and other islands, China treats the South China Sea as inland waters. Customary law and the United Nations Convention on the Law of the Seas (UNCLOS) however, consider these waters to be international waters, just like the Mediterranean or the Black Sea. This is based on the *freedom of the seas*, as it was first expressed by Grotius, in his treaty *Mare Liberum* in 1609. The freedom of the seas can only partially be restricted by legal claims in accordance with UNCLOS, which was also ratified by China. Overlapping claims of Brunei, China, Malaysia, Philippines, Taiwan and Vietnam complicate jurisdiction issues. In addition to these claims, more than 41.000 ships, accounting for more than half of the world shipping volume 5000 billions of dollars' worth of trade, sail through the sea annually.[19]

16.6 Maritime Strategy in the South China Sea

The views on sphere of influence, sovereignty and jurisdiction differ largely between China and the United States and its allies. The US navy maintains a continuous presence in the South China Sea, even with carrier battle groups operating from Guam, Japan and South Korea. The US Freedom of Navigation program is to maintain the legacy of Grotius. Though ironically, the US didn't ratify UNCLOS, due to procedural conservative opposition in Congress, it is one of the strongest proponents of freedom of navigation. According to UNCLOS, freedom of navigation exist within the Exclusive Economic Zone (EEZ) of a coastal state, which means that even a foreign (war)ship can approach the coastal state up to its territorial waters, 12 nautical miles from its shores.[20] This led to incidents, like the incident with USNS Impeccable, an unarmed Navy ocean surveillance ship in 2009. USNS Impeccable was harassed by five Chinese fishing vessels in Chinese EEZ in

[18] Qian 2016.

[19] Jenner and Thuy 2016, p. 173.

[20] UNCLOS III, Article 58, *Rights and duties of other States in the exclusive economic zone*.

the South China Sea. They attempted to snag the Impeccable's acoustic array sonars from beneath the ship, and manoeuvred in front of the ship, coming to a complete stop and almost causing a collision. The Chinese vessels at times came within 25 feet of the ship.[21] The mere fact that USNS Impeccable was in the Chinese EEZ did not justify the attempts to chase it. From the Chinese point of view however, it was an infringement in their sphere of influence without peaceful purposes.

The Chinese are fully aware that at present they cannot win a direct confrontation with the US Navy. Therefore they adopt an indirect approach. Sun Tzu already advocated: "He who knows the art of the direct and indirect approach will be victorious. Such is the art of manoeuvring".[22] The PLAN probably won't equal the US Navy in the next decade or even decades, therefore it will not be possible to gain full sea control of the South China Sea. The strategy aims at sea denial of the US Navy in the waters close to the Chinese mainland. This sea denial strategy opposes the US Freedom of Navigation programme in and South China Sea. The US response to this is a comprehensive Air-Sea Battle concept. As the weaker party, China chooses asymmetric means as a deterrent: submarines and carrier-killer ballistic missiles. But the actual warfare maybe different, according to Sun Tzu: "In war, numbers alone confer no advantage. Do not advance relying on sheer military power".[23] So for instance the actual war may be fought electronically. A vast network of electronic warfare devices is being installed on the newly occupied (artificial) island in the South China Sea.[24]

16.7 Unrestricted Warfare in the South China Sea

How does China apply unrestricted warfare in the South China Sea? It is clear that it has been gradually extending its influence in this area. Though Sun Tzu favours achieving goals without the use of force, Chinese naval vessels fought a short war with Vietnam in 1974 over the Paracel Islands and China has been occupying them ever since. But how does China "subdue the enemy without fighting?" In 2014, a standoff around China National Offshore Oil Corporation's (CNOOC) Haiyang Shiyou 981 drilling rig took place, the most serious clash between China and Vietnam in the disputed South China Sea since the battle in the Paracels in 1974 and a skirmish near Johnson South Reef in the Spratlys in 1988. But this time fishing vessels from both countries were playing the key roles on the frontline instead of warships. From May 2014, nearly 200 ships from China and Vietnam, including naval, coastguard and civilian vessels, gathered in the area around the CNOOC platform. The Vietnamese vessels were attempting to disrupt drilling work

[21] Footage provided by Navy Visual News Service 2009.

[22] Sun Tzu 1963, p. 106.

[23] Sun Tzu 1963, p. 122.

[24] Thomas-Noone 2016.

in the disputed area, while the Chinese ones formed a protective cordon around the rig until it withdrew in July.[25]

At present, China's territorial extension is mainly based on civilian or para-military activities. Like Fisher Militia. These militia extend China's presence, without the need to deploy the PLAN or the coastguard, which however, are often in the background to back up the militia's activities. These activities amongst others consist of passing information on foreign naval ships to the PLAN or Chinese coastguard. Their fishing vessels are also chasing fishing vessels from other countries and they are a means of transport of building materiel for construction works on the disputed Islands. Low tide elevations are thus being transformed to true artificial Islands, with runways, radars, ports and military installations. According to Reuters, the fishing fleet based in the tiny port town of Tanmen on Hainan Island is getting everything, from military training and subsidies to even fuel and ice, as China creates an increasingly sophisticated fishing militia to sail into the disputed South China Sea. The training and support includes exercises at sea and requests to fishermen to gather information on foreign vessels, provincial government officials, regional diplomats and fishing company executives said in recent interviews. "The maritime militia is expanding because of the country's need for it, and because of the desire of the fishermen to engage in national service, protecting our country's interests," said an advisor to the Hainan government who did not want to be named. State-controlled fishing companies dominate the fleets that go regularly to the Spratlys and are recipients of much of the militia training and subsidies, according to industry sources.[26]

To highlight the importance of South China Sea, President Xi Jinping visited Tanmen in Hainan within a month after becoming president. In the centre of the small town on the east coast of Hainan province, a large billboard celebrates the April 2013 visit of President Xi Jinping, who was photographed with local fishermen a month after assuming the presidency. Xi's first stop in town was the Tanmen Marine Militia Company. Since its establishment in 1985, the militia had recruited local fishermen, provided military training in their spare time, and organised voyages to "actively collect maritime information, assist island and reef construction, and contribute to sovereignty defence in the South China Sea" when summoned, the official *Hainan Daily* reported. The president shook hands with militiamen, praised them for their combat capability, discipline and patriotism, encouraged them to master modern equipment, and promised to "improve their work and living conditions", the newspaper said. Xi had set the goal of building China into a "maritime power" after he became Communist Party general secretary in late 2012.[27]

[25] Zhen 2016.
[26] Rajagopalan 2016.
[27] Zhen 2016.

16.8 All Warfare Is Based on Deception[28]

According to international law, outside the territorial waters, but inside the Exclusive Economic Zone, "no State may validly purport to subject any part of the high seas to its sovereignty".[29] China seems to create a self-fulfilling prophecy by repeating that it has "historic sovereign rights" over 80% of the South China Sea, based on the nine-dash-line which figured on a chart from 1947. According to the 12th July 2016 Award of the Permanent Court of Arbitration, China's position turns out to be deceptive, because the Unites Nations Convention on the Law of the Seas (UNCLOS) leaves no room for it. Coastal states are entitled to territorial waters, a contiguous zone, an Exclusive Economic Zone of 200 nautical miles and the subsoil to 350 Nautical miles if the continental shelf is extending that far. All these zones depart from the base line, which is the astronomical low tide line of the mainland. Under strict circumstances, islands may generate jurisdiction too. And if there are overlapping claims, these have to be negotiated on the basis of equity. China however does not abide by these rules, but claims the South China Sea as inland waters, just like e.g. the Suez Canal is inland waters of Egypt. UNCLOS, of course, is not applicable on inland waters.

In 2012 China seized Scarborough Shoal, which in fact is lying on the continental shelf of the Philippines. Philippine fishermen were expelled from the area by Chinese fishermen. Under the rule of its former president Aquino, in 2013, the Philippines turned to The Hague's Permanent Court of Arbitration over the South China Sea jurisdiction disputes with China. China did not recognize the Court and boycotted its hearings. China's main argument was that it considers that it has sovereignty in the South China Sea and therefore the disputes cannot be subject to ruling by the Court. On 12th July of 2016 the Court ruled that China cannot make historic sovereignty claims and virtually all Philippine griefs against China were ruled in favour of the Philippines.

China was upset by the outcome of the tribunal, but it has managed to weaken the effect. It seems that another strategy of the old master Sun Tzu is being applied: "Thus, what is of supreme importance in war is to attack the enemy's strategy (...). Next best is to disrupt his alliances."[30] Based on this principle China already tried to play apart the members of the Association of Southeast Asian Nations (ASEAN).[31] In its disputes over the South China Sea, as the dominant player, it largely prefers to deal with other states bilaterally. Now that the Arbitration Court award has been given, the Philippines are being incited to make "a pivot to China." This seems to work, as the present president of the Philippines, Duterte, tries to detach the Philippines from its ally the United States and seeks for closer ties with China. Interestingly, the recent visit of the Philippines president to Beijing led to the

[28] Sun Tzu 1963, p. 66.

[29] UNCLOS III, Article 89—Invalidity of claims of sovereignty over the high seas.

[30] Sun Tzu 1963, pp. 77–78.

[31] Nguyen 2016.

promise of large investments in infrastructure in the Philippines and the opening of the Scarborough Shoal to fishing vessels from the Philippines.[32]

16.9 Conclusion

The old writings of Sun Tzu and the modern writings of Liang and Xiangsui, together with the analysis of the tensions in the South China Sea demonstrate that the strategies being used by China do not always concur with Western ways of thinking. China didn't launch a large scale offensive against its neighbours to control the South China with its natural resources and sea lines of communication. Over the last decades it has been patiently broadening its influence in the area, often by using civilian or paramilitary means. It now occupies most of the Spratlys, Paracels and Scarborough Shoal. Because of its economic and diplomatic dominance China is persuasive behind the scenes in bilateral negotiations with neighbouring countries. The military, symmetric, deterrent response of the West, by sailing through these waters with warships may be effective, but may not be sufficient to secure the status of international waters and freedom of navigation in the long term. In dealing with China in the disputes over the South China Sea, Westerners need to acquire knowledge about Chinese strategic thinking, in order to have the most adequate response. Or, to end with a quote of their most famous strategist Sun Tzu: "If ignorant both of your enemy and of yourself, you are certain in every battle to be in peril".[33]

References

Jenner CJ, Thuy TT (eds) (2016) The South China Sea. Cambridge University Press, Cambridge

Liang Q, Xiangsui W (1999) Unrestricted Warfare. PLA Literature and Arts Publishing House. http://www.cryptome.org/cuw.htm Accessed 1 March 2017

Martinson RD (2016) Panning for gold. Naval War College Review 69(3):23–44

Navy Visual News Service (2009) https://www.youtube.com/watch?v=hQvQjwAE4w4. Accessed 1 March 2017

Nguyen P (2016) ASEAN Learning to Navigate a New Age in Great Power Politics. Southeast Asia from Scott Circle 7(13):1–3

NRC Handelsblad (2010) Overval in de Rotterdamse haven (31 December 2010) http://www.nrc.nl/nieuws/2010/12/31/overal-in-de-rotterdamse-haven-duiken-chinezen-op-11984262-a1038271 Accessed 1 March 2017

NRC Handelsblad (2016) Duterte kondigt "scheiding" Filippijnen van VS aan (21 October 2016) http://www.nrc.nl/nieuws/2016/10/21/duterte-kondigt-scheiding-filippijnen-van-vs-aan-4919776-a1527665 Accessed 1 March 2017

[32] NRC Handelsblad 2016.

[33] Sun Tzu 1963, p. 84.

Pehrson CJ (2006) String of Pearls: Meeting the challenge of China's rising power across the Asian littoral. US Army Strategic Studies Institute, Carlisle. http://www.strategicstudiesinstitute.army.mil/pdffiles/PUB721.pdf Accessed 1 March 2017

Qian W (2016) Defense Ministry's regular press conference (25 Feb 2016) http://www.eng.mod.gov.cn/Press/2016-02/25/content_4644801.htm Accessed 1 March 2017

Rajagopalan M (2016) China trains 'fishing militia' to sail into disputed waters. Reuters (30 April 2016) http://www.reuters.com/article/us-southchinasea-china-fishingboats-idUSKCN0XS0RS Accessed 1 March 2017

Schulte Nordholt H (2016) What are Beijing's motivations? Atlantisch Perspectief 40(3):34–38

Sun Tzu (1963) The Art of War (translated by Griffith SB). Oxford University Press, Oxford

Thomas-Noone B (2016) Electronic warfare in the South China Sea. ASPI The Strategist (24 August 2016) http://www.aspistrategist.org.au/electronic-warfare-south-china-sea/ Accessed 1 March 2017

UNCLOS III (1982) United Nations Convention on the Law of the Sea

US Department of Defense (2015) A Cooperative Strategy for 21st Century Seapower (March 2015) http://www.navy.mil/local/maritime/150227-CS21R-Final.pdf. Accessed 1 March 2017

Van der Putten F-P, Meijnders M (2015) China, Europe and the Maritime Silk Road. Netherlands Institute of International Relations Clingendael, The Hague

Woudstra NA (2011) China en maritieme strategie. Marineblad 121(7):24–27

Woudstra NA (2013) Honderd jaar maritieme strategie en zeetactiek. Militaire Spectator 182 (11):480–486

Zhen L (2016) Fisher militia? Maritime units put humble fishermen on front line in South China Sea. South China Morning Post (24 May 2016) http://www.scmp.com/news/china/diplomacy-defence/article/1951614/fisher-militia-maritime-units-put-humble-fishermen. Accessed 1 March 2017

Author Biography

Captain (RNLN) N.A. Woudstra, LL.M. is Associate Professor of Maritime Strategy and Operations at the Department of War Studies, Netherlands Defence Academy. He holds a Master degree of Law and International Relations from the Dutch Open University.

Chapter 17
Maritime Economic Enforcement Measures: 'Winning Without Killing' in Modern Conflicts at Sea

Martin Fink

> *The economic purpose of the war, ultimately fundamental to all war-makers, was more obvious to the maritime powers and more openly admitted by them...[...]...Their fleets might just occasionally fight others ...[but]... The real war at sea was commercial: choking the enemy's trade channels until he surrendered and then appropriating his trade to oneself.*
>
> Best 1996, pp. 35–36.

Abstract This chapter deals with the non-kinetic side of naval operations. In particular, it deals with the use of naval forces in applying economic maritime enforcement measures. There has always been a solid connection between the use of the navy and the economy. Traditionally, the relationship between naval forces and the economy has come to be expressed in two principles where during conflict the navy aims to destroy the opponent's economy, and during both peacetime and conflict the navy aims to protect the State's own economic interests. At an operational level, three models of enforcing economic maritime enforcement measures can be identified that translate these principles to today's naval operations: traditional economic warfare operations, maritime embargo operations and maritime security operations. These measures are also often considered to be at the low end of the use of force scale.

Keywords Naval Forces · Maritime Embargo Operations · Maritime Interception Operations · Economic Sanctions · Maritime Enforcement · Maritime Security Operations

M. Fink (✉)
Administrative Law Branch, Ministry of Defence, Amsterdam Centre of International Law (ACIL), Amsterdam, The Netherlands
e-mail: md.fink.01@mindef.nl

P.A.L. Ducheine and F.P.B. Osinga (eds.), *Netherlands Annual Review of Military Studies 2017*, NL ARMS, DOI 10.1007/978-94-6265-189-0_17

Contents

17.1 Introduction

Next to seeking, challenging and destroying the opponent's naval forces through kinetic means, a traditional task for navies has always been to ground the opponent's economy. This task primarily aims to force the opponent to cease its fighting by targeting the opponent's economy through maritime economic enforcement measures, such as blockades and contraband operations. These activities may serve as an example of how the military, and naval power in particular, can play a role in supporting the different strands of a political strategy. When expressed within the model, the four instruments of national power, Diplomacy, Information, Military, Economy (DIME), through maritime economic enforcement measures naval forces not only contribute to the military power instrument of the State's political strategy, but also support the economic power instrument. Furthermore, maritime economic enforcement measures are often considered as non-kinetic in nature, or at least situated at the low end of the use of force spectrum. This chapter first briefly emphasises the role of naval forces within the economic dimension of conflict and will then discuss three operational level models of maritime economic enforcement measures that can be distinguished, through which naval forces in the modern era of today's conflicts can support political ends.

17.2 Naval Forces and the Economy

Naval forces and the economy have always been seen as a logical combination in the thought process of applying State power. Obviously, the historical roots of commerce at sea, the rise of empires and the use of naval forces in the protection of seaborne trade and denying the trade of others, is at the basis of this long-lasting marriage.[1] With regard to this relationship between the navy and economic interests, a widely accepted concept exists in which the core tasks of the navy are shaped upon two general principles: during armed conflict, naval forces must ensure that the opponent cannot use the sea for its purposes. During peacetime, it should protect the State's seaborne trade.[2]

With regard to the first principle, a State taking maritime economic enforcement measures during armed conflict is more traditionally regarded as practising economic warfare. Essentially, economic warfare does not aim to win decisive military battles, but aims to stop the opponent's ability to sustain its warfighting capability by cutting the economic sea lines of communication, pressurizing its internal political legitimacy and to minimize manoeuvrability of the opponent by denying the use of the world's oceans. One important reason for achieving either command, control or denial of the sea is chiefly to destroy or prevent the opponent's seaborne trade.[3] The establishment of naval blockades and the conduct of contraband operations[4] are the manner in which the strategic aim of 'targeting' the opponent's economy is operationalized.

The second principle aims at the protection of the State's own economic interests. The famous 19th century naval strategist Rear Admiral Alfred T. Mahan opined that apart from keeping a navy purely as a branch of the military establishment, the necessity of a navy is based on the existence of commercial shipping that must be protected in times of peace and war.[5] Today, in the 21st century, it is not the State's overseas colonial adventures on which the State's wealth is based and which makes the sea an important area, but more likely the dependency on a global economy. This dependency makes the focus on the protection of economic interests beyond the State's border, however, no less important than in the days of Mahan. The sea lines of communication are, therefore, still essential to the State's economic interests.

The task of protecting the State's economic interests is not limited to situations of armed conflict, but applies at all times. It aims to protect all economic security interests that a State might have. Although the principle of protecting seaborne trade has always existed and has been widely accepted, for instance in the task of protecting merchants against piracy, during the last decade the number of issues that have fallen under the umbrella of this principle has grown extensively. Next to the classic examples of the protection of the sea lines of communication against piracy and the issue of illegal fishing, also threats to the economy such as terrorist acts, migrants, human and drug trafficking, protecting the routes of oil and other essential

[1] Till 2004, pp. 96–97.

[2] See also Gretton 1965, pp. 182–183.

[3] See e.g. Corbett 1911, pp. 99–100.

[4] Older manifests on maritime strategy often mention the term "commerce raiding" in this context. This term, however, has a negative connotation. I therefore prefer the more factual term of contraband operations.

[5] Mahan 1890, p. 26.

commodities and ensuring the non-proliferation of weapons of mass destruction are all part of this. Furthermore, the understanding that protecting the economy encompasses a wide range of issues (or threats against maritime security, as they are often called) coincided with the growing acceptance of governments and international security organisations that the enhancement of the security of economic interests should also be supported through military means. The main focus today is on protecting economic interests, rather than conducting economic warfare during armed conflict. The reasons for this focus are numerous. Mentioned here are the evolution of conflict in which the difference between war and peace has melted away, a greater emphasis on the protection of the interests of a State rather than the submission of another State, the fact that the character of conflict has moved to non-international armed conflicts, a changing political strategy on the global commons that has moved from freedom to awareness and control of the global commons, and lastly, the acceptance of the use of military (naval) power regarding a broad range of issues that go beyond the purely military use and beyond armed conflict. The pendulum, therefore, that swings between the two above-mentioned principles has shifted from the idea that naval forces are a branch of the military establishment, to the idea that the military can support a State's interests in the widest sense. All this means that naval forces have become an important actor outside armed conflicts and in situations that are considered to be less kinetic in character than regular warfare.

17.3 Three Models of Economic Enforcement Measures at Sea

Based on the above mentioned background and recent practice of economic enforcement at sea, the operational level conduct of maritime economic enforcement measures can be divided into three models. First is the classic economic warfare-model that applies during international armed conflict. Second is the maritime embargo operations-model that aims to support economic sanctions taken by the UN-Security Council. And third is the maritime security operations-model, in which naval operations aim to enhance maritime security and good order at sea to protect a State's security interests. Although two of these models have come into existence in parallel with the evolution of conflict during the last 25 years, none of the models have become obsolete by the appearance of the other. They could even be used as a tool within the same conflict. The three models will be discussed in the following paragraphs in more detail. In general terms, however, what these models have in common is firstly that they are generally considered to be at the low end of the use of force-spectrum. And secondly, that maritime enforcement measures have legally challenging frameworks, which therefore (and thirdly) will depend heavily on a whole-of-government-approach. These points of general similarity will also be briefly addressed below.

17.3.1 Economic Warfare Measures During International Armed Conflict

Traditional economic warfare is operationalized by the conduct of contraband and blockade operations. Such operations are both a means to choke the opponent's economy and create manoeuvrability for one's own forces.[6] Belligerent warships can stop and search suspected neutral and enemy merchant supporters of the opponent and can confiscate contraband, enemy merchant vessels or any vessel that breaches a blockade. Economic warfare is applied during international armed conflict and is played by the rules of the laws of naval warfare. These laws are a specialized subset of the laws of armed conflict, specifically intended for naval warfare. As mentioned earlier, economic warfare does not aim to win decisive military battles, but aims to stop the opponent's ability to sustain its warfighting capability, to pressurize its internal political legitimacy and to minimize manoeuvrability of the opponent by cutting the use the world's oceans. The interesting part of these enforcement measures is that although the strategic aim focuses on the opponent, the actual activity at sea is generally focused on neutrals, or non-belligerents. Neutrals are searched for possible contraband and are forbidden to enter or leave blockaded enemy ports. Neutral and enemy merchant vessels and goods are taken when the laws of naval warfare provide this authority to combatants and governments have decided to use such authority. Essentially, therefore, this means that economic enforcement measures are taken against civilians, which are often also civilians from a third State. Classic examples of a political grand strategy of which economic warfare was part of are the British blockades of Germany during the First and Second World War. On a lesser scale, current-day examples are the Israeli naval operations against the Gaza-strip since 2009,[7] and against Hezbollah in Lebanon, during the conflict in 2006.[8] Although it must be said that from a legal perspective, it is arguable whether or not the laws of naval warfare applied in these recent examples,[9] it is clear that States at least de facto applied economic warfare measures during these conflicts.

---

[6] Doctrinally, blockades can be divided into fleet (or naval) blockades and economic blockades. The first focuses at the denial of manoeuvrability of the opponent. The second supports the economic warfare aims. Obviously, the two types of blockades may overlap in one blockade, for instance when the enemy fleet is held up in a port which is also an important economic port. See Corbett 1911, pp. 185–210.

[7] Operation Cast Lead.

[8] Operation Change of Direction. See on this operation Ducheine and Pouw 2009, pp. 51–96.

[9] The legal question underlying these conflicts is whether they can be characterized as either international or non-international armed conflicts. In the latter case, the laws of naval warfare do not apply.

17.3.2 *Maritime Embargo Operations in Support of the UN-Security Council*

A second model of economic enforcement measures at sea emerged since the end of the Cold War. The increase of UN-peacekeeping operations also saw an increase of the use of naval assets to support economic sanctions taken by the UN-Security Council. Naval forces were now also contributing to the decisions of the Council at sea. An arms or oil embargo, or the stopping of mercenaries can for instance be authorized by the Council in order to take away factors that may fuel the conflict or may able the target State to sustain its unwanted actions. In the practical sense, maritime economic enforcement measures aim to stop and confine the conflict by means of inspecting vessels in a certain maritime area[10] and closing a State's ports for ingress of material or persons that may result in continued resistance of a State against the will of the UN-Security Council. The authority to hold and to confiscate vessels and items flows from the extant UN-resolutions.

Maritime embargo operations are a tool for the Council to restore international peace and security. Like economic warfare, maritime embargo operations are not the primary instrument to win a military conflict. Similarly, it is aimed against the opponent, but the actual activities are against all others. Instead, apart from containing the conflict, they are a means used to gradually step through the collective security system. It serves, therefore, as a pressuring tool through military means, without immediately resorting to military force against a State against whom sanctions are imposed. Apart from the international community taking a gradual response to a developing crisis, politically, States may also be more easily committed as these operations are usually at the low end of the use of force spectrum and neither requires 'boots on the ground' nor to cross state boundaries. Also, maritime embargo operations are quickly put in place without having to deal with issues such as territorial sovereignty.

Since the end of the Cold War, the UNSC has resorted seven times to maritime embargo operations.[11] Interestingly, NATO's very first 'out of area'-operation, Operation *Maritime Monitor*, was the start of what became a typical maritime enforcement operation in support of the decisions of the Council when it evolved into operation *Sharp Guard*. Maritime embargo operations were used most recently during the Libya-conflict in 2011, during the NATO operation *Unified Protector*.[12] Although this particular NATO operation ended in October 2011, after Qaddafi's death, the UN-weapons embargo remained in place. It is now revived through UN-Resolution 2292 (2015) that authorizes States to implement the weapons embargo imposed against Libya. Apart from frustrating the business-model of

---

[10] Different names exist for this area, such as the area of maritime operations (AMO) which is obviously close to the target State.

[11] During the crises in Southern Rhodesia, Iraq, the Former Yugoslavia, Haiti, Sierra Leone, Lebanon and Libya.

[12] See Fink 2011a, pp. 237–260.

migrant smuggling and trafficking and training of the Libyan coastguard and navy, the EU-operation *Sophia* has now taken on this role.

17.3.3 Maritime Security Operations (MSO): Policing the Seas in Peacetime Circumstances

The third model of economic enforcement measure at sea, the so called maritime security operations-model, is a result of the reaction to the current focus on threats against a State's maritime security interests. Since 9/11, the issue of enhancing maritime security against threats that may arise in the global commons became a prominent subject in the maritime dimension. This discussion was fuelled by earlier attacks against tankers such as the *MV Limburg* and the possibility that terrorists, or non-state actors, would use the maritime dimension to proliferate and use weapons of mass destruction. Important to note is that, as said, enhancing maritime security ventures into subjects beyond the classic type of military security issues, which are, therefore, outside the situation of armed conflict. In many descriptions of what maritime security operations should mean, a particular legal dividing-line is noticeable, in which operations that are executed in the context of enhancing maritime security are considered to be below the threshold of an armed conflict.

The aim of enhancing maritime security is not to win an armed conflict or to restore international peace and security, but to secure the good order of the oceans. It brings three thought processes together. First is the idea that to protect one's own interests, one must 'police' the seas. This idea has been a rather unnoticed and gradual, but nonetheless unique change in thinking about governing the maritime environment. The point of departure has traditionally been that the freedom of the use of the sea would thrive by non-interference. In this traditional context, States do not accept that other States have any form of policing rights at sea, apart from a number of very limited subjects, such as piracy. The current thinking on enhancing maritime security has changed this. The second idea is that in order to police the seas, maritime security awareness needs to be established at sea to understand what is happening in the global commons. So called maritime security awareness should enable States to act timely against whatever may threaten a State.[13] Third is the idea that States and international organisations, such as NATO and the EU, have embraced the possibility that naval forces can play a role in enhancing maritime security beyond their traditional role in armed conflict and restoring international peace and security. These roles are not military in nature, but more likely fall within the context of law enforcement, often also called a constabulary role.

Enhancing maritime security is operationalized through maritime security operations (MSO). Warships engage with the maritime community in certain areas

[13] In NATO terms abbreviated as MSA. In the US terminology MSA is more often called maritime domain awareness (MDA).

Table 17.1 Summary of the three models of maritime economic enforcement measures

	Economic warfare	Economic enforcement measures i.s.o. UNSC	Maritime security operations
Political end state	Winning the armed conflict	Restore international peace and security	Good order at sea
Military strategic aim	Derail opponent State's economy	Pressurize target State to conform to Council's decisions	Protect economic interests of own State
Character of conflict	International armed conflict	Depends on the situation. Can be in all spectrums of the character of conflict.	Law enforcement (peacetime circumstances)
Target audience	Neutrals (civilians),	Complete maritime community (civilians)	Criminal suspects (civilians)
Tactical level mission	Stop and search suspected supporters of the opponent	Maritime community to conform to measures taken by the UNSC	Secure sea lines of communication and arrest suspected criminals

of the oceans in order to raise their awareness, gather information and, if needed, act against possible threats. In this model, warship are actively 'policing' the seas. Apart from the well-known anti-piracy operations off the Horn of Africa[14] and the migrant operations in the Central Southern Mediterranean Sea,[15] the multinational[16] maritime initiative of Combined Maritime Forces (CMF) is important to note.[17] According to its website, CMF:

> is a multi-national naval partnership, which exists to promote security, stability and prosperity across approximately 3.2 million square miles of international waters, which encompass some of the world's most important shipping lanes. CMF's main focus areas are defeating terrorism, preventing piracy, encouraging regional cooperation, and promoting a safe maritime environment.[18]

Apart from CMF, also NATO has recently moved more actively into MSO-type naval operations. In October 2016, NATO decided to terminate its (only) Article-5 operation *Active Endeavour* (OAE) and changed it into Operation *Sea Guardian* (OSG). With regard to the latter operation, NATO's website announces that:

> Sea Guardian is a flexible maritime operation that is able to perform the full range of maritime security tasks, if so decided by the North Atlantic Council. It is currently

[14] NATO's operation Ocean Shield, the EU-Operation Atalanta and several individual anti-piracy initiatives.

[15] NATO's SNMG2-operation in the Aegean Sea, and the EU-operations Sophia and Triton in the Southern Central Mediterranean.

[16] Currently 31 States are participating in CMF.

[17] See Kuijper and Both 2010, pp. 600–617, for a good background and purpose of the Combined Maritime Forces organisation.

[18] See Combined Maritime Forces (undated).

performing three tasks in the Mediterranean Sea: maritime situational awareness, counter-terrorism at sea and support to capacity-building.[19]

With this change from OAE to OSG, NATO developed its Article 5-operation, which was primarily meant as a measure to support actions against Al Qaida after 9/11, to a standing peacetime naval operation that has a broad focus on maritime security (Table 17.1).

17.4 The 'Non-Kinetic-ness' of Maritime Economic Enforcement Measures

Naval strategist and historian Admiral Sir Herbert Richmond wrote that one quote from Mahan must not be forgotten: 'Less familiar, but no less worthy of being widely known, is that picture of an economic warfare. The contestants 'met in a deadly strife in which no weapon was drawn'.'[20] This quote that Richmond wants us to remember underlines the feeling of non-kinetic-ness of taking economic enforcement measures. This holds true for all three mentioned models. They have in common that the conduct of economic enforcement measures is often considered to be at the low end of the use of force spectrum. Several reasons can be found for this, of which four are mentioned here.

First, the aim is not to fight, but to display other coercive types of measures with less kinetic force, such as control, contain and deny by stopping, searching, confiscating and capturing. Its success is not measured by the destruction of the opponent, but by whether in the overall schemes of things the maritime dimension is rendered un-accessible for the opponent and its supporters. This situation should ultimately have its effects on the opponent warfighting ability and internal political legitimacy.

Second, the physical nature of economic enforcement measures, which includes the boarding of a vessel to check paperwork and cargo, and possibly seizing cargo and arresting persons, does usually not include the need to apply force. The tactical 'endgame' is not to defeat the opponent, but to deny goods, vessels or persons into certain geographical areas, or to capture goods and vessels and arrest persons. This is, however, not to say that there may not be any use of force involved in taking economic enforcement measures (surely, persons could resist in their arrest), but the application of force in these operations is not a first resort means.

Third, enforcement measures are not directed against a military opponent, but against civilians. The target audience is not the opponent's armed forces, but civilians from both neutral and enemy States. These civilians have in one way or another made themselves subject to these measures. For example, because they are in a certain geographical area, or because they are suspected of criminal activities,

---

[19] NATO 2016.

[20] Richmond 1947, p. 13.

carrying contraband or suspected of carrying goods that are sanctioned by the Security Council.

Fourth, when applying force in enforcing economic measures, it will often be within a legal paradigm that fits more closely to law enforcement operations than to a situation of armed conflict. Consequently, legally this urges to consider the application of force as a last resort measure.

As maritime economic enforcement measures are at the low end of the use of force spectrum in which no direct destruction of life or object takes place, it may also explain why, in contrast to kinetic targeting of economic targets, such as oil refineries or banks of which its destruction has basically the same effect, there is no discussion that in the naval dimension of conflict the economy can still be targeted. It is not, however, without complete critique. One example is the closing of Yemeni ports during Saudi Arabia-led operation *Decisive Storm* against Houthi-rebels. As part of that operation in the end of March 2015, the coalition force stated it was going to impose a blockade of Yemeni ports to stop the influx of arms getting into the hands of the Houthi's.[21] The established coalition naval measures were, however, also viewed as seriously deteriorating the humanitarian situation on land. Almost as an immediate response to the establishment of the blockade, growing humanitarian disaster appeared, and the UN-humanitarian relief coordinator urged to lift the naval measures. Interestingly, also during the *Mavi Marmara*-incident in 2009 humanitarian considerations we brought into the discussion of the legality of the blockade. Although the law of blockade does, in principle, not concern itself with humanitarian principles, it is maturing in the sense that there is an increasing willingness to embrace humanitarian principles.[22]

17.5 The Legal Complexity of Maritime Enforcement Measures

Another consequence of the non-kinetic-ness of economic enforcement measures is that it comes with a complex legal environment in which these operations take place. With kinetic operations, generally, and when performed within the applicable legal boundaries, it ends there and then after the engagement. Maritime economic enforcement measures generally, however, have a 'wet' and a 'dry' dimension. The wet dimension concerns the actual activities at sea. For example, the confiscation of goods or vessels, or the detention of persons. The dry dimension concerns the process that comes after the capture, arrest or confiscation. Suspected pirates or criminals need to be taken to a court of law. Captured contraband must be adjudicated in a prize court to see whether ownership of the captured goods has legally passed to the capturing State. This part of the action is sometimes called the 'legal

[21] The Guardian 2015.

[22] See elaborately on this issue Fink 2011b, pp. 191–215.

finish'. The practical challenges that flow from the legal finish are numerous: how does one get persons timely before an investigative judge, how must one sail an enemy vessel back to the State for prize adjudication, which State will offer a diversion port for inspection and further legal process during embargo operations? Apart from these practical challenges, the legal authority can be based on many different legal frameworks that range from ad hoc consent from flag-states, the laws of naval warfare and UN-resolutions, to the many treaties that exist to deal with criminal matters at sea. Such as the treaties regarding drug-or human trafficking, terrorist acts, the protection of fixed platforms at sea or treaties that deal with illegal fishing. Commanders who are confronted with taking maritime enforcement measures should, therefore, have a good understanding in what kind of legal environment they are performing their task, not have their initial success at sea undone in the dry dimension of the action.

Because of the dry and wet dimensions in maritime enforcement and the legal complexities that flow from it, there is also a need to take a more whole of government approach to such operations. Instead of destroying the enemy at sea, which is the sole responsibility of the navy, maritime enforcement measures conducted in the complete range conflict-types urges for more governmental agencies to be involved. To name one example, with regard to piracy the navy is involved in physically capturing the suspect-pirate, Foreign Affairs will be involved in facilitating the suspect to be transported to a court of law on land and Justice will have to deal with the suspect in court, and Internal Affairs are involved when issues of asylum may arise. A successful process of non-kinetic action, therefore, urges to have more governmental agencies involved, than only the Ministry of Defence.

17.6 Conclusion

Naval forces and the economy of a State have always been very much related to each another. In today's conflicts, which concerns the whole spectrum from armed conflicts to peacetime crises, this relationship has become no less important. The two principles of targeting seaborne trade during armed conflict and protecting the State's economy are still valid today, where its current focus is on the latter. On an operational level, the conduct of maritime economic enforcement measures has resulted in three different maritime enforcement models, each with their own characteristics and purpose. These models have however in common that they are considered to be at the low end of the kinetic spectrum and are legally complex due to its 'double dimension'—a wet and dry dimension—of process. Maritime economic enforcement measures, in either three forms, are a useful instrument in the fight against the threats of modern conflict, in which military means are applied, and in particular naval forces, without killing.

References

Best G (1996) War and Law since 1945. Clarendon Press, Oxford

Combined Maritime Forces (undated) http://combinedmaritimeforces.com/about/ Accessed 2 March 2017

Corbett JS (1911) Principles of maritime strategy, 2004 Dover edition. Longmans, Green and Co, London/New York

Ducheine PAL, Pouw EH (2009) Operation Change of Direction - A short survey on the legal basis and applicable legal regimes. In: De Weger M, Osinga F, Kirkels H (eds) NL ARMS. Netherlands Defence Academy, Breda pp 51–96

Fink M (2011a) UN-Mandated Maritime Arms Embargo Operations in Operation Unified Protector. Military Law and the Law of War Review 50(1–2):237–260

Fink M (2011b) Contemporary views on the law of naval blockades. Aegean Review on the Law of the Sea 1(2):191–215

Gretton P (1965) Maritime strategy - A study of British Defence problems. Cassell & Company, London

Kuijper D, Both SJJ (2010) Combined maritime forces - Het belang van deelname aan een unieke coalitie. Militaire Spectator 179(12):600–617

Mahan A (1890) The influence of sea power upon history, 1660-1783. Dover Publications Inc., New York

NATO (2016) Operation Active Endeavour. http://www.nato.int/cps/en/natohq/topics_7932.htm (27 October 2016). Accessed 2 March 2017

Richmond H (1947) Sea power in the modern world. G. Bell & Sons, London

The Guardian (2015) US Backed Saudi Naval Blockade Leaves 20 m Inside Yemen Facing Humanitarian Disaster (5 June 2015) http://www.mintpressnews.com/us-backed-saudi-naval-blockade-leaves-20m-inside-yemen-facing-humanitarian-disaster/206286/ Accessed 2 March 2017

Till G (2004) Sea Power - A Guide for the twenty-first century. Frank Cass, London/Portland

Author Biography

Commander (RNLN) Dr. M.D. Fink, LL.M. is Section Head of the Administrative Law Branch, Netherlands Ministry of Defence, and Senior Researcher Law of Armed Conflict and Peace Operations (ACIL/LACPO). He holds a law degree from Leiden University and obtained his Ph.D. on the Legal Framework for Maritime Interception Operations at the University of Amsterdam in 2016.

Chapter 18
Employing Comprehensive Intelligence: The UN Experience in Mali

Sebastiaan Rietjens, Erik de Waard and Paul C. van Fenema

Abstract Understanding the nature of the war, and the broader strategic and operational environment is of great importance for effective decision-making. However, most of today's military missions are multidimensional having military, political, economic, and judicial aims. As a result, military organizations need comprehensive intelligence on a broad range of issues, including but not limited to security, governance, socio-economic and criminal challenges. Drawing upon situational awareness literature, this chapter analyses the efforts of the United Nations mission MINUSMA in Mali to acquire such comprehensive intelligence and effectively use it to inform decision-making. The main findings of the chapter include that (1) a comprehensive scope jeopardizes the effectiveness of the intelligence operations in terms of overloading scarce resources—both sensors and analytical capabilities, (2) aligning the intelligence capabilities with the operational capabilities proves crucial for mission success, (3) organizing intelligence in a mission network is challenging (in the case of MINUSMA due to e.g. complex command and control structure, incompatible technical systems, as well as the changing security situation) and (4) the combination of high and low tech resources and working methods is of great importance.

Keywords Comprehensive intelligence · United Nations · Understanding

Contents

S. Rietjens (✉) · E. de Waard · P.C. van Fenema
Faculty of Military Sciences, Netherlands Defence Academy, Breda, The Netherlands
e-mail: sjh.rietjens.01@mindef.nl

P.A.L. Ducheine and F.P.B. Osinga (eds.), *Netherlands Annual Review of Military Studies 2017*, NL ARMS, DOI 10.1007/978-94-6265-189-0_18

18.1 Introduction

Operations in Afghanistan and Iraq since 2001 have offered a stark reminder of the truism already offer by Sun Tzu and Clausewitz that military success requires deep understanding of the nature of the war, and the broader strategic and operational environment. Doctrinal publications and supporting studies published as a reflection of the challenges encountered during the counterinsurgency and stabilizations missions reflect this awareness. As the UK JPD-04—titled Understanding—argues, 'understanding provides the context for the decision-making process which informs the application of national power. The purpose of understanding is to equip decision-makers at all levels with the insight and foresight required to make effective decisions as well as manage the associated risks and second and subsequent order effects'.[1] These studies harp on the need for cultural understanding, advocate social network analysis to make sense of political and societal power relations, and offer suggestions for developing effective strategic communication with local population as a means to create influence.[2]

This is reflected in the demand for changing intelligence processes. As McChrystal admonished after his command experience in Iraq and Afghanistan, contemporary military deployment has shown that actionable intelligence has become a prerequisite for decisiveness.[3] During recent missions, where armed forces got involved in activities such as protecting humanitarian aid, disarming belligerent factions, monitoring cease-fires, and negotiating among reluctant players, while the risk of military escalation was always present, the need for adequate and timely intelligence proved overwhelming.[4] Typically, many of these missions were not solely kinetic, but multidimensional in nature, having military, political, economic, and judicial aims.[5] As such, intelligence on merely security-related aspects did not suffice. Military organizations needed understanding on a broad range of issues, including but not limited to governance, socio-economic and

[1] UK MOD DCDC 2010, p. 1-1.

[2] See for instance US Army 2013 (Army Manual FM 3–13); Larsen et al. 2009.

[3] McChrystal 2015.

[4] Cammaert 2003; Smith 1994.

[5] Koops et al. 2015, p. 613.

criminal challenges.[6] For example, NATO and Coalition of the Willing alliances during their missions in Afghanistan and Iraq learned that population-centric intelligence was essential to progress.[7] As Flynn et al. stress:

> The lesson of the last decade is that failing to understand the human dimension of conflict is too costly in lives, resources, and political will … a new [intelligence] concept should seek to explain how populations understand their reality, why they choose either to support or resist their governments, how they organize themselves socially and politically, and why and how their beliefs transform over time.[8]

In addition to the need for comprehensive intelligence, a second development emphasizes the contribution that intelligence makes to the decision-making process. This is often referred to as intelligence-driven operations[9] and means that operations are "driven in timing and objectives by intelligence, including operations to gain intelligence".[10] To this end, intelligence capabilities need to gather information and produce actionable intelligence for the leadership of a mission. This development resembles the broader trend towards results-based management[11] that stresses the outcome or effectiveness of the effort that is undertaken, in this case the intelligence effort.

In Afghanistan and Iraq the US made great strides in this respect by creating new organizational structures, which tied different US military and national intelligence agencies and special operations units together in so called Cross-Functional Teams. This allowed fusing of a variety of intelligence sources, benefiting from the wide range of expertise and types of sensors that this cooperation produced. It also facilitated rapid analysis and dissemination of information, which in turn facilitated rapid decision-making and responsive new actions in the form of raids, tracking movements of individuals, monitoring the cell phone communication of specific identified members of terrorist and criminal organizations or the arrest of them.[12]

Within the Multidimensional Integrated Stabilization Mission in Mali (MINUSMA) that started in 2013, the UN incorporated both developments (i.e. comprehensive intelligence and intelligence-driven operations) in its mission design as it increased its intelligence capacity in an unprecedented manner. Several Western European countries contributed highly professional capabilities to MINUSMA's intelligence chain. These capabilities were kinetic such as a special forces unit[13] and a helicopter detachment as well as non-kinetic. A prime example

[6] Bruls and Dorn 2014.

[7] See e.g. Flynn et al. 2010; Kitzen 2012; Perugini 2008.

[8] Flynn et al. 2012, p. 13.

[9] Norheim-Martinsen and Ravndal 2011.

[10] Dorn 2009, p. 806.

[11] See e.g. Try and Radnor 2007.

[12] For the evolution in intelligence operations, see, for instance, Lamb and Munsing 2011; Naylor 2015.

[13] For a detailed analysis on the contribution of the special forces to MINUSMA's intelligence process, see Rietjens and Zomer 2017 (forthcoming).

of the latter and also the flagship of MINUSMA's intelligence capacity is a unit that is coined ASIFU, the All Sources Information Fusion Unit. ASIFU is intended to "contribute especially to traditionally non-military intelligence analysis, such as illegal trafficking and narcotics-trade; ethnic dynamics and tribal tensions; corruption and bad governance within Mali and MINUSMA area of interest."[14] The intelligence products that ASIFU delivers should fuel the decision-making process of MINUSMA's military and civilian leadership. To this end, ASIFU has control of many high-tech sensors, well-educated intelligence personnel and a state-of-the-art IT, including databases and command systems.[15] All these aspects make ASIFU a unique asset within the context of UN missions.

Within this chapter we focus at ASIFU. More specifically, the objective of this chapter is to increase understanding on comprehensive intelligence within peace support operations, by analysing the intelligence contribution of ASIFU. To this end, the chapter uses a theoretical framework on situational awareness that is widely applied to military as well as civilian domains. Since situational awareness is a well-developed theoretical domain that addresses the way actors obtain comprehension of their external environment and the way they use this to inform decision-making, it is very applicable to study the way in which ASIFU gained comprehensive intelligence and how this contributed to MINUSMA's decision-making process. Section 2 provides an overview of the literature on situational awareness. Section 3 subsequently introduces the conflict in Mali and the mission design of MINUSMA in general and ASIFU specifically. Section 4 analyses the functioning of ASIFU and the contribution it made to the decision-making process. Section 5 provides a discussion and concludes the chapter.

18.2 Situational Awareness

The term situational awareness (SA) has its origin in the world of military aviation, where achieving high levels of SA was found to be both critical and challenging early in aviation history.[16] These military pilots were confronted by a constantly evolving 'picture' of the state of their environment. As a result, they depended heavily on what became to be known as situational awareness in their decision-making processes. In this context, SA means appreciating all you need to know about what is going on in relation to the full scope of your task. For example, for a fighter pilot, SA means knowing the threats and intentions of the enemy forces but at the same time knowing the capabilities, responses and intentions of one's own aircraft.

---

[14] UN Department of Peacekeeping Operations 2015.

[15] Van Dalen 2015.

[16] Endsley and Jones 2011.

Despite the wide use of the concept of SA, there is no universally accepted definition.[17] Sarter and Woods go as far as to refer to SA as an "ill-defined phenomenon".[18] Nevertheless, the most cited definition of SA, and one that appears to have stood the test of time,[19] is suggested by Endsley, namely: "the perception of the elements in the environment within a volume of time and space, the comprehension of their meaning and the projection of their status in the near future".[20] While defining SA in this way, Endsley also emphasizes what SA is not. First, she states that SA is not an action or performance. As such, the understanding of a situation is quite distinct from the manual or vocal actions taken in response to that situation.[21] Second, Endsley stresses that SA is different from long-term memory. Wickens adds to this point by stressing that the construct of SA is primarily applicable to dynamic situations where variables are changing.[22] This aligns with the definition of SA offered by Durso et al.: "comprehension or understanding of a *dynamic* environment" (emphasis added).[23] Hence, static knowledge about the state of a system is less directly relevant to the concept of SA.[24] To this end, various SA researchers and practitioners are cautious in assuming that merely presenting more information will lead to increased SA.[25] For example, Bolia et al. argue against the proposition that the increased availability of information means that operators will make better decisions.[26] Rather, factors such as information overload, poor interpretation, presence of non-relevant information in the light of a task, and false data are likely to degrade rather than enhance SA in this scenario.[27]

Although several different SA models have been developed, Endsley's three-level model[28] is by far the most cited in the literature.[29] Her model is generic and presents a relatively simple and intuitive description of SA, which has since been applied in different domains.[30] The model uses a linear information processing chain and, in drilling down, Endsley draws a careful distinction between three levels of SA which, from the bottom up, she refers to as: 'perception' (including 'noticing')—Level 1; 'comprehension'—Level 2; and 'projection'—Level 3.

---

[17] Salmon et al. 2008.
[18] Sarter and Woods 1991.
[19] Wickens 2008.
[20] Endsley 1995, p. 36.
[21] Wickens 2008.
[22] Wickens 2008.
[23] Durso et al. 2007, p. 164.
[24] Wickens 2008.
[25] Stanton et al. 2010.
[26] Bolia et al. 2007.
[27] Stanton et al. 2010.
[28] Endsley 1995.
[29] Stanton et al. 2010.
[30] Salmon 2008.

Endsley's model is presented in Fig. 18.1. It shows that SA acquisition and maintenance is influenced by various factors including individual factors (e.g. experience, training, workload) and task and system factors (e.g. system complexity, interface design).[31]

A key point in Endsley's model is that the higher levels of SA depend on success at the lower levels. Thus, an air traffic controller may first notice a change in an aircraft's trajectory (level 1), then comprehend that this means that two aircraft are on converging trajectories (level 2) and, finally, understand that an accident may take place and how serious it would be (level 3). In this respect, Wickens notes that "the diagnostic distinction between these three levels is important, not only because they point to different perceptual/cognitive operations but also because breakdowns in each may have different consequences for addressing them, through training or system design".[32]

The first step in achieving SA involves perceiving the status, attributes, and dynamics of task-related elements in the surrounding environment.[33] Different sensors may contribute to the perception of information, including visual, auditory, taste or a combination of these. At this stage, the data is merely 'perceived' and no further processing of the data takes place. In many systems, a strong emphasis is placed on the electronic displays and read-outs that are provided, but Endsley and Jones argue that much of the SA at level 1 comes from the individual directly perceiving the environment with human senses.[34] Endsley and Jones identify several factors that contribute to uncertainty at level 1 SA and which may lead to errors.[35] These include missing information, reliability of the data (or lack thereof), incongruent or conflicting data, timeliness of the data and ambiguous or noisy data.

The second step in Endsley's model focuses on understanding what the data and cues perceived mean in relation to relevant goals and objectives. During the acquisition of level 2 SA "the decision maker forms a holistic picture of the environment, comprehending the significance of objects and events".[36] At this stage, it involves integrating multiple pieces of data to form information, and prioritizing that combined information's importance and meaning as it relates to achieving the present goals.[37]

The third and highest level of SA involves predicting the future states of the system and elements in the environment.[38] By using both level 1 and 2 SA-related knowledge, and merging this with experience in the form of mental models,

[31] Endsley 1995.

[32] Wickens 2008, p. 398.

[33] Endsley 1995.

[34] Endsley and Jones 2011.

[35] Endsley and Jones 2011.

[36] Endsley 1995, p. 37.

[37] Endsley and Jones 2011.

[38] Endsley 1995.

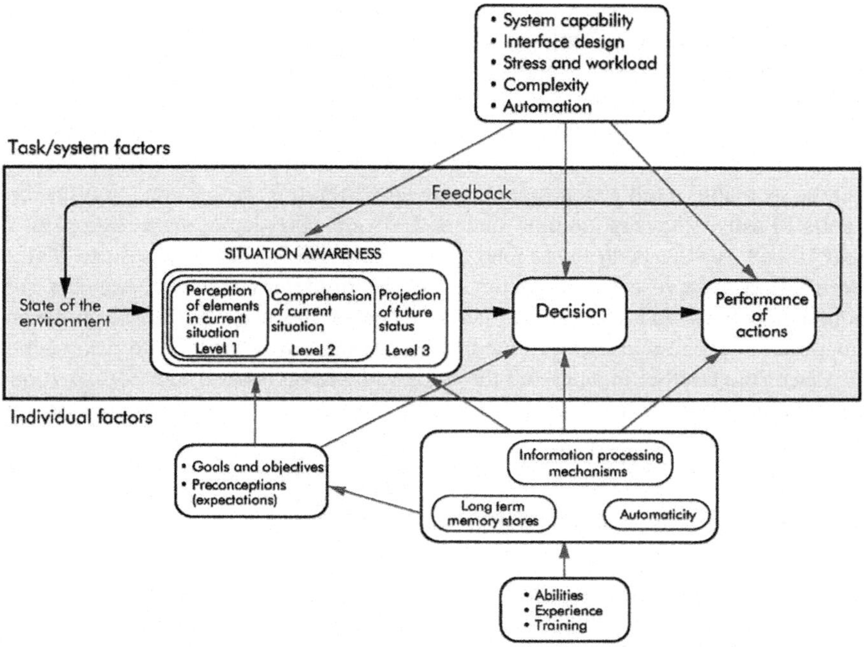

Figure 18.1 Three-level model of situation awareness. Adapted from Endsley 1995; Salmon 2008

individuals can forecast likely future states in a particular situation.[39] Endsley and Jones suggest several causes of failure to achieve such accurate predictions.[40] These include workload fatigue issues such as information processing overload and insufficient domain-specific knowledge.[41]

How can professionals and researchers use SA? As Endsley states, SA is not a construct that exists solely for its own sake.[42] From a top-down perspective, SA should inform the decision makers. They can then select activities that will align the perceived environment with their plans and goals based on that understanding. From a bottom-up perspective, patterns in the environment may be recognized that will indicate that different goals are necessary.

---

[39] Salmon 2008.

[40] Endsley and Jones 2011.

[41] The three levels that Endsley identifies closely resemble the levels that are recorded in the UK Joint Doctrine Publication 04 (JDP-04) on Understanding: see UK MOD DCDC (2010). Within the JDP-04, these levels are defined as (1) situational awareness ("perception of a particular area of interest, problem or situation bounded by time and space in the context of a mission or task", p. 1-5), (2) insight ("the capacity to gain an accurate and deep understanding of something", p. 2-1) and (3) foresight ("the ability to anticipate future events or requirements", p. 2-1).

[42] Endsley 1995.

The theoretical perspective described above was used as a lens to analyse the way ASIFU obtained situational awareness and how this contributed to the decision-making process within MINUSMA. To acquire empirical data the researchers on this project conducted 93 semi-structured interviews with key MINUSMA personnel (civilian as well as military) who had deployed between March 2014 and December 2015. We also analysed documents from the mission, such as available intelligence reports, standard operating procedures, and meeting reports. Lastly, the first author attended several pre-deployment exercises of ASIFU, and together with the second author made a 2.5-week field visit to Mali in late 2015 to observe the mission first-hand. From these various sources and subsequent feedback and validation,[43] an analysis of the way ASIFU acquired and situational awareness and contributed to MINUSMA was made. The next section introduces the conflict in Mali and the design of MINUSMA. Then, Sect. 4 zooms in on the way ASIFU acquired and shared situational awareness.

18.3 Context and the Design of MINUSMA

The establishment of MINUSMA by the UN Security Council in Resolution 2100 on 25th April 2013 was the result of several intertwined events. The northern regions of Mali had become unstable due to developments such as illicit trafficking, a lack of state control, and heavily armed Tuareg fighters returning from Libya. This explosive cocktail had led to mutinies throughout the country, a military coup in March 2012, and a marginalization of the Armed Forces of Mali (FAMA), which constantly lacked ammunition and reinforcements.[44] In broad terms, MINUSMA's task was to promote stabilization of key population centres, protect civilians and guide the political reform process.[45] To achieve this, MINUSMA consisted of approximately 500 international civilians, 120 UN volunteers, over 1,000 policemen and close to 9,000 soldiers.[46] These military troops originated from 41 different countries (e.g. Bangladesh, France, Germany, Niger, Senegal, The Netherlands) and were commanded by the Force Headquarters (FHQ) in the capital Bamako, as well as three sector headquarters (SHQs) that covered dedicated operational areas.[47]

---

[43] To ensure internal validity, four officers and one civilian analyst from The Netherlands who were all closely involved in the mission reviewed draft versions of the paper. The draft versions were met with responses of recognition, as well as reactions indicating that important issues had been revealed.

[44] For a detailed account on the conflict in Mali, see Boutellis 2015.

[45] Lotze 2015.

[46] MINUSMA 2017.

[47] SHQ-West covered the province of Timbuktu, whereas SHQ-East covered the eastern province of Gao. A SHQ-North was created in 2014, based in Kidal and covering a smaller but very turbulent region.

MINUSMA's force design contained the typical military intelligence units (designated by the number 2, per standard military staff convention) within its battalions (S2), Sector headquarters cells (G2) and Force Headquarters cell (U2). These units were supposed to provide MINUSMA's commanders with current intelligence, especially relating to security. To address intelligence deficiencies that were observed in previous UN missions,[48] the UN deployed an additional intelligence unit that was coined the All Sources Information Fusion Unit (ASIFU). As referred to earlier, ASIFU's mission was to provide intelligence on a wide range of topics. This was often referred to as X-PMESII, indicating that information was to be gathered and analysed on Political, Military, Economic, Social, Infrastructure and Information domains (following NATO conventions).[49] The X (cross) implied that these domains were interconnected and could not be seen separately. Doing this, ASIFU's role is

> To improve the processing and production of MINUSMA broad information and intelligence in order to have accessible and useable information on time. This will support the decision-making processes on the operational (force headquarters) and tactical (sector headquarters) level. But ASIFU should also be able to support the strategic level: the special representative of the secretary-general through the [Joint Mission Analysis Centre] JMAC and [UN Department of Safety and Security] UNDSS.[50]

ASIFU was attached to the UN headquarters in the capital of Bamako. Its intelligence staff initially consisted of 30 military officers from seven different countries (the Netherlands, Denmark, Norway, Sweden, Estonia, Finland, and Germany) and 1 Dutch civilian advisor. In time ASIFU's capacity would increase to about 70. ASIFU also had two ISR (intelligence, surveillance and reconnaissance) companies under its command which focused on intelligence gathering. The first company consisted of 55–65 mostly Dutch soldiers[51] and was deployed in the province of Gao from March 2014 onwards. As such it worked within MINUSMA's Sector East. The company had several distinct capabilities including human intelligence (HUMINT), civil-military interaction and Unmanned Aerial Vehicles (UAVs). The second unit was the Swedish ISR Task Force. Its intelligence capacity was approximately twice the size of the Dutch unit and started to operate a year later, in March 2015. It was situated in the western province of Timbuktu and attached to MINUSMA's SHQ West. The capabilities of the Swedish Task Force included, amongst others, military reconnaissance personnel, a weapons intelligence team, and small UAVs.

The next section presents findings from our study on ASIFU's effort to obtain SA in war-thorn Mali. To this end we used Endsley's model as a frame of reference. Some of the findings clearly support specific interrelationships hypothesized in the

[48] See e.g. Rietjens and De Waard 2017.

[49] Van Dalen 2015.

[50] First Commander ASIFU, Colonel Keijsers, cited in Karlsrud and Smith 2015, p. 11.

[51] Several other countries contributed soldiers to the ISR Company, including Belgium, Denmark, Estonia and Switzerland.

model. Others add new insights to the model's perceived logic. But also, some contradicting dynamics were uncovered that offer insight for other missions.

18.4 Results

18.4.1 Goal-Setting and Situational Awareness

In support of the model, the ASIFU case has clearly shown the linkage between goal-setting and achieving SA. This connection may seem straightforward. Yet against the reality of today's security dilemmas—such as nation-building, fighting terrorism, and hybrid warfare—formulating clear and unambiguous objectives is extremely difficult. Mali has for many years remained outside the security scope of most Western countries. After the collapse of the Gadhafi regime and the parallel rise of Islamic terrorism in the Middle East, Mali suddenly became a hotspot that needed urgent international attention. However, knowledge of the country's political and social structures and power relations was very limited and widely distributed. Despite this lack of understanding the international community still needed realistic scenarios on Mali's security and societal development patterns. This would enable development of accurate expectations about the way the country could be assisted in building sustainable peace. Thus, ASIFU followed a learning-by-doing approach, using the previous ISAF deployment to Afghanistan as a frame of reference. To be more precise, ISAF's X-PMESII framework was embraced to consciously unravel MINUSMA's complex intelligence demand into clear and manageable parts. At the same time, the intelligence process of ISAF, which was a NATO mission, had a strong hierarchical grounding, where the selection of priority information requirements (PIRs), specific information requirements (SIRs), and essential elements of information (EEIs) cascaded down the hierarchy, fuelling the operational units with clear task-settings. Instead, within MINUSMA initially a bottom-up approach dominated. In the beginning, MINUSMA's leadership was so busy with the here and now of getting the entire organization into gear, that articulating long-term intelligence goals for ASIFU was not a priority. Thus, ASIFU had to translate the Force Commander's general intent into comprehensible intelligence goals, and basically, determine its own performance indicators. The following citation of one of ASIFU's analysts describes this situation quite well.

> We formulated the Information Requirements ourselves. In theory the Force Commander should task us via an OPORDER [Operational Order], but the OPORDER has never officially been published. There was also no mention of ASIFU, which meant that officially we did not have a task in the command structure of the Force Commander and the SRSG [Special Representative of the Secretary General]. Therefore we mainly operated according to the vision of ASIFU's Commander, who formulated a task description and several (Priority) Information Requirements that were later approved by the Force Commander. Instead of a traditional top-down this was more a bottom-up, self-directing process.

Interesting to mention is that the bottom-up approach further affected the relationship between goal-setting and SA. Especially during the first year of MINUSMA's deployment, the intelligence goals were not formulated by the Force Commander but by peers of earlier rotations, which implied that they could be more easily disputed. Each time an ASIFU detachment was replaced by a new rotation the PIRs, SIRs, and EEIs were brought up for debate and were changed according to the views of the newcomers. So, the intended process, in which the intelligence goals would be gradually refined over the rotations, did not materialize. Overall, ASIFU's intelligence scope remained quite broad, because successive rotations kept coming up with new ideas and priorities. Having said that, it is important to stress that excessive pride in oneself is not the main reason for these fluctuations, but they seem to be far more a logical consequence of the task complexity at hand. The great variation in intelligence requirements and approaches, basically, emphasizes that reaching consensus on complex societal questions is extremely challenging and can never be an exact science. For example, MINUSMA's intelligence collection plan included questions such as: Under what circumstances will participants in the peace implementation abandon the process and opt for a military solution? Or, what is the threat posed by criminal, terrorist, and other non-compliant armed groups throughout Mali?[52] What makes the exercise of answering such questions even more difficult is that changing circumstances in Mali may well result in new or additional questions surfacing at the expense of already existing questions in the intelligence collection plan.

18.4.2 System Complexity and Situational Awareness

Another confirmation of Endsley's model this study has found, refers to the relationship between system complexity and SA. In this respect, system complexity has to do with MINUSMA's organizational set-up. It concerns a UN operation leading to a complex collaboration effort between civilian and military partners. In practice, the structural differentiation between the collection of current and mid- to long-term intelligence strained the realization of SA of MINUSMA's military force. In designing MINUSMA's military component it was decided to make the regular intelligence branches (S2, G2 and U2) responsible for the current intelligence process and to create a separate organizational entity specialized in the gathering of information with a longer time horizon—ASIFU. Yet, in practice personnel of the regular intelligence branches lacked experience, were often generalists rather than intelligence experts and did not have the adequate technical means at their disposal to process the information. In addition, most of these branches hardly contained analysis capacity. As a result, the current intelligence picture remained far too vague and scattered for the Force Commander to get a comprehensive command

[52] MINUSMA 2015.

and control process going. In addition, the security situation in Mali seriously deteriorated when armed groups started to attack UN supply convoys with IEDs (improvised explosive device) to thwart MINUSMA. Since the attacks cost many human lives,[53] the Force Commander was anxious to better anticipate these threats. His call for actionable current intelligence was so urgent, that there was no time to wait for the professional development of the operational battalions. Instead the Force Commander increasingly asked ASIFU and its state-of-the-art intelligence capabilities for immediate assistance. However, ASIFU was fully geared to collect mid- to long-term intelligence and initially refused to commit valuable resources for chasing suspects or keeping convoys under surveillance, arguing that with its intended task-setting it could already contribute to countering the IED danger. After all, rather than just fighting the symptoms of a problem, it is just as important to come to the root cause of the problem. This requires investing energy and capabilities in higher order issues, such as why certain groups in Mali are so discontent that they resort to violence. Eventually, ASIFU was forced to follow a dual approach of trying to both satisfy the Force Commander's critical current intelligence needs and to stick to its original long term intelligence commitment. This trade-off complicated the organizational system even more. The fact that there was a misbalance between the country's size and the available intelligence sensors[54] made matters grew worse when ASIFU had to start shifting its attention between current and more generic intelligence task assignments. The fragmentation of operational intelligence capacity made ASIFU's SA on security-related threats as well on more multi-faceted societal developments quite shallow and disputable.

18.4.3 Levels of Situational Awareness

A third empirical result deals with the interrelationship between the three levels of SA. ASIFU had many different sensors that contributed to obtaining level 1 SA—perception of the elements. To subsequently process all the incoming information, ASIFU contained analysis fusion cells,[55] one at ASIFU HQ and the other at the Dutch ISR Company. Both cells consisted of 12–16 persons and included collators,

[53] MINUSMA fatality statistics: 6 (2013), 39 (2014); 29 (2015), with most fatalities from malicious acts. Statistics available at http://www.un.org/en/peacekeeping/resources/statistics/fatalities.shtml; BBC News 2015.

[54] Sector North, East and West respectively measure 80,000, 170,00 and 400,000 km^2, corresponding roughly with twice, four times, and ten times the size of Switzerland. One can imagine that covering these vast regions and collecting strong evidence to unravel complex societal dynamics is a major challenge for just a limited number of sensors.

[55] Within the Dutch ISR Company this cell was coined the All Sources Intelligence Cell (ASIC), while at ASIFU HQ such a cell was named the All Fusion Cell (AFC). The Swedish ISR Taskforce had a more traditional layout. To analyse the information its sensors gathered the taskforce had a Military Sources Intelligence Cell (MSIC), consisting of four military analysts.

technicians and different kinds of analysts including military, geospatial and human terrain analysts. The aim of both fusion cells was to comprehend the environment in which they were operating (i.e. level 2 SA) and if possible to project the future status of the environment (i.e. level 3 SA). The Commander of the Dutch-led ISR Company illustrated this as follows:

> We started to collect as much information as possible on the PMESII factors, everything we could find about the situation in Gao because for Bamako this was still one big question mark. Every piece of information we were able to collect was useful because there was nothing there yet. We focused on the PMESII factors, this was already decided upon in advance, and additionally we looked into threat to the force and threat to the mission but that was of secondary importance. Of course, I would like to know when a bomb is going to explode, but what is more important for me is to find out who makes that bomb, who tasks that person, who creates the plan. We want to know in advance what is going on in people's minds, and to integrate a predictive element to our analyses.

The most prominent intelligence report that ASIFU made and that aimed at predicting the future status of the Mali environment, was the so-called "Quarterly Outlook".[56] Based on an extensive scenario analysis, ASIFU produced these intelligence reports every three months. The reports tended to be very comprehensive in nature, including not only information about the armed groups, but also about tribal tensions, smuggling routes, and the perception of the Malian population towards MINUSMA. ASIFU made assessments of the likely places of "greatest potential for violence" and of civil unrest. Furthermore, ASIFU reports included some creative scenario-building, e.g., for the possible outcomes of peace negotiations held in Algiers.

Despite these attempts, getting to higher levels of SA proved difficult for various reasons. Many analysts of ASIFU displayed only little awareness of the complexity of the conflict, the history of Mali and the ethnic sensitivities. Thus, they had difficulties to unravel the dynamics of the environment and address the information requirements they were tasked with. Other reasons included amongst others the lack of interpreters,[57] the difficulties in coordinating with the operational battalions of the main force[58] and the lack of technical systems that supported information sharing.[59]

Where Endsley's model hypothesizes a sequential development from perception via comprehension to projection, this study has uncovered that SA does not have to evolve in an advancing order per se. Although, in line with the ruling logic, this study supports the idea that higher levels of SA depend on the success of lower levels, it also complements the main argument by discovering that individual as well as system factors can stimulate a reverse movement from level 2 back to level 1. To be more specific, the study has shown that the system of detachments rotating

---

[56] MINUSMA ASIFU 2015.

[57] Maas 2016.

[58] Rietjens and De Waard 2017.

[59] Rietjens and Baudet 2017.

in and out every four months hinders progress in the level of SA. Basically, each rotation starts its own SA sequence when arriving in the mission area. A time span of four to six months is, however, rather limited to reach level 2—comprehension of the situation. When a detachment just begins to comprehend the general societal and security dynamics it is replaced by a new one. Although for each rotation a formal hand-over, take-over process is organized, in which the unit leaving helps its successors to settle, there still is insufficient time to get to know and understand all the real practical details. As such, each new deployment starts its own level 1 phase —perception of the elements. Re-evaluating the existing intelligence collection plan of PIRs, SIRs, and EEIs, which was discussed earlier, is an essential part of this renewed perception. Also, scarcity of assets causes a re-assessment of the situation by a new detachment to make its own judgment on how to best balance goals, ambitions, and available means. For instance, the first rotation of the Dutch-led ISR Company lacked force protection elements, and therefore was forced to aim all its intelligence sensors at the nearest town of Gao and its surrounding area. When additional force protection units arrived, the second rotation could expand the area of operation, also directing sensors to rural areas further away. Yet, the next detachment argued that expanding the operational reach had diluted the intelligence picture, and thus, decided to concentrate all the available means on Gao city again. Recapitulating, where Endsley's model suggests a direct relationship between system complexity and SA, the ASIFU case also points to an indirect correlation with system complexity leading to a re-assessment of existing goals and ambitions and, then, the initiation of a new SA sequence.

18.4.4 Situational Awareness Is (Not) Action

A fourth, more fundamental, finding opposes Endley's key assumptions that SA is not action or performance, arguing that understanding a situation [SA] differs from the action taken into response of it. This study has found evidence for a decision-making process that Isenberg has termed "fighting empirically".[60] In his vision the growing complexity and ambiguity of military operations require a behavioural pattern in which creating situation awareness and action are intertwined. More specifically Isenberg states: "the most effective officers may take action with a very incomplete understanding of the battle situation [...and...] will define and solve problems, and implement the solutions in a continually reiterating cycle".[61] Isenberg strongly bases his reasoning on Weick's enactment theory. Enactment refers to the idea that only by acting organizations can make sense of their environment.[62] Basically, enactment sets the stage for understanding and

[60] Isenberg 1985, p. 178.
[61] Isenberg 1985, p. 178.
[62] Weick 1979.

adaptability. The enactment perspective denies that the environment and the organization are separate entities. On the contrary, it stresses that both are in a symbiotic relationship, where organizational measures actively shape the environment and people dynamically invent rather than passively consider what is happening around them. It is because of mutual interaction between organization and environment that Weick propagates, for organizations to progress, that people should put more energy in reflecting upon the things they do and undertake instead of viewing the environment as given and confronting it as an objective bystander. For ASIFU this was, basically, the modus operandi. The operational task-setting was so multi-faceted and equivocal that action was the only practical way to probe the environment and learn by doing. Everyone realized that an ex-ante holistic understanding of the situation was simply unobtainable. By sending out reconnaissance patrols, tasking UAVs to monitor certain areas, or street-level engagements with local people, etcetera information was gathered and combined to gain understanding and decide on follow-up measures.

18.4.5 Situational Awareness, Decision-Making and Execution

A final finding deals with the relationship between SA, decision-making and the execution of activities. According to its design, ASIFU's intelligence effort had to inform the decision-making processes of MINUSMA's leadership at the tactical (Sector Headquarters), operational (Force Headquarters) as well as the strategic level (Special Representative). However, as was raised earlier, MINUSMA Force Commander as well as his Sector Commanders, lacked adequate current and security-related intelligence, mostly because of disfunctioning regular intelligence branches. Meanwhile, ASIFU was tasked to provide comprehensive intelligence at a mid- to long-term. During the research period (March 2014-December 2015) ASIFU employed several initiatives to provide current and security-related intelligence, but largely maintained its focus at comprehensive and mid- to long-term intelligence. However, as the security situation seriously deteriorated during this period, current and security-related intelligence had the highest priority at the tactical and operational levels. ASIFU's comprehensive and predictive analyses, most notably its Quarterly Outlook, therefore only played a marginal role in the decision-making processes. The difficulties that ASIFU met in drafting these analyses (see section "Levels of SA") contributed to this as well.

At a strategic level, ASIFU was to support the special representative of the secretary-general (SRSG) through the Joint Mission Analysis Centre (JMAC) and the UN Department of Safety and Security (UNDSS). The interaction between ASIFU, JMAC and UNDSS was, however, poor for a number of reasons. These included technological (e.g. organisations using different technical systems), organisational (e.g. while ASIFU was directed by the Force Commander, UNDSS and JMAC were directed by the SRSG and his chief of staff) as well as political (e.g. there was large rivalry between the three organizations).[63] Largely as a result of this, the contribution of ASIFU to the strategic decision-making process was limited. A notable exception to this was the first half year of ASIFU's deployment. During that period the SRSG was a Dutch diplomat, Mr. Koenders, who had close ties with the Dutch commander of ASIFU. This enhanced ASIFU's contribution to the strategic decision-making process, but further fuelled anxiousness of JMAC and UNDSS towards ASIFU.

In addition to the strained relation between SA and decision-making, the relation with the execution was further challenged by the functioning of MINUSMA's main force. It appeared that MINUSMA's battalions, most of them consisting of African troops,[64] had a hard time sustaining and protecting themselves. By and large, they were insufficiently organized, trained and equipped to perform their tasks such promoting stabilization of key population centres and protection of civilians. Moreover, they had little to no experience working with NATO-style intelligence. As a result, the battalions made only very limited use of ASIFU's intelligence products to inform their operations.

18.5 Generalizing Insights and Conclusion

The objective of this chapter was to increase understanding on comprehensive intelligence within peace support operations, by analysing the intelligence contribution of ASIFU. Our study generates several insights that can leverage on lessons learned during the MINUSMA mission. First, while we introduced two intelligence developments—comprehensive intelligence and intelligence driven operations—our study shows that these two are closely related and that this relation is somewhat problematic. The intelligence driven operation represented a dedicated effort to support a comprehensive mission, which concerns kinetic and non-kinetic aspects of (actions in) the theatre. Hence, requirements from the comprehensive mission translated into the need for an intelligence driven operation that included a range of dimensions (X-PMESII). However, this extensive scope clearly also jeopardizes the effectiveness of the intelligence driven operation in terms of overloading scarce resources—both sensors and analytical capabilities. Scoping thus appears an essential process to organize the intelligence operation in conjunction with the dynamics of a comprehensive operation. In addition, aligning the intelligence capabilities with the operational capabilities proved crucial for mission success. In the MINUSMA case Western European countries put large effort in optimizing the

[63] For a more detailed account on this see Rietjens and De Waard 2017; Rietjens and Baudet 2017; Zomer 2016.

[64] In December 2015 Sector North contained two infantry battalions: one from Chad and one from Guinee. In Sector East three infantry battalions were deployed originating from Niger, Bangladesh and Senegal. Finally, Sector West had two infantry battalions from Burkina Faso and from Togo.

intelligence organization. However, in doing so they overlooked the operational capacities that MINUSMA had in place. As most UN troops had difficulties sustaining and protecting themselves, they had only limited possibilities and interest to carry out intelligence-driven operations. This clearly decreased the value of the intelligence effort.

Second, organizing intelligence in a mission network deserves more attention. SA literature has originally been built around individual operators acting in a centralized organization. Researchers have been struggling with the individual-collective unit of analysis and positivist-constructivist philosophical dimension in relation to SA. The unit of analysis deals with the question whether collective (team, organization, or network) SA simply aggregates individual SA or represents a different phenomenon and the extent to which a collective can have the same SA. This dovetails with the philosophical dimension of SA: does SA represent a reified 'thing' (e.g. a report or visual) that can be shared (positivist view), or is SA an ongoing construction of diverse individuals (constructivist view advocated by for instance Weick). Current research tends to promote a view that combines collectiveness with decentralization or distributedness. Examples include distributed cognition,[65] leadership,[66] planning,[67] and distributed situation awareness (DSA).[68] Such theorizing acknowledges the uniqueness of individuals and their own specialized knowledge and dynamic understanding. At the same time, interactions of knowing individuals and network structures and technologies can lead to compatible SA as a 'glue that holds the distributed system together'.[69] The direction DSA takes varies with the mission, and this influences the required expertise and mode of involvement of specialized actors. In our case, collective (MINUSMA level) intelligence depended originally on regular intelligence branches providing current intelligence and ASIFU providing long-term insights. In practice, this mix did not work out because of for instance the complex command and control structure, incompatible technical systems, as well as the changing security situation.

Third, ASIFU operated in an environment that combined high and low tech resources and working methods. Units such as ASIFU can develop their embeddedness in a technological, information, and expertise sense. Yet in a partially low-tech environment, they may lack social embeddedness, which has been defined as 'the relational embeddedness and the structural embeddedness of organizational actors, which generates the likelihood of shared understandings between and among organizations'.[70] While some researchers and practitioners advocate the use of advanced technologies (e.g. analytics, algorithms, various ways of social media use,

[65] Hollan et al. 2000.

[66] Ancona and Backman 2010.

[67] Johnson et al. 2012.

[68] Stanton et al. 2006.

[69] Stanton et al. 2006.

[70] Jones and Lichtenstein 2008, p. 238.

and simulations) and 'digital peacekeeping',[71] others point at new questions and the complex impact of technologies.[72] Critical research on and practical experimentation with advanced technologies is therefore necessary to optimize their added value.

To conclude, much like the challenges ISAF experienced in implementing effectively the concept of the comprehensive Approach, as Brocades Zaalberg demonstrates elsewhere in this volume, ASIFU's experience illustrates the challenge of designing and adapting an intelligence operation within a comprehensive mission like MINUSMA. Apart from the quality and innovation of intelligence practices, the network context deserves more attention. This will scope, focus and resource intelligence operations in a manner that undergirds operational success. Moreover, learning as a network proves paramount to be able to adapt when ex ante design does not work out, or when circumstances change. The latter is highly likely in current and future comprehensive missions.

References

Ancona DG, Backman E (2010) Distributed Leadership: Going from Pyramids to Networks. Leadership Excellence 27(1):11–12

BBC News (2015) World's most dangerous peacekeeping mission (20 November 2015) http://www.bbc.com/news/world-africa-34812600 Accessed 10 March 2017

Bolia RS, Vidulich MA, Nelson WT, Cook MJ (2007) A history lesson in the use of technology to support military decision-making and command & control. In: Cook MJ, Noyes JM, Masakowski Y (eds) Decision making in complex systems. Ashgate, Aldershot, pp 191–200

Boutellis A (2015) Can the UN Stabilize Mali? Towards a UN Stabilization Doctrine? Stability: International Journal of Security & Development 4(1):1–16

Bratic V (2016) Peacebuilding in the Age of New Media (Editorial). Media and Communication 4 (1):1–3

Bruls F, Dorn AW (2014) Human Security Intelligence: Towards a Comprehensive Understanding of Complex Emergencies. In: Hobbs C, Moran M, Salisbury D (eds) Open Source Intelligence in the Twenty First Century: New Approaches and Opportunities. Palgrave Macmillan, London, pp 123–144

Cammaert PC (2003) Intelligence in Peacekeeping Operations: Lessons for the Future. In: De Jong B, Platje W, Steele RD (eds) Peacekeeping Intelligence: Emerging Concepts for the Future. OSS International Press, Oakton, pp 11–30

Dorn AW (2009) Intelligence-led Peacekeeping: The United Nations Stabilization Mission in Haiti (MINUSTAH), 2006–07. Intelligence and National Security 24(6):805–835

Dorn AW (2016) Smart Peacekeeping: Toward Tech-Enabled UN Operations. International Peace Institute, New York

Durso F, Rawson K, Girotto S (2007) Comprehension and situation awareness. In: Durso F, Nickerson R et al (eds) Handbook of Applied Cognition, 2nd edn. Wiley, Hoboken, USA, pp 163–194

Endsley MR (1995) Toward a theory of situation awareness in dynamic systems. Human Factors 37(1):32–64

[71] Dorn 2016.

[72] Bratic 2016.

Endsley MR, Jones DG (2011) Designing for situation awareness: An Approach to User-Centered Design. Taylor & Francis, London

Flynn MT, Pottinger M, Batchelor PD (2010) Fixing Intell: A Blueprint for Making Intelligence Relevant in Afghanistan. Center for a New American Security, Washington DC

Flynn MT, Sisco J, Ellis DC (2012) Left of Bang: The Value of Sociocultural Analysis in Today's Environment. Prism 3(4):13–21

Hollan JE, Hutchins E, Kirsch D (2000) Distributed Cognition: Toward a New Foundation of Human-Computer Interaction Research. ACM Transactions on Computer-Human Interaction 7 (2):174–196

Isenberg DJ (1985) Some Hows and Whats of Managerial Thinking. In: Hunt J, Blair J (eds) Leadership of the Future Battlefield. Pergamon, New York

Johnson LB, Kopeikin AN, Choi HL, How JP (2012) Distributed Planning Strategies to Ensure Network Connectivity for Dynamic Heterogeneous Teams. IEEE Journal on Selected Areas in Communications 30(5):861–869

Jones C, Lichtenstein BB (2008) Temporary Inter-Organizational Projects: How Temporal and Social Embeddedness Enhance Coordination and Manage Uncertainty. In: Cropper S, Ebers M et al (eds) The Oxford Handbook of Inter-Organizational Relations. Oxford University Press, Oxford, pp 231–255

Karlsrud J, Smith AC (2015) Europe's Return to UN Peacekeeping in Africa? Lessons from Mali. Providing for Peacekeeping No. 11. International Peace Institute, New York

Kitzen MWM (2012) Close encounters of the tribal kind: The implementation of cooption as a tool for de-escalation of conflict; the case of the Netherlands in Afghanistan's Uruzgan province. Journal of Strategic Studies 35(5):713–734

Koops JA, MacQueen N, Tardy T, Williams PD (eds)(2015) The Oxford Handbook of United Nations Peacekeeping Operations. Oxford University Press, Oxford

Lamb CJ, Munsing E (2011) Secret Weapon: High-value Target Teams as an Organizational Innovation. Strategic Perspectives No. 4. Institute for National Strategic Studies, Washington DC

Larson EV, Darilek ER, Gibran D et al (2009) Foundations of Effective Influence Operations, A Framework for Enhancing Army Capabilities. RAND Corporation, Santa Monica

Lotze W (2015) United Nations Multidimensional Integrated Stabilization Mission in Mali (MINUSMA). In: Koops JA, MacQueen N, Tardy T, Williams PD (eds) The Oxford Handbook of United Nations Peacekeeping Operations. Oxford University Press, Oxford, pp 854–864

Maas B (2016) The Interrelation between Servicemen and Interpreters in Peace Support Operations: Case Study of MINUSMA, Master thesis. Netherlands Defence Academy, Breda

McChrystal S (2015) Team of Teams; New Rules of Engagement for a Complex World. Random House, Portfolio/Penguin, New York

MINUSMA (2015) Version 1.4 MINUSMA Force Intelligence Collection Plan. MINUSMA, Bamako

MINUSMA ASIFU (2015) ASIFU information brief. PowerPoint presentation (November 2015) MINUSMA, Bamako

MINUSMA (2017) MINUSMA Facts and Figures http://www.un.org/en/peacekeeping/missions/minusma/facts.shtml Accessed 15 April 2015

Naylor S (2015) Relentless Strike, The Secret History of Joint Special Operations Command. St. Martin's Press, New York

Norheim-Martinsen PM, Ravndal JA (2011) Towards Intelligence-Driven Peace Operations? The Evolution of UN and EU Intelligence Structures. International Peacekeeping 18(4):454–467

Perugini N (2008) Anthropologists at War: Ethnographic Intelligence and Counter-Insurgency in Iraq and Afghanistan. International Political Anthropology 1(2):213–227

Rietjens SJH, Baudet FH (2017) Stovepiping within multinational military operations: The case of Mali. In: Goldenberg I, Soeters JMML, Dean W (eds) Information Sharing in Military Operations. Springer, Berlin, pp 201–220

Rietjens SJH, De Waard E (2017) UN peacekeeping intelligence: The ASIFU experiment. International Journal of Intelligence and Counterintelligence 30(4) (accepted for publication 2017)

Rietjens SJH, Zomer J (2017) In search of intelligence: The Dutch Special Forces in Mali. In: Glicken Turnley J, Michael K, Ben-Ari E (eds) Special Operations Forces (SOF) Around the World: Perspectives from the Social Sciences. Routledge, London
Salmon PM (2008) Distributed situation awareness: Advances in theory, measurement and application to teamwork, unpublished PhD thesis. Brunel University, London
Salmon PM, Stanton NA, Walke GH et al (2008) What really is going on? Review of situation awareness models for individuals and teams. Theoretical Issues in Ergonomics Science 9 (4):297–323
Sarter NB, Woods DD (1991) Situation Awareness: A Critical But Ill-defined Phenomenon. The International Journal of Aviation Psychology 1(1):45–57
Smith H (1994) Intelligence and UN Peacekeeping. Survival 36(3):174–192
Stanton NA, Salmon PM, Walker GH, Jenkins DP (2010) Is situation awareness all in the mind? Theoretical Issues in Ergonomics Science 11(2):29–40
Stanton NA, Stewart R, Harris D et al (2006) Distributed situation awareness in dynamic systems: theoretical development and application of an ergonomics methodology. Ergonomics 49(12–13):1299–1311
Try D, Radnor Z (2007) Developing an understanding of results-based management through public value theory. International Journal of Public Sector Management 20(7):655–673
UK MOD DCDC (2010) Joint doctrine Publication 04 Understanding. UK MOD, Shrivenham
UN Department of Peacekeeping Operations (2015) UN Multi-dimensional Integrated Mission in Mali MINUSMA. PowerPoint presentation at the Conference on Peacekeeping in Mali. US Army Peacekeeping Institute, Carlisle, USA, 28 January 2015
US Army (2013) US Army Manual FM 3-13: Inform and Influence Activities. Headquarters Department of the Army, Washington DC
Van Dalen H (2015) ASIFU: Baanbrekend inlichtingenexperiment in Mali [ASIFU: A pioneering intelligence experiment in Mali]. Militaire Spectator, 184(7–8):306–320
Weick KE (1979) The Social Psychology of Organizing, 2nd edn. McGraw-Hill, New York
Wickens CD (2008) Situation Awareness: Review of Mica Endsley's 1995 Articles on Situation Awareness Theory and Measurement. Human Factors 50(3):397–403
Zomer J (2016) The integrated intelligence structure of MINUSMA: One for all and all for one? Master thesis, University of Leiden

Author Biographies

Dr. S.J.H Rietjens, an engineer by training, is Associate Professor of Information Management, Department of Military Management Studies at the Faculty of Military Sciences, Netherlands Defence Academy. He obtained his master degree and Ph.D. on civil-military interaction during peace support operations from Twente University in 2006. His research focusses on effectiveness of (intelligence) operations.

Dr. E.J. de Waard is Associate Professor of Strategic Organizing and Change, Department of Military Management Studies at the Faculty of Military Sciences, Netherlands Defence Academy. He holds a Master of Science degree from Radboud University Nijmegen and obtained his Ph.D. at the Erasmus University Rotterdam in 2010.

Professor Dr. P.C. van Fenema is Professor of Military Logistics, Department of Military Management Studies at the Faculty of Military Sciences, Netherlands Defence Academy. He obtained his Master of Science degree in Law and Economics from Utrecht University, and his Ph.D. in Business Administration at the Erasmus University in 2002, both cum laude.

Printed by Printforce, the Netherlands